T0134820

Studies in Computational Intelligence

Volume 647

Series editor

Janusz Kacprzyk, Polish Academy of Sciences, Warsaw, Poland
e-mail: kacprzyk@ibspan.waw.pl

About this Series

The series "Studies in Computational Intelligence" (SCI) publishes new developments and advances in the various areas of computational intelligence—quickly and with a high quality. The intent is to cover the theory, applications, and design methods of computational intelligence, as embedded in the fields of engineering, computer science, physics and life sciences, as well as the methodologies behind them. The series contains monographs, lecture notes and edited volumes in computational intelligence spanning the areas of neural networks, connectionist systems, genetic algorithms, evolutionary computation, artificial intelligence, cellular automata, self-organizing systems, soft computing, fuzzy systems, and hybrid intelligent systems. Of particular value to both the contributors and the readership are the short publication timeframe and the worldwide distribution, which enable both wide and rapid dissemination of research output.

More information about this series at http://www.springer.com/series/7092

Liming Chen · Supriya Kapoor
Rahul Bhatia
Editors

Emerging Trends and Advanced Technologies for Computational Intelligence

Extended and Selected Results
from the Science and Information Conference
2015

 Springer

Editors
Liming Chen
School of Computer Science
De Montfort University
Leicester
UK

Rahul Bhatia
The Science and Information
(SAI) Organization
Bradford, West Yorkshire
UK

Supriya Kapoor
The Science and Information
(SAI) Organization
Bradford, West Yorkshire
UK

ISSN 1860-949X   ISSN 1860-9503 (electronic)
Studies in Computational Intelligence
ISBN 978-3-319-81491-9   ISBN 978-3-319-33353-3 (eBook)
DOI 10.1007/978-3-319-33353-3

Printed on acid-free paper

This Springer imprint is published by Springer Nature
The registered company is Springer International Publishing AG Switzerland

Editor's Preface

The Science and Information (SAI) Organization is an international professional organisation dedicated to promoting research, technology and development by providing multiple platforms for professionals and researchers to exchange and disseminate the latest research results and findings. It has covered a wide range of topics with special emphasis being placed on the general areas of computer science and information technologies. This includes the emerging technology trends, e.g. cloud computing, big data and ambient intelligence; communication systems, e.g. 3G/4G network evolution and mobile ad hoc networks; electronics, e.g. novel sensing and sensor networks; security, e.g. secure transactions, cryptography and cyber law; machine vision, e.g. virtual reality and video analysis; and intelligent data management, e.g. artificial intelligence, neural networks, data mining and knowledge management and e-learning and e-business. Science and Information (SAI) Conference is a premier event organised by SAI for researchers and industry practitioners to share their new ideas, original research results and practical development experiences from all of the aforementioned areas.

In addition to providing researchers with an opportunity for presenting their findings and views, SAI2015 also helps increase public engagement with science and community. The conference features a number of interviews, publicity of presentations and keynotes in social platforms such as YouTube, which are made available to the general public to raise awareness of the advancements and application of science and information technologies.

SAI Conference 2015 has attracted huge attention from researchers and practitioners around the world. During the three-day event, 260 scientists, technology developers, young researcher including Ph.D. students, and industrial practitioners from 56 countries have engaged intensively in presentations, demonstrations, open panel sessions and informal discussions. The inspiring keynote speeches and the state-of-the-art lectures have deeply motivated attendees and envisioned future research directions. The conference has greatly facilitated knowledge transfer and synergy, bridged gaps between different research communities/groups, laid down foundation for common purposes and helped identify opportunities and challenges for interested researchers and technology and system developers.

To further the dissemination of high-quality research and novel technologies presented in SAI Conference 2015, 21 papers, which received highly recommended feedback from peer review, have been selected for this special edition of Springer book on SAI Conference 2015. All papers have gone through substantial extension and consolidation and are subject to another round of rigorous review and additional modification. We believe that these papers represent the state of the art of the cutting-edge research and technologies in related areas and can help inform relevant research communities and individuals of the future development in science and information.

Contents

A Plantar Inclinometer Based Approach to Fall Detection in Open Environments

Jianfei Sun, Zumin Wang, Liming Chen, Baofeng Wang,
Changqing Ji and Shuai Tao

Abstract In this paper, we report a threshold-based method of fall detection using plantar inclinometer sensor, which provides us the information of angle variations during walking, and of angle status after a fall. The angle variations and status are collected in three-dimensional space. We analyzed the normal range of angle variations during walking, and selected the thresholds by testing the distribution of plantar angles of falls. In the experiments, thresholds were selected from plantar angles of fall status in four directions: forward, backward, left and right. Using the selected thresholds, we detected falls of five subjects in different situations for five hundred times and obtained the average detection rate of 85.4 %.

1 Introduction

Fall accident is not only a severe threat to public health, but even fatal for poor physical elderly people with walking difficulties. With the social aging problems getting worse these days [6], the number of elderly people living alone gradually increased and most of them could not get good health services [9, 18]. The elderly people are usually living alone, which gives rise to risk of fall accident. According to statistics, more than 33 % of people aged 65 years or older have one fall per year [14]. Sometimes, it is not the fall but the subsequent injury is the threat to the elderly people [10], and almost 62 % of injury-related hospitalizations for seniors are the results from falling [15]. Sometimes the subsequent injuries like cerebral hemorrhage happen and they will be lethal if timely rescues are unavailable [7]. Therefore, an accurate and real-time fall detection method will make timely rescue possible that avoids the subsequent injury and danger.

J. Sun · Z. Wang · B. Wang · C. Ji · S. Tao (✉)
Information and Engineering College, Dalian University, Dalian, China
e-mail: taoshuai@dlu.edu.cn

L. Chen
School of Computer Science and Informatics, De Montfort University, Leicester, UK
e-mail: liming.chen@dmu.ac.uk

© Springer International Publishing Switzerland 2016
L. Chen et al. (eds.), *Emerging Trends and Advanced Technologies
for Computational Intelligence*, Studies in Computational Intelligence 647,
DOI 10.1007/978-3-319-33353-3_1

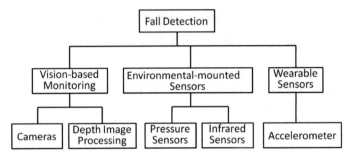

Fig. 1 Three types of method based on different principles for fall detection

At present, there are several types of method based on different principles can realize fall monitoring, which were shown but not limited to the contents in Fig. 1.

First of all, the vision-based monitoring type, for example uses a MapCam (omni-camera) to capture images and performs image processing over the images [13], of which fall monitoring success rate closed to 80 %. Wong obtained the rate of 86.19 % in detecting the human faint event using thermal cameras in indoor environment [21]. An unsupervised method was proposed in [16] for detecting abnormal activity using the fusion of some simple features. The detection rate of falls was about 66.67 %. Williams used a distributed network of smart cameras to detect and localize falls with a high accurate of 97.5 % [20]. But these methods can be used only for indoor fall monitoring. Furthermore, due to the application of HD panoramic camera as the sensor [17], it is likely to involve privacy problems. Even without the privacy problem, the camera will bring more or less discomfort to monitored people.

In our previous study, we used an infrared sensor network locating on the ceiling to monitor the action of people, with the higher monitoring accuracy (up to a maximum of 95.14 %) compared with the HD panoramic camera based methods, which could also solve the personal privacy problems (without camera equipments). But there are still some limitations, such as the indoor application only and its system is relatively complex, the setup and maintenance is more difficult.

The third type of fall monitoring method is based on the use of wearable devices, such as the fall sensor fixed on the waist in early time with small volume and high sensitivity [4, 5]. But the wearing discomfort and other shortcomings severely limit its development; the fall detection function was embedded into the smart bracelet and even vest [1, 2], which could break through the indoor limitation. But the limits of this type of method are that the elderly people usually forget to wear the equipments and it is difficult to maintain the monitoring precise in daily life.

Take the most commonly used smart wristband as an example, it is very easy and convenient for the stylish and curious young people to accept new things and wear such smart wristband, but for the majority of elder people, watch-like portable devices can bring inconvenience and some of the elderly due to loss of memory, inevitably forget to wear some unnecessary devices. To solve this kind of problems, some research groups tried to place pressure sensor and three-axis acceleration sensor on the head and the waist to monitor fall and they obtained a sensitivity of 95.71 % and a specificity of 97.78 % [3], but wearing this equipment is very cumbersome for elder people and everyday wearing will be a burden.

In this research, we applied the plantar axial inclinometer based human fall monitoring method, which could effectively reduce limitations of present methods, provide an efficient and stable system with no environmental dependency and a wide application without violation of privacy issues.

In summary, our fall monitoring system and method the advantages of simple structure and low computational complexity, so we can easily integrate it into shoes or insoles. The elderly people will remember to wear them as long as they simply take on shoes. So compared with other fall detection systems, ours will not only preserve user privacy but also with low influence and easy applicable for elderly people.

2 A Plantar Inclinometer Based Approach to Fall Detection

2.1 Plantar Inclinometer Sensor

In the plantar inclinometer, we use the single axis sensor *SCA60C-N*1000060, which is a high-precision, reliable and low-price analog sensor. The photograph of the sensor module is shown in Fig. 2. The sensor of *SCA60C-N*1000060 is applied widely in the research of intelligent control and abnormal alarm [12, 19]. Due to the character of single axis, in this study, for correctly reflecting the plantar angle, we need to use two *SCA60C-N*1000060 in the sensor module (shown in Fig. 2). In the sensor module, two sensors are orthogonal to each other for detecting the plantar angle data in two axes.

Fig. 2 The photograph of the sensor module

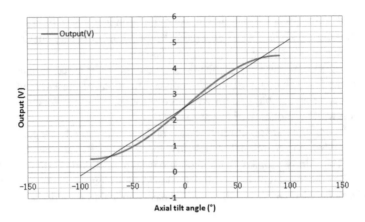

Fig. 3 The relationship between digital value and analog output of the sensor module

The sensor module can detect the plantar angles from $-90°$ to $90°$, the corresponding output digital quantities are from -0.5 to 4.5 V. The relationship between digital value and analog output is shown in Fig. 3.

In Fig. 3, we see that the relationship between digital value (voltage) and analog output (angle) is nonlinear, and similar to sinusoidal vibration, which is shown as

$$\alpha = arcsin(\frac{V_{out} - Offset}{Sensitivity}), \tag{1}$$

where α is the axis angle of the sensor module, V_{out} is the digital output. When the angle of the module is $0°$ (horizontal direction), the value of *Offset* is 2.5 V. The *Sensitivity* under room temperature is 2 V/g. Therefore, formula (1) can be rewrite as

$$\alpha = arcsin(0.5V_{out} - 1.25). \tag{2}$$

By collecting the value of V_{out}, and calculating by formula (2), we obtain the axis angles of the module.

Before data collection, we need to connect the sensor module and data processing center (including A/D converter, communication and host system), calibrate the system for the first time. For the convenience of data processing, we transformed the angle range of $-90°–90°$ to $0°–180°$. If the angle values stay in the range of $90° \pm 1°$ and remain stable when we place the sensor module in horizontal, the system is calibrated successfully. Then we mounted the sensor module onto a pad, which is shown in Fig. 4a. The pad has a degree of flexibility for winding. After mounting the sensor module, we calibrated the system for the second time. If the angle values stay in the range of $90° \pm 1°$ and remain stable when we place the pad in horizontal, the system is calibrated successfully.

(a) **(b)** **(c)**

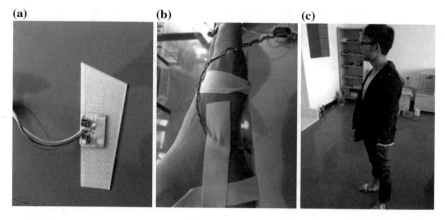

Fig. 4 The sensor module mounted onto a pad, the pad sticked under the feet, a volunteer wearing the pad

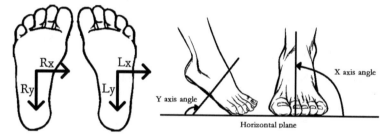

Fig. 5 The x-axis and y-axis of *left* and *right foot*

Then we sticked the pad under the feet, which is shown in Fig. 4b, and calibrated the system for the third time. A volunteer with the pad walked 10 m with a normal speed. We collected the angle values. After walking, when the volunteer stays still which is shown in Fig. 4c, if the angle values stay in the range of $90° \pm 1°$ and remain stable, the system is calibrated successfully. After three times calibration, the experiments could be started.

We set the x-axis and y-axis of left and right foot shown in Fig. 5. Therefore, when a person stands upright, the angles of x-axis and y-axis will both be 90°.

2.2 Data Collection During Falls

The subject in this test is a student in our laboratory, who is healthy and can walk normally. We detected the right-foot angles during walking and found that the angles are almost 90° when the person is standing, and fluctuate rapidly during walking.

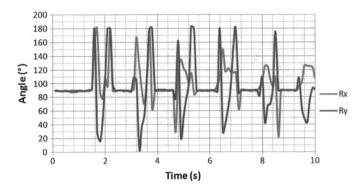

Fig. 6 The variation trend of right-foot angles during walking in 10 s

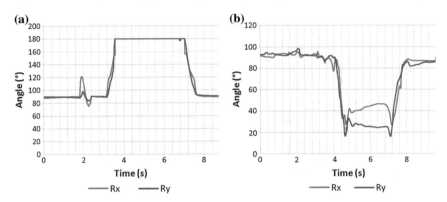

Fig. 7 The variation of right-foot angles (left-foot is symmetrical) when the subject falls forward and backward. **a** Fall forward. **b** Fall backward

The y-axis has a larger fluctuate range. The variation trend of angles during walking in 10 s is shown in Fig. 6.

We also investigated the variation of right-foot angles (left-foot is symmetrical) when the subject falls forward and backward. From Fig. 7, we see that when falls happen, the angles of both x-axis and y-axis have strenuous vibrations, which are very different from the situation of walking. Then plantar angle data were collected after falls in four different situations: forward, backward, left and right for determining the thresholds.

2.2.1   Forward Falls

The volunteer falls forward to an air bed, stays still naturally, the plantar angle data are then collected. The distribution of plantar angles after forward falls 120 times is shown in Fig. 8a, b.

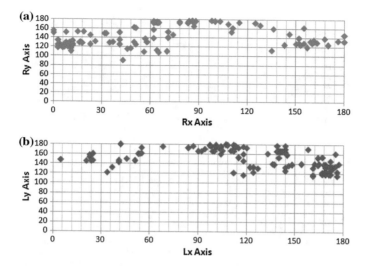

Fig. 8 The distribution of plantar angles after forward falls 120 times

From Fig. 8, we see that the variation range of plantar angle is limited after falls happening. For this reason, we could select some thresholds to recognize falls. When a person falls forward, there will be a wide variation range of x-axis, but a narrow variation range of y-axis.

2.2.2 Backward Falls

The volunteer falls backward to an air bed, stays still naturally, the plantar angle data are then collected. The distribution of plantar angles after backward falls 120 times is shown in Fig. 9a, b.

From Fig. 9, we see that when a person falls backward, there will be a wide variation range of x-axis, but a narrow variation range of y-axis. The y-axis angle direction of backward falls (0°–90°) is contrary to forward falls (90°–180°).

2.2.3 Left Falls

The volunteer falls to left side to an air bed, stays still naturally, the plantar angle data are then collected. The distribution of plantar angles after left falls 120 times is shown in Fig. 10a, b.

From Fig. 10, we see that the variation range of plantar angle is limited after falls happening. When a person falls to left side, most of the variation range of x-axis is 0°–120°, most of the variation range of y-axis is 0°–120°.

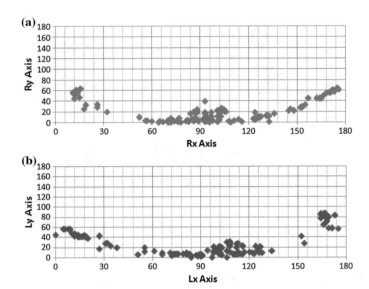

Fig. 9 The distribution of plantar angles after backward falls 120 times

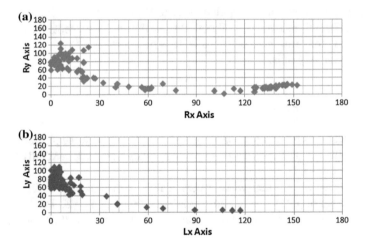

Fig. 10 The distribution of plantar angles after left falls 120 times

2.2.4 Right Falls

The volunteer falls to right side to an air bed, stays still naturally, the plantar angle data are then collected. The distribution of plantar angles after right falls 120 times is shown in Fig. 11a, b.

From Fig. 11, we see that the change range of plantar angle is limited after falls happening. When a person falls to right side, most of the variation range of x-axis is 60°–180°, most of the variation range of y-axis is 0°–120°.

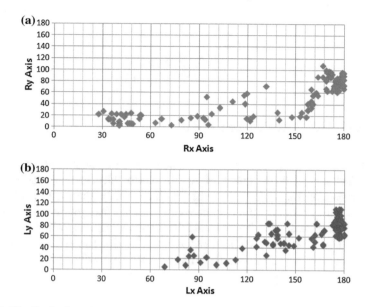

Fig. 11 The distribution of plantar angles after right falls 120 times

The limitation to the range of plantar angles of falls caused by physiological structure [8, 11] of people is an important basis for fall detection in this research.

2.3 Fall Detection Method

By observing the collected data of plantar angles, we see that, for x-axis or y-axis under one of the feet, at least one of them is very large or small. Therefore, based on this rule, we selected some thresholds for fall detection. The thresholds should satisfy most of the angle values for fall detection. By statistical analysis, we found that all the angles of at least one axis is in the interval of [57°, 131°]. Therefore, the thresholds of plantar angle for fall detection is set to [57°, 131°]. We detect falls by the formula below.

$$State = [(57° \geq R_x \vee R_x \geq 131°) \vee (57° \geq R_y \vee R_y \geq 131°)]$$
$$\wedge [(57° \geq L_x \vee L_x \geq 131°) \vee (57° \geq L_y \vee L_y \geq 131°)] \qquad (3)$$

In formula (3), when state is 1, a fall happens, if it is 0, there is no fall.

To avoid some false alarms caused by walking or some abnormal legs movements, we improved the method by adding a tally function which is shown in formula (4). When fall status is detected for x times, we recognize it as a real fall. In the experiment in next section, we set x to 15.

$$Fall = \begin{cases} 1, & \sum_{n=1}^{x} State_n = x \\ 0, & \sum_{n=1}^{x} State_n < x \end{cases} \qquad (4)$$

3   Experiments of Fall Detection

In the experiment, there are five subjects. The information of subjects are shown in Table 1. Each subject did five groups of experiments for fall detection in the horizontal plane (forward, backward, left side, right side and free fall). There are 16 falls in each group and totally 80 falls. In addition, We did another five groups of experiments for fall detection on a slope. There are 4 falls in each group and totally 20 falls. The number of falls of each subject in the experiment is 100. The experiment result of each subject is shown in Fig. 12. The average detection results of 500 falls of five subjects are shown in Fig. 13.

Table 1 The information of five subjects

Subject	Sub. 1	Sub. 2	Sub. 3	Sub. 4	Sub. 5
Gender	Male	Male	Male	Male	Female
Age	20	22	21	22	24
Height (cm)	180	170	169	168	170
Weight (kg)	67	65	67.5	62	54.5
Detection rate (%)	78	85	88	88	88

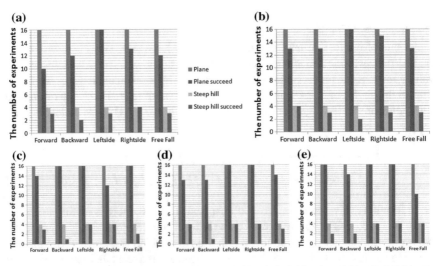

Fig. 12 The experiment results of five subjects. **a** Subjct A. **b** Subjct B. **c** Subjct C. **d** Subjct D. **e** Subjct E

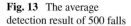

Fig. 13 The average detection result of 500 falls

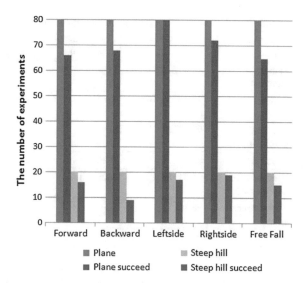

From Fig. 13, we see that there are 427 falls detected successfully. The detection rate was 85.4%. However, the detection rate of backward falls is only 77%. The most possible reason is that slope brings bad effects to the detection. It could cause a low detection rate in the situation of backward fall due to the effect on plantar angles.

4 Discussion

In this paper, we report a method of fall detection using plantar inclinometer sensor, which provides us the information of angle variations during walking, and of angle status after a fall. We analyzed the normal range of angle variations during walking, and selected the thresholds by testing the distribution of plantar angles. We detected falls in different situations by five subjects for 500 times and obtained the average detection rate of 85.4%.

By comparison with some similar researches in Table 2, it is not difficult to find: the best success rates were yielded by distributed multi-cameras [20], thermal infrared matrix solutions [17] and pressure/acceleration sensor solutions [3], the success rates were above 95%; while the thermal infrared matrix fall detection belongs to environment detection scheme, which is similar to the camera fall detection [13] that is with larger hardware scale, more complicated algorithm and it is difficult to be integrated on small scale leading to the restriction to interior application.

The integrated waist/hand acceleration sensors are currently popular fall detection solutions on the market [1–5], which are characterized by less hardware complexity, easy to be integrated on small scale and no interior limitation.

The success rate of most of the products is not high except several solutions applying multiple sensor types. In order to solve this problem, developers always

Table 2 The detection results compared with some similar research

Fall detection scheme	Detection rate (%)	The complexity of hardware	The difficulty of algorithm	Environment
Plantar inclinometer	85.4 [this study]	Low	Low	Indoor/outdoor
Camera	79.8 [13], 66.67 [16]	Medium	High	Indoor
Distributed multi-cameras	97.5 [20]	High	High	Indoor
Thermal camera	86.19 [21]	Medium	High	Indoor
Pressure and acceleration	95.71 [3]	Medium	High	Indoor/outdoor
Infrared sensor array	95.14 [17]	High	High	Indoor

use more efficient and complex algorithms, however, which will increase the cost of hardware due to higher sensitivity, and the complexity of software system. Compared with the solutions above, the main advantages of our system are as follows: high accuracy, simple hardware, easily to be integrated and without the external equipment and environment support. However, there are some deficiencies in the system: lower sensor accuracy and anti-interference ability; the impedance in sensor is insensitive to vibrations and it is not possible to achieve fast dynamic detection; extensible function is limited; the cost of the hardware is relatively high.

Although we obtained a high detection rate of falls, our threshold-based method is relatively naive, and the subjects are all young people in the experiment. To solve the problems mentioned above, we will apply machine learning techniques to build classified models of different activities of elders, and to learn users' pattern/threshold values. In addition, the difference of angle variation between young people (the subjects in this study) and elders will be investigated. We expect to increase the proformance by using more advanced methods with the same device of plantar inclinometer sensors in future study.

Acknowledgments This work was supported by the National Natural Science Foundation of China (Grant No. 61501076)

References

1. Angelini, L., Caon, M., Carrino, S., Bergeron, L., Nyffeler, N., Jean-Mairet, M., Mugellini, E.: Designing a desirable smart bracelet for older adults. In: Proceedings of the 2013 ACM Conference on Pervasive and Ubiquitous Computing Adjunct Publication, pp. 425–434. ACM (2013)

2. Bourke, A.K., Van de Ven, P.W., Chaya, A.E., OLaighin, G.M., Nelson, J.: Testing of a long-term fall detection system incorporated into a custom vest for the elderly. In: Engineering in Medicine and Biology Society, 2008. EMBS 2008. 30th Annual International Conference of the IEEE, pp. 2844–2847. IEEE (2008)
3. Chen, D., Zhang, Y., Feng, W., Li, X.: A wireless real-time fall detecting system based on barometer and accelerometer. In: 2012 7th IEEE Conference on Industrial Electronics and Applications (ICIEA), pp. 1816–1821. IEEE (2012)
4. Chen, G.C., Huang, C.N., Chiang, C.Y., Hsieh, C.J., Chan, C.T.: A reliable fall detection system based on wearable sensor and signal magnitude area for elderly residents. In: Aging Friendly Technology for Health and Independence, pp. 267–270. Springer (2010)
5. Chen, J., Kwong, K., Chang, D., Luk, J., Bajcsy, R.: Wearable sensors for reliable fall detection. In: 27th Annual International Conference of the Engineering in Medicine and Biology Society, 2005. IEEE-EMBS 2005, pp. 3551–3554. IEEE (2006)
6. Cockerham, W.C.: This aging society (1991)
7. Dunning, K., Moomaw, C., Flaherty, M.L., Osborne, J., James, M.L., Worrall, B.B., Woo, D.: Abstract t p279: falls after intracerebral hemorrhage. Stroke **46**(Suppl 1), ATP279–ATP279 (2015)
8. Elveru, R.A., Rothstein, J.M., Lamb, L.R.: Goniometric reliability in a clinical setting. Subtalar and ankle joint measurements. Phys. Ther. **68**(5), 672–677 (1988)
9. Gao, L., Cao, X., Zhang, M.: The study on community health education of empty nest elderly. Engineering **5**(10), 137 (2013)
10. Guimaraes, R., Isaacs, B.: Characteristics of the gait in old people who fall. Disabil. Rehabil. **2**(4), 177–180 (1980)
11. Leardini, A., O'Connor, J.J.: C.F.G.S.: A geometric model of the human ankle joint. J. Biomech. **32**(6), 585591 (1999)
12. Liu, X.Q., Kuang, L.J., Li, J.C., Huang, J.Z.: The design and production of panels control system based on the angle sensor sca60c. Mech. Electr. Eng. Technol. **8**, 011 (2012)
13. Miaou, S.G., Sung, P.H., Huang, C.Y.: A customized human fall detection system using omni-camera images and personal information. In: 1st Transdisciplinary Conference on Distributed Diagnosis and Home Healthcare, 2006. D2H2, pp. 39–42. IEEE (2006)
14. Noury, N.: A smart sensor for the remote follow up of activity and fall detection of the elderly. In: Biology 2nd Annual International IEEE-EMB Special Topic Conference on Microtechnologies in Medicine & amp, pp. 314–317. IEEE (2002)
15. P.H.A.: Report on Seniors' Falls in Canada [electronic Resource]. Division of Aging and Seniors, Public Health Agency of Canada (2005)
16. Rougier, C., Meunier, J., St-Arnaud, A., Rousseau, J.: Monocular 3d head tracking to detect falls of elderly people. In: 28th Annual International Conference of the IEEE Engineering in Medicine and Biology Society, 2006. EMBS'06, pp. 6384–6387. IEEE (2006)
17. Tao, S., Kudo, M., Nonaka, H.: Privacy-preserved behavior analysis and fall detection by an infrared ceiling sensor network. Sensors **12**(12), 16920–16936 (2012)
18. Tian, Q., Meng, C.: An empirical investigation of rural empty-nesters in chongqing and a construction of service system. Asian Agric. Res. **5**(02) (2013)
19. Wenlong, Z., Qing, G., Baoshan, L.: Design of landslide warning system. In: 2011 Third International Conference on Measuring Technology and Mechatronics Automation (ICMTMA), vol. 1, pp. 974–977. IEEE (2011)
20. Williams, A., Ganesan, D., Hanson, A.: Aging in place: fall detection and localization in a distributed smart camera network. In: Proceedings of the 15th international conference on Multimedia, pp. 892–901. ACM (2007)
21. Wong, W.K., Lim, H.L., Loo, C.K., Lim, W.S.: Home alone faint detection surveillance system using thermal camera. In: 2010 Second International Conference on Computer Research and Development, pp. 747–751. IEEE (2010)

Using Fuzzy Evidential Reasoning for Multiple Assessment Fusion in Spondylarthropathic Patient Self-management

Giovanni Schiboni, Wolfgang Leister and Liming Chen

Abstract This paper proposes an approach for an ICT-supported medical assessment, by merging measures of signs and symptoms from heterogeneous sources. The disease status estimate of patients that suffer from spondylarthropathy is evaluated with different types of uncertainties using a fuzzy rule-based evidential reasoning (FURBER) approach. The approach treats measures of signs and symptoms in order to define the disease status. We take in consideration the Bath indices and the ASDAS index, described by using fuzzy linguistic variables. A fuzzy rule-base designed on the basis of a belief structure is exploited to capture uncertainty and non-linear relationships between these parameters and the disease status. The inference of the rule-based system is implemented using an evidential reasoning algorithm. An expected utility-based health score is used to assess disease activity over time and to measure the response to treatment. Our tool may be particularly helpful in monitoring the response of treatments and in interpreting the response to therapeutic interventions in clinical trials. A case study is used to illustrate the application of the proposed approach.

Keywords Spondyloarthritis · Medical assessment · Multiple attribute decision analysis · Fuzzy rule-base · Utility · Evidential reasoning · Uncertainty modelling

G. Schiboni (✉) · L. Chen
School of Computer Science and Informatics, De Montfort University, Leicester, UK
e-mail: giovanni.schiboni@uni-passau.de

L. Chen
e-mail: liming.chen@dmu.ac.uk

W. Leister
Norsk Regnesentral/Norwegian Computer Centre, Oslo, Norway
e-mail: wolfgang.leister@nr.no

© Springer International Publishing Switzerland 2016
L. Chen et al. (eds.), *Emerging Trends and Advanced Technologies
for Computational Intelligence*, Studies in Computational Intelligence 647,
DOI 10.1007/978-3-319-33353-3_2

1 Introduction

The spondyloarthropathies (SpA) are a group of related conditions that includes ankylosing spondylitis (AS), reactive arthritis, psoriatic arthritis (PsA), undifferentiated spondyloarthropathy and inflammatory, bowel-disease-associated arthritis. Collectively, these arthropathies are characterized by inflammatory arthritis, extra-articular inflammation, preceding bacterial infection, seronegativity for rheumatoid factor and a strong HLA-B27 association [10]. Usually, there is no definite direction for SpA assessment. Outcome measurement in SpA, particularly AS, has been a rapidly growing field over the last decade, with enormous progress being made in patient-reported outcomes, clinical assessments, physical measurements and composite scoring of disease state, and response to treatment. The purpose is to perform a process for determining a *SpA disease status estimate* from a subject through signs and symptoms of the disease. Signs are directly discovered by the physician or by health monitoring systems [28] and symptoms are obtained from a patient's experience and feelings. The acquisition of signs and symptoms knowledge implies uncertain measures. Acquisition systems need to be tolerant against sensor failures while being able to deal with a high degree of uncertainty arising from the open world setup. Further, patients cannot describe exact conditions. Consequently, the medical decision making process is dominated by uncertainty [20]. Thus, inevitably, the main problem for constructing an accurate and complete mathematical model for a disease status assessment is representation of uncertainty.

Rationally, reliably, and correctly handling uncertainties in medical diagnosis and treatment decisions are major challenges that have been researched more than four decades [21]. The first well-known medical rule based expert system [37] was used to diagnose bacterial infections. Several frameworks have been proposed [4, 9, 14, 22, 30, 36, 43] but all of them have a lack of procedures to address uncertainty, relying heavily on supporting statistical information that may not be available. Note well this list is not exhaustive.

One way to deal with information that may be imprecise, ill-defined, and incomplete, for which traditional quantitative approaches (e.g., statistical approach) do not give an adequate answer, is the incorporation of subjective and/or vague terms, i.e., linguistic assessments instead of numerical values. Fuzzy logic approaches [55], employing fuzzy IF-THEN rules (where the conditional part and the conclusions contain linguistic variables [58]), can model the qualitative aspects of human knowledge and the reasoning process without employing precise quantitative analysis. This provides a tool for working directly with the linguistic variables, when a communication process is involved between medical expert and patient.

A linguistic variable for a fuzzy set [59] is a variable for whose values are words (or linguistic terms) rather than numbers. Each linguistic variable represents a universal set and the different linguistic terms are fuzzy sets in the universal set. The different terms or linguistic values are also characterized by membership functions defined on the universe of discourse [1].

The fuzzy encoding of signs and symptoms of a SpA disease status estimate well suits with the approximate reasoning for medical assessment. The motivations are clear from the characteristics of linguistic variables:

- linguistic variables may be regarded as a form of information compression called granulation [56];
- they serve as a means of approximate characterization of phenomena that are either too ill-defined, too complex, or both, to permit a description in sharp terms [57];
- they provide a means for translating linguistic descriptions into numerical and computable ones. Therefore, the duality between symbolic and numerical processing becomes natural instead of antagonistic [12].

In view of the complexity of a mathematical model for such purpose, the knowledge representation power of fuzzy rule-based systems is severely limited if only fuzziness is taken into account in representing uncertain knowledge. In a SpA disease status estimate, intrinsically vague information may coexist with conditions of *lack of specificity* originating from evidence not strong enough to completely support a hypothesis but only with degrees of belief or credibility [2]. This could come from erroneous measures of sensor failures or from a degree of ignorance of an expert. Dempster-Shafer (D-S) theory of evidence [6, 34], based on the concept of belief function, is well suited to modeling subjective credibility induced by partial evidence [41]. The D-S theory describes and handles uncertainties using the concept of the degrees of belief, which can model incompleteness and ignorance explicitly. It also provides appropriate methods for computing belief functions for combination of evidences [29]. Besides, the D-S theory also shows great potentials in multiple attribute decision analysis (MADA), where an evidential reasoning (ER) approach for MADA has been developed, on the basis of a distributed assessment framework and the evidence combination rule of the D-S theory [46, 49–51, 53, 54]. The combination may become substantial when a lack of specificity in data is prevalent. In these cases, experts may have difficulty in structuring and articulating causal relationships [23].

A Belief Rule Base (BRB) for a Clinical Decision Support Systems (CDSS) architecture to make clinical decision, disease suspicion or diagnosis has been developed recently. This BRB CDSS employed three layers architecture with BRB framework, which allows the handling of various types of uncertainty found in clinical domain knowledge as well as clinical and medical data. They applied a rule-base inference using the evidential reasoning approach, i.e., RIMER [47] and fuzzy logic [46, 49, 51]. The same RIMER approach has been applied for the design, development and application of an expert system to diagnose influenza under uncertainty [35].

In this paper, we describe an approach for modeling a SpA disease status that is based on fuzzy logic and the evidential reasoning approach. In order to deal with fuzziness and incompleteness in the assessment, we use a fuzzy rule-based evidential reasoning (FURBER) approach [25]. In our approach, signs and symptoms parameters are described by using fuzzy linguistic variables and the outcomes of a fuzzy rule-base, or consequents, with belief structure, describe the disease status.

The fuzzy rules with belief degrees are used to capture uncertainty and causal relationships between the signs and symptoms and the disease status. To the best of our knowledge, this is the first time that a fuzzy BRB is used, in conjunction with the evidential reasoning, in order to perform a SpA assessment, addressed as a MADA problem. This BRB with fuzzy logic approach gives us the ability to handle both vague information and ignorance or incompleteness due to evidence which is not strong enough to make simple true or false judgments, but with degrees of belief. In fact, the approach brings the following advantages both in:

- *Encoding gathered informations.* It allows to take in consideration vagueness or fuzzy uncertainty in the description of SpA signs and symptoms, using subjective and vague linguistic terms which may overlap in their meanings instead of independent crisp sets. For example, the assessment grades 'Moderate' and 'Severe' are difficult to be expressed as clearly distinctive crisp sets, but quite natural to be defined as two dependent fuzzy sets. In other words, the intersection of the two fuzzy sets may not be empty.
- *Decoding a consistent estimate as result of the assessment process.* In real-life diagnosis, a doctor cannot judge one patient disease status to be 100 % consistent with the actual level of the disease. Opinion or judgment with belief levels must be expressed. For example: "Severe" outcome from the Bath indices is consistent with a high degree of belief of SpA disease with a 80 % probability; "Moderate Disease Activity" from ASDAS outcome is a medium degree of belief of SpA disease with a 60 % probability. Hence, the probability of a degree of SpA disease for the patient can be High, Medium or Low with different degrees of belief. From the above, it can be inferred that the uncertain knowledge that exists with the SpA assessment should need to be processed by using refined knowledge representation schema and inference mechanism.

Our expert system is designed to be the assessment module of a personalized Chronic Patient Self-Management System (CPSMS) for SpA [32].

2   A Fuzzy Evidential Reasoning Based Approach to Multiple Assessment Fusion

The CPSMS is an ontology-based decision support system with the goal of improving the quality of life (QoL) of the patients by facilitating patients with chronic disease to carry out personalized self-management. Based on personalized rules for training and using a disease and training diary, the patients will be given recommendations to handle their disease in self-management. It is done via knowledge description and inference for both patients and their conditions by providing a set of non-pharmacological treatment plans or suggestions.

To support the CPSMS, it is necessary to predict an individual's SpA disease status based on a combination of physiological, behavioral and psychological features. A quantitative clinical analysis of the status of patients with muskuloskeletal disorder throughout their stay in a domestic environment has to be performed. The disease states are correlated to non-invasive objective (physiological, behavioral) and subjective (psychological) assessment measurements. Trends in disease states through the course of treatment can then be tracked. The combination of different complementary metrics, based on the measures, would be in the study of response to treatment in clinical spondyloarthrosis. While behavioral symptoms are all highly discriminative in predicting a state of suffering, they have traditionally been very qualitative in nature. Using sensor and mobile technology, we would be able to precisely and quantitatively measure these symptoms, in order to develop objective diagnostic assessments. In addition, to have a more reliable measure of the patient status, the physiological/contextual measures will be sustained by subjective outcomes from clinical psychological measures in the form of questionnaires. To this end, this paper develops a fusion model which integrates physiological and behavioral measures, with the ultimate purpose of accurately determining the disease status of a SpA patient as well as trending the effects of treatment. Using these objective scales, in fact, it's possible to have clinically significant outcomes related to a subject's state, marked at different spaced intervals, throughout the course of the treatment. The performance of a treatment and the quantification of a subject's progress over time would be then feasible, by exploiting an automated tracking of the patient's response to treatment over time. Thus, quantitatively measuring the degree of change in the measures, allows to evaluate the degree of response in treatment.

The process of modeling our SpA medical assessment comprises the following four elements, each described in detail in their corresponding section: (1) identification of measures of symptoms; (2) definition of the input and fuzzy variables; (3) the fuzzy rule-based belief structure, and (4) the rule-based inference mechanism.

2.1 Identification of Measures of Symptoms

In order to obtain a SpA disease status estimate, we consider analysis of Bath indices and ASDAS index outcomes, demonstrating the procedure involved in the inference process. In Table 1 the list of measures of symptoms considered is presented.

The Bath indices [13] present outcome measures from SpA patients and comprise of four indices:

- BASMI, the Bath AS Metrology Index;
- BASFI, the Bath AS Functional Index;
- BASDAI, the Bath AS Disease Activity Index;
- BAS-G, the Bath AS Patient Global Score.

The BASMI quantifies the mobility of the axial skeleton in AS patients and allows objective assessment of clinically significant changes in spinal movement. The index is determined by clinical measures of cervical rotation, tragus to wall distance, lumbar flexion, lumbar side flexion, and intermalleolar distance.

Table 1 Measures of symptoms and their characteristics

Test	Measure	Content	Response	Scale	References
BASMI	Objective	Sensor based	Numerical	0–10	Jenkinson et al. [18]
BASFI	Self-reported	Questionnaires	Likert	0–10	Calin et al. [3]
BASDAI	Self-reported	Questionnaires	Likert	0–10	Garrett et al. [11]
BAS-G	Self-reported	Questionnaires	Likert	0–10	Jones et al. [19]
ASDAS (CRP)	Hybrid	Questionnaires/lab analysis	Numerical	$0–\infty$	Lukas et al. [26]

The BASFI defines and monitors physical functioning of patients with AS. The index is determined by eight items concerning activities referring to the functional anatomy of the patients (bending, reaching, changing position, standing, turning, and climbing steps), and two items assessing the patients' ability to cope with everyday life.

The BASDAI measures patient-reported disease activity in patients with AS. The index is determined by patient-reported levels of back pain, fatigue, peripheral joint pain and swelling, localized tenderness, and the duration and severity of morning stiffness.

The BAS-G gives a global assessment of the well-being of the person with AS over a given time period. The index is determined by two visual analog scales to measure the effect of AS on the respondent's well-being, the first estimated over the last week, the second over the last 6 months.

The subjective measures have a numeric response scale (0–10) anchored by adjectival descriptors as a Likert scale.

The Ankylosing Spondylitis Disease Activity Score (ASDAS) measures disease activity in AS based on a composite score of domains relevant to patients and clinicians, including both self-reported items and objective measures. The index is determined by patient-reported assessments of back pain, duration of morning stiffness, peripheral joint pain and/or swelling, general well-being, and a serologic marker of inflammation (erythrocyte sedimentation rate [ESR] or C-reactive protein [CRP]). ASDAS has a continuous scale from zero with no defined upper end. A clinically important change is present when a difference of 1.1 units or greater appears; a major change is defined as a change of 2.0 units or more [27]. More information about these measures of symptoms can be found in literature [60].

2.2 Definition of the Input and Fuzzy Variables

A granularity of linguistic terms sets, used for describing each fundamental attribute, i.e., measures of symptoms, has to be defined, according to its own qualitative characteristics, i.e., linguistic terms or assessment grades, and quantitative characteristics, i.e., numerical scales. Subjective assessments are more appropriate for SpA analysis by using these parameters, as they are always associated with great uncertainty, since they come from imprecise source of knowledge, as questionnaires. Each has its own set of fuzzy assessment grades, fuzzy grades for short, characterized by linguistic variables, instead of precise numbers in probabilistic terms, and by fuzzy membership functions (FMFs). A total of $N = 5$ variables must be defined in terms of fuzzy grades. In general, attributes e_i, with $i = 1\ldots 5$ are described by j_i fuzzy grades $\{A_{i,j}, j = 1\ldots j_i\}$, respectively. Thus, a type of FMF has to be selected for each attribute description. There is some flexibility defining FMFs depending on experts considerations. The application of categorical judgements is suitable in terms of the nature of our source of informations.

As far as BASMI (e_1), BASFI (e_2), BASDAI (e_3) and BAS-G (e_4) are concerned, we follow the granularity of the respective questionnaires, i.e., 'Very Severe' (VS), 'Severe' (SE), 'Moderate' (MO), 'Mild' (MI), and 'None' (NO), as well agreed in the medical community.

In compact notation:

$$\{A_{i,j}, i = 1\ldots 4 \text{ and } j = 1\ldots j_i\} = \{NO, MI, MO, SE, VS\}, \qquad (1)$$

where $j_i = 5$.

Figure 1a shows an example for the FMFs definition. The choice of the FMFs has to be done by expert-knowledge.

As far as the ASDAS index (e_5) is concerned, the choice of the fuzzy grades comes from [27], where the following disease activity states are described: 'Very High Disease' (VD), 'High Disease Activity' (HD), 'Moderate Disease Activity' (MD), 'Inactive Disease Activity' (ID). In compact notation:

$$\{A_{5,j}, j = 1\ldots j_5\} = \{ID, MD, HD, VD\}, \qquad (2)$$

where $j_5 = 4$.

The scale on which the FMFs are distributed is relevant during modeling. In Fig. 1b, for example, the scale is different from the other attributes, according to the

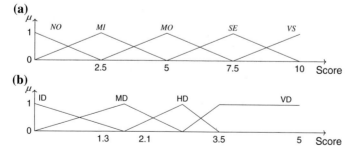

Fig. 1 **a** FMFs for assessment grades and e_1, e_2, e_3 and e_4. **b** FMFs for assessment grades for e_5

definition of the ASDAS index. Note that the straight-line FMFs are just chosen for their simplicity. More complex, e.g., non-linear FMFs could be used, according to different kinds of information sources considered and to the different requirements. If numerical values from the measures scores above are not available at all, then the modeling can also be carried out only based on subjective judgements, following this granularity. Furthermore, the belief distribution assessment scheme, proposed in Sect. 3, permits inference even in case no FMF is available.

Now, let us assume that the *SpA disease status estimate* of a patient, i.e., output of the assessment, can be classified into several categories (grades) of values. Our choice, for describing linguistically the assessment uses the values as consequent variable: 'High' (Hi, C_4), 'Medium' (Me, C_3), 'Low' (Lo, C_2) and 'None' (No, C_1).

In compact notation:

$$\{C_j, j = 1 \ldots j_c\} = \{No, Lo, Me, Hi\}, \tag{3}$$

where $j_c = 4$.

2.3   Construct a Fuzzy Rule-Base with the Belief Structure

Let us define a rule-based system for receiving information from different assessments and inferring conclusions. This step is the knowledge-based core of the approach since it merges the various heterogeneous assessment measures coming from the different sensorial assessments. The set of rules has to be defined by domain experts and observation facts. It is a reasoning scheme provided to the model as a priori.

The key constituents are the IF-THEN rules, extended with belief degrees, in which each antecedent attribute takes referential values and each possible consequent is associated with belief degrees [46]. The knowledge representation parameters, i.e., the rule weights, antecedent attribute weights and belief degrees with consequents, are extended by incorporating the concept of belief rule into fuzzy IF-THEN rules for a SpA medical assessment:

$$R_k : \text{ if } (e_1 \text{ is } A_1^k) \wedge \cdots \wedge (e_N \text{ is } A_N^k),$$
$$\text{then } \{(C_1, \bar{\beta}_{1,k}) \ldots (C_{j_c}, \bar{\beta}_{j_c,k})\}$$
$$\left(\sum_{j=1}^{j_c} \bar{\beta}_{j,k} \leq 1\right), \text{ with a rule weight } \theta_k \tag{4}$$
$$\text{and attribute weights } \delta_{k,1} \ldots \delta_{k,N}$$
$$k \in \{1 \ldots L\}$$

Fig. 2 Evaluation hierarchy

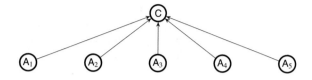

where A_i^k represents the value of the ith antecedent attribute e_i in the kth rule. $\bar{\beta}_{j,k}$ is the belief degree to which C_j is believed to be the consequent if, in the kth packet rule, the input satisfies the packet antecedents $A^k = \{A_1^k \ldots A_N^k\}$. If $\sum_{j=1}^{j_c} \bar{\beta}_{j,k} = 1$ the kth rule is said to be complete, otherwise it is incomplete.

The schema of this belief rule base is depicted in Fig. 2. It represents the logic framework of the reasoning process. A bottom-up paradigm is used to solve the inference. In the first layer, the antecedents A_i, with $i = 1 \ldots N$, represent pieces of evidence for the rules. Such information is then aggregated using the ER approach in Sect. 3.4, obtaining the consequents outcome C.

3 The Fuzzy Evidential Inference Mechanism for Multiple Assessment Fusion

Once a rule-base is established, the knowledge contained can be used to perform inference for the given input. The inference procedure is composed of different components that are: (1) input transformation, (2) rule activation weight calculation, (3) rule update mechanism, (4) aggregation of the rules, (5) expected utility definition. In Sect. 3.1, we describe the way in which the input for an antecedent can be decoded into a distribution representation of the linguistic terms using belief degrees. In Sect. 3.2, we describe how to compute the total degree to which the input matches to the packet antecedents. In other words, how to quantify the contribution of each rule, through the weights of activation, for the subsequent aggregation process. In Sect. 3.3, we describe how to handle incomplete information due to a lack of data. In Sect. 3.4, we describe the evidential reasoning algorithm. Eventually in Sect. 3.5, we describe how to extract a health score from the distributed assessment, output of the ER, exploiting the concept of expected utility.

3.1 Input Transformation

Inputs are discretized into a distributed representation of linguistic terms in antecedents using belief degrees. The general input form corresponding to the antecedent attribute in the kth rule is:

$$(A_1^*, \varepsilon_1) \wedge \cdots \wedge (A_5^*, \varepsilon_5) \tag{5}$$

where A_i^* is the actual input for the attribute e_i, ε_i is the degree of belief assigned by an expert to the association of A_i^*. A distribution assessment approach [25, 47, 49] permits to assess the input (A_i^*, ε_i) for an antecedent e_i to a distribution representation of the linguistic terms using belief degrees:

$$S(e_i) = S(A_i^*, \varepsilon_i) = \{(A_{i,j}, \alpha_{i,j}), \; j = 1 \ldots j_i\} \tag{6}$$

for $i = 1 \ldots N$, where $A_{i,j}$ is the jth linguistic value; $\alpha_{i,j}$ is the belief degree to which the input (A_i^*, ε_i) for e_i belongs to the linguistic value $A_{i,j}$ with $\alpha_{i,j} \geq 0$ and $\sum_{j=1}^{j_i} \alpha_{i,j} \leq 1$.

Consider that $\alpha_{i,j}$ in Eq. 6 can be generated in different ways depending on the nature of the antecedent attribute and the availability of data [25, 38]. Due to the possible uncertainty in the inputs, for the purpose of a disease status modeling, it is assumed that each input parameter (e_i with $i = 1 \ldots N$) could be presented in terms of FMF (if it is available) in any of the following forms based on available information, history data or expert experience:

- A single deterministic value with 100 % certainty.
- A closed interval defined by an equally likely range.
- A triangular distribution defined by a most likely value, with lower and upper least likely values.
- A trapezoidal distribution defined by a most likely range, with lower and upper least likely values.

If the measures of symptoms are the same as identified in Sect. 2.1, the main scenario is the one in which a numerical score x_i is provided for each e_i. In this case there is the need to calculate the distributed assessment of x_i, relative to the ith attribute, on the fuzzy grades $A_{i,j}$ [52]. This can be done by defining its degrees of belief for which it belongs to the fuzzy grades by normalizing its membership degrees to these grades. Thus, firstly, the membership degrees of x_i must be computed on the assessment grades (see Fig. 1b) correspondent to attribute i. This is simply done applying the formula:

$$\mu_{i,j} = \begin{cases} 0, & x_i \leq a \\ \frac{x_i - a}{b - a}, & a \leq x_i \leq b, \\ \frac{c - x_i}{c - b}, & b \leq x_i \leq c, \\ 0, & c \leq x_i, \end{cases} \tag{7}$$

$$\text{for } j = 1 \ldots J_i$$

where $\mu_{i,j}$ is the degree of membership of the triangular FMF relative to the jth term of the ith attribute and a, b, c are the parameters of the FMF. To obtain the distributed assessment of an original score x_i on the ith attribute, these membership degrees are normalized to generate its degrees of belief as follows:

$$\alpha_{i,j} = \frac{\mu_{i,j}}{\mu_{i,j} + \mu_{i,j+1}}$$

$$\alpha_{i,j+1} = \frac{\mu_{i,j+1}}{\mu_{i,j} + \mu_{i,j+1}} \tag{8}$$

$$\text{for } j = 1 \ldots j_i - 1$$

Note that if the requirement of mutual complementarities for the FMFs is satisfied $\alpha_{i,j} = \mu_{i,j}$.

If the input parameters, instead, correspond to forms 2, 3 or 4, $\alpha_{i,j}$ can be formulated as follows [25]:

$$\alpha_{i,j} = \frac{\tau(A_i^*, A_{i,j})\varepsilon_i}{\sum_{j=1}^{j_i}[\tau(A_i^*, A_{i,j})]}, \tag{9}$$

$$\text{for } i = 1 \ldots N \text{ and } j = 1 \ldots j_i$$

where (A_i^*, ε_i) is the actual input corresponding to the ith antecedent, τ is a matching function, $\tau(A_i^*, A_{i,j}) = \tau_{ij}$ is a matching degree to which A_i^* belongs to $A_{i,j}$. Note that, even if A_i^* completely belongs to the jth linguistic expression, i.e., $\tau(A_i^*, A_{i,j}) = 1$, $\alpha_{i,j} \neq 1$ if $\varepsilon_i \neq 1$.

Other matching functions can be employed [47]. For qualitative parameters, for example, a subjective numerical scale may be used against which the range of parameters is mapped. A medical judgement, coming from the doctor, or physiologist, experience and records, could be included in the assessment process.

3.2   Rule Activation Weight Calculation

Consider an input as in Eq. 5 corresponding to the kth rule as defined in Eq. 4

$$e_1 \text{ is } (A_1^k, \alpha_1^k) \wedge \cdots \wedge e_5 \text{ is } (A_5^k, \alpha_5^k)$$

where $A_i^k \in \{A_{i,j}, \ j = 1 \ldots j_i\}$ and $\alpha_i^k \in \{\alpha_{i,j}, \ j = 1 \ldots j_i\}$ is the individual belief degree that the input belongs to A_i^k of the individual antecedent e_i appearing in the kth rule. The total degree α_k to which the input matches to the packet antecedent A^k in the kth rule is defined by combining the individual degrees α_i^k for $i = 1 \ldots N$ [25]. The weight of activation of the kth rule w_k is computed by exploiting the flowing formula [33]:

$$w_k = \frac{\theta_k \alpha_k}{\sum_{j=1}^{L} \theta_j \alpha_j} = \frac{\theta_k \prod_{i=1}^{N}(\alpha_i^k)^{\bar{\delta}_{ki}}}{\sum_{j=1}^{L} \theta_j \left[\prod_{l=1}^{N}(\alpha_i^j)^{\bar{\delta}_{jl}} \right]} \tag{10}$$

with

$$\bar{\delta}_{ki} = \frac{\delta_{ki}}{\max_{1=1...N} \{\delta_{ki}\}}. \tag{11}$$

Here $\bar{\delta}_{ki}$ is the relative weight of A_i^k used in the kth rule, obtained by dividing the weight of A_i^k with maximum weight of all the antecedent attributes of the kth rule. This normalized value $\bar{\delta}_{ki}$ implies that the range of its value is between 0 and 1. Finally, consider that $\alpha_k = \prod_{i=1}^{N} (\alpha_i^k)^{\bar{\delta}_{ki}}$ is the combined matching degree that is computed by using multiplicative aggregation function. It can be used since the rules are composed only by $\wedge$ operators.

3.3   Rule Update Mechanism

The rule update mechanism is incorporated in an update of the original belief degree $\bar{\beta}_{ik}$. When the kth rule is activated, the consequent of a rule could be incomplete due to the lack of data of the antecedents. If there is an incomplete input for an attribute, it will produce an incomplete output in each of the rules where the attribute is used. So the original belief degree $\bar{\beta}_{j,k}$ in the jth consequent C_j of the kth rule must be updated based on the actual input information available [33, 40]. The original belief degree is updated as follows:

$$\beta_{j,k} = \frac{\bar{\beta}_{j,k}\left[\sum_{i=1}^{N} \xi(i,k) \sum_{t=1}^{j_i} \alpha_{i,t} \right]}{\sum_{i=1}^{N} \xi(i,k)} \tag{12}$$

$$\text{where } \xi(i,k) = \begin{cases} 1, & \text{if } R_k \text{ is used} \\ 0, & \text{otherwise} \end{cases}$$

where $\bar{\beta}_{j,k}$ is given in Eq. 4. Just note that $0 \leq \beta_{j,k} \leq 1$ for all k and $1 - \sum_{j=1}^{j_c} \beta_{j,k}$ denotes both the ignorance incurred in establishing R_k and the incompleteness that may exist in the input information.

3.4   Aggregation of the Rules

Let us consider a fuzzy rule-base with the belief structure that is given by $R = \{R_1, R_2 \ldots R_L\}$. The kth rule in Eq. 4 is expressed as:

R_k: if e is A^k then SpA disease status estimate is C with belief degree β_k

where e is the antecedent attribute vector $(e_1 \ldots e_N)$, A^k is the packet antecedents $\{A_1^k \ldots A_N^k\}$, C is the consequent vector $(C_1 \ldots C_{j_c})$ and β_k is the vector of the belief degrees $(\beta_{1,k} \ldots \beta_{j_c,k})$ with $k \in \{1 \ldots L\}$. The distributed assessment, referred to as a *belief structure* is represented by:

$$S(A^k) = \{(C_i, \beta_{j,k};\ j = 1 \ldots j_c)\} \tag{13}$$

where $\beta_{j,k}$ is the degree to which C_i is the consequent if the input activates the antecedent A^k in the kth rule, that is given by Eq. 12 with $0 \leq \sum_{j=1}^{j_c} \beta_{j,k} \leq 1$ with $k = 1 \ldots L$.

The ER approach is used to aggregate all the packet antecedents of the L rules, to obtain the degree of belief of each referential values of the consequent attribute. First, the degrees of belief $\beta_{j,k}$ for all $j = 1 \ldots j_c$ have to be transformed into basic probability masses [54]:

$$m_{j,k} = w_k \beta_{j,k},\ j = 1 \ldots j_c$$
$$m_{C,k} = 1 - \sum_{j=1}^{j_c} m_{j,k} = 1 - w_k \sum_{j}^{j_c} \beta_{j,k} \tag{14}$$
$$\bar{m}_{C,k} = 1 - w_k \text{ and } \tilde{m}_{C,k} = w_k \left(1 - \sum_{j=1}^{j_c} \beta_{j,k}\right).$$

with $m_{C,k} = \bar{m}_{C,k} + \tilde{m}_{C,k}$ for all $k = 1 \ldots L$ and $\sum_{j}^{L} w_j = 1$.

All the packet antecedents of the L rules now can be aggregated to generate the combined degree of belief in each possible consequent term C_j in C. Let us consider $m_{j,I(k)}$ the combined degree of the belief in C_j by aggregating the first k packet antecedents $(A^1 \ldots A^k)$ and $m_{C,I(k)}$ the remaining degree of belief unassigned to any output term. The overall combined degree of belief β_j in C_j is computed with the evidential algorithm in Eq. 15.

$$\{C_j\}:\ m_{j,I(k+1)} = K_{I(k+1)}[m_{j,I(k)}m_{j,k+1} + m_{j,I(k)}m_{C,k+1} + m_{C,I(k)}m_{j,k+1}]$$
$$m_{C,I(k)} = \bar{m}_{C,I(k)} + \tilde{m}_{C,I(k)},\ k = 1 \ldots L$$
$$\{C\}:\ \tilde{m}_{C,I(k+1)} = K_{I(k+1)}[\tilde{m}_{C,I(k)}\tilde{m}_{C,k+1} + \bar{m}_{C,I(k)}\tilde{m}_{C,k+1} + \tilde{m}_{C,I(k)}\bar{m}_{C,k+1}]$$
$$\{C\}:\ \bar{m}_{C,I(k+1)} = K_{I(k+1)}[\bar{m}_{C,I(k)}\bar{m}_{C,k+1}]$$
$$K_{I(k+1)} = \left[1 - \sum_{j=1}^{j_c} \sum_{t=1,t \neq j}^{j_c} m_{j,I(k)}m_{t,k+1}\right]^{-1},\ k = 1 \ldots L - 1 \tag{15}$$
$$\{C_n\}:\ \beta_j = m_{j,I(L)}/(1 - \bar{m}_{C,I(L)}),\ j = 1 \ldots j_c$$
$$\{C\}:\ \beta_C = \bar{m}_{C,I(L)}/(1 - \bar{m}_{C,I(L)}).$$

We have that β_C is the belief degree unassigned to any C_j. It can be proved that $\sum_{j=1}^{N} \beta_j + \beta_C = 1$ [54].

The final distributed assessment, which is obtained by aggregating the L rules activated by the actual input vector $A^* = \{A^{*k}, k = 1 \ldots L\}$ for e_i with $i = 1 \ldots N$, is expressed as:

$$S(A^*) = \{(C_j, \beta_j), j = 1 \ldots j_c\} \tag{16}$$

The final result is still a belief distribution on disease status estimate, which gives a global view about the SpA disease activity level for a given input.

3.5 Expected Utility Definition

Since the aim is to judge the quality of the treatment, i.e., monitoring the status of the patient by assessing if the condition, related to disease activity, deteriorates, remains stable or improves, the trend of the outcomes of assessments needs to be analyzed. Thus, some kind of numerical score, equivalent to the distributed assessment in a sense, must be defined. Expected utility concept can be employed to define such score. Let us consider an utility $u(C_j)$ of the grade C_j with

$$\begin{aligned} u(C_{j+1}) &> u(C_j) \\ \text{if } C_{j+1} &\text{ is preferred to } C_j, \end{aligned} \tag{17}$$

where $u(C_j)$ may be estimated using the probability assignment method [44] or by constructing regression models using partial rankings or pairwise comparison [45]. It is worthy to mention that in [31], they conduct a statistical mapping analysis between a standard investigator assessment of disease severity in AS and domain responses in a standard index of health utility. They implemented the above mapping in an optimized algorithm to estimate utility.

If the assessments are complete and precise, i.e., $\beta_C = 0$, the expected utility of a distributed assessment outcome Φ may be considered a *health score* that can be used for ranking assessments. It is calculated as:

$$u(\Phi) = \sum_{j=1}^{j_c} \beta_j u(C_j). \tag{18}$$

An assessment Φ_1 is preferred to an assessment Φ_2 if and only if $u(\Phi_1) > u(\Phi_2)$. In case the assessment is incomplete, some procedure exists [54] in order to deal with.

4 A Case Study in SpA Treatment

The primary target of a SpA treatment is remission of the disease. When it fails, minimizing inflammatory activity of signs and symptoms is the priority, protecting the patient's quality of life by preventing structural damage and functional disability. Therefore, persistence of activity indicates the need for a change of treatment. Our tool may be particularly helpful to monitor patients in a self-monitoring setting. These patients will monitor their health condition, create a log of their condition, and be alerted when changes in their condition does not meet the target, that is their condition is about to worsen. Other applications, such as monitoring the impact of changes in treatment, e.g., in clinical trials are also possible. Since SpA patients react individually to their health condition, one-fit-all recommendations for interpreting the disease activity are not available. Despite this fact, there is a general agreement in the medical community on recommendations for monitoring patients. In particular, the experts agreed that the BASDAI plays a major role in the follow-up of AS [7].

We demonstrate the procedure involved in the FURBER inference for the SpA disease status assessment, by using the following simplified use scenario, similar to the one proposed in [32]. Mark is a SpA patient, male, 25 years old. His SpA disease was confirmed 1 year ago, and now it is under stable condition. He has a regular appointment for treatment at the local clinic every second week. The most obvious symptom for him is back pain and morning stiffness. Usually doctor prescribes pharmacological treatment. With the assistance of the home-based self-management system, he is able to monitor his disease condition on his own. The CPSMS provides Mark with his condition information and both pharmacological and non-pharmacological treatment plans. The CPSMS self-management system tracks the trend of the SpA disease treatment in a period of 8 weeks. The measurements are made every 2 weeks by using mobile phones and sensors and by filling out the Bath and ASDAS questionnaires. In Table 2 outcomes of these measures relative to entire period are presented.

Five levels of linguistic variables are used for Bath indices and four levels for ASDAS index. The definition of their linguistic terms corresponds to the ones in Eqs. 1 and 2 and the corresponding FMF are given in Table 3 and in Fig. 1. Four linguistic expressions are used as well for the consequent variable as the ones in Eq. 3.

4.1 Rule-Base with Belief Structure Construction

According to the number of linguistic terms used for describing the antecedents, a rule-base with a total number of 27 fuzzy rules with belief structures is used in the case study, presented in Table 4. The belief degrees that characterize the rule-base are chosen arbitrarily only for illustration purposes, as the actual degrees of belief depend on the context of applications and an expert knowledge is required for their definition. It is worth noting that this is an extremely simplified belief rule-base since

Table 2 Original measures of symptoms related to a patient and their transformed distributed values

Test	Week 1	Week 3	Week 5	Week 8
BASMI (e_1)	6	5.3	5.6	4.9
	{(MO, 0.6), (SE, 0.4)}	{(MO, 0.88), (SE, 0.12)}	{(MO, 0.76), (SE, 0.24)}	{(MI, 0.04), (MO, 0.96)}
BASFI (e_2)	5.5	4.9	5.5	4.5
	{(MO, 0.8), (SE, 0.2)}	{(MI, 0.04), (MO, 0.96)}	{(MO, 0.8), (SE, 0.2)}	{(MI, 0.2), (MO, 0.8)}
BASDAI (e_3)	5.8	4.7	5.3	4.4
	{(MO, 0.68), (SE, 0.32)}	{(MI, 0.12)(MO, 0.88)}	{(MO, 0.88), (SE, 0.12)}	{(MI, 0.24), (MO, 0.76)}
BAS-G (e_4)	5.1	4.5	4.8	4.6
	{(MO, 0.96), (SE, 0.04)}	{(MI, 0.2), (MO, 0.8)}	{(MI, 0.08), (MO, 0.92)}	{(MI, 0.16), (MO, 0.84)}
ASDAS (e_5)	2.1	1.5	1.8	1.3
	{(MD, 0.64), (HD, 0.36)}	{(ID, 0.12), (MD, 0.88)}	{(MD, 0.91), (HD, 0.09)}	{(ID, 0.24), (MD, 0.76)}

Only non-zero values are presented

Table 3 Membership functions of the fuzzy assessment grades

Linguistic terms	Bath (e_1, e_2, e_3, e_4)				
	None (*NO*)	Mild (*MI*)	Moderate (*MO*)	Severe (*SE*)	Very severe (*VS*)
Grade FMF	(0, 0, 2.5)	(0, 2.5, 5)	(2.5, 5, 7.5)	(5, 7.5, 10)	(7.5, 10, 10)
Linguistic terms	ASDAS (e_5)				
		Inactive (*ID*)	Moderate (*MD*)	High (*HD*)	Very high (*VD*)
Grade FMF		(0, 0, 1.7)	(0, 1.7, 2.8)	(1.7, 2.8, 3.5)	(2.8, 3.5, 5, 5)

in a real-case scenario a higher number of rules would be necessary in order to have a reliable model. Nevertheless, the case study demonstrates that these rules provide a flexible and rational way to construct knowledge base.

4.2 Input Trasformation

The inputs in Table 2 are transformed into the distributed representation of linguistic terms in the antecedent using Eqs. 7 and 8 based on the corresponding FMF in Table 3 and in Fig. 1. In the rule-base, 27 rules have been defined, of which only a part is fired depending on inputs of different assessments. The fired rules for each assessment are all list in Table 5. The activation weights w_k of each rule in the fire sub-rule-base are calculated with Eq. 10 where we considered the attribute weights $\delta_i = 1$ and the rule weight $\theta_k = 1$. For an instance, the relative assessment on rule #7 is expressed as:

$$S(A^7) = \{(No, 0.4), (Lo, 0.55), (Me, 0.05), (Hi, 0)\}$$

with activation weight $w_7 = 0.0178$ for week 3 global assessment.

4.3 Fired Rule Combination Using the ER Algorithm

Exploiting the ER algorithm in Eq. 15 as described in Sect. 3.4 one can aggregate the fired rules and generate a SpA disease status estimate. The distributed assessments, result of the disease status estimate over a period of 8 weeks, are as follows:

Table 4 Rule-base with belief structure

Rule	Belief				
	Antecedent attributes	Disease status estimate			
		None	Low	Medium	High
R#1	(NO, NO, NO, NO, ID)	1			
R#2	(NO, NO, NO, NO, MD)	0.80	0.20		
R#3	(NO, NO, NO, MI, MD)	0.65	0.35		
R#4	(NO, NO, MI, MI, MD)	0.5	0.5		
R#5	(NO, MI, MI, MI, MD)	0.3	0.7		
R#6	(MO, MI, MI, MO, ID)	0.1	0.5	0.4	
R#7	(SE, MO, MO, MI, MD)	0.4	0.55	0.05	
R#8	(SE, MI, MI, MO, ID)		0.45	0.55	
R#9	(MI, MO, MI, MO, ID)	0.15	0.35	0.5	
R#10	(MO, MI, MO, MI, MD)		0.45	0.55	
R#11	(MI,MI,MO,MO,ID)		0.6	0.4	
R#12	(MI, MI, MI, MI, MD)	0.1	0.9		
R#13	(MI, MI, MI, MI, HD)		0.95	0.05	
R#14	(MI, MI, MI, MO, HD)		0.8	0.2	
R#15	(MI, MI, MO, MO, HD)		0.7	0.3	
R#16	(MI, MI, MO, NO, MD)	0.1	0.5	0.4	
R#17	(MI, MO, MO, MO, HD)		0.55	0.45	
R#18	(MO, MO, MO, MO, HD)		0.35	0.65	
R#19	(MO, MO, MO, MO, VD)		0.15	0.85	
R#20	(MO, MO, MO, SE, VD)			0.9	0.1
R#21	(MO, MO, SE, SE, VD)			0.8	0.2
R#22	(MO, SE, SE, SE, VD)			0.75	0.25
R#23	(SE, SE, SE, SE, VD)			0.6	0.4
R#24	(SE, SE, SE, VS, VD)			0.55	0.45
R#25	(SE, SE, VS, VS, VD)			0.45	0.55
R#26	(SE, VS, VS, VS, VD)			0.25	0.75
R#27	(VS, VS, VS, VS, VD)				1

4.4   Trend Analysis of the Disease Activity During Treatment

The driven paradigm is based on the SpA treat-to-target principles [42]. A therapy plan can be divided into *target change* and *treatment change*. If both non-pharmacological and pharmacological treatments are administered and an assessment result is, after a certain period of time, not on target, then, the treatment plan must be changed accordingly to patient conditions. An instance of target can be that the ASDAS index does not get significantly worse. Thus, it's necessary a health

Table 5 Fuzzy rule expression matrix of the fired rules for each assessment

Input	Disease status estimate			
	None	Low	Medium	High
Assessment week 1				
$A^{18}(0.2005)$		0.35	0.65	
$A^{19}(0.1128)$		0.15	0.85	
$A^{20}(0.0047)$			0.9	0.1
$A^{21}(0.0022)$			0.8	0.2
Assessment week 3				
$A^{6}(0.0004)$	0.1	0.5	0.4	
$A^{7}(0.0178)$	0.4	0.55	0.05	
$A^{8}(0.0005)$		0.45	0.55	
$A^{10}(0.0055)$		0.45	0.55	
Assessment week 5				
$A^{18}(0.4479)$		0.35	0.65	
$A^{19}(0.0443)$		0.15	0.85	
Assessment week 8				
$A^{6}(0.0093)$	0.1	0.5	0.4	
$A^{9}(0.0015)$	0.15	0.35	0.5	
$A^{10}(0.0177)$		0.45	0.55	
$A^{11}(0.0012)$		0.6	0.4	
$A^{12}(0.0002)$	0.1	0.9		

The values in the parentheses are the weights of the packet antecedent attributes A^k generated using Eq. 10, where we assumed that the weights of rules and the weights of antecedents are equal. Just note that, since the inputs are complete, no update is necessary for the degrees belief

score that measures clinical disease activity at specific time-points for supporting decisions about entry into clinical trials, for supporting treatment changes and for defining therapeutic goals (Fig. 3).

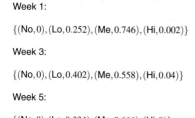

Week 1:

$\{(No,0),(Lo,0.252),(Me,0.746),(Hi,0.002)\}$

Week 3:

$\{(No,0),(Lo,0.402),(Me,0.558),(Hi,0.04)\}$

Week 5:

$\{(No,0),(Lo,0.334),(Me,0.666),(Hi,0)\}$

Week 8:

$\{(No,0.023),(Lo,0.471),(Me,0.506),(Hi,0)\}$

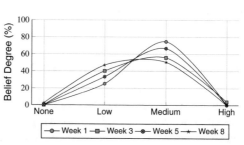

Fig. 3 Distributed assessments of the SpA disease status in a period of 8 weeks

Fig. 4 Expected utility
defined for assessment
grades

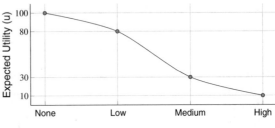

Fig. 5 Health score trend
related to SpA disease status
in a period of 8 weeks

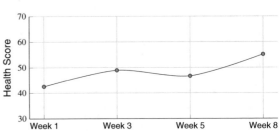

Following what is described in Sect. 3.5, the expected utilities of individual assessment grades need to be defined first. We defined such expected utility values for different assessment grades, see Fig. 4. More specifically, we assigned an utility score of 100 to grade No, 80 to Lo, 30 to Me and 10 to Hi. In this way, a distributed assessment can be transformed to a health score using Eq. 18. Note that a deeper study, made by expert knowledge, should be done in order to define more reliable values.

Finally, applying Eq. 17, health scores can be obtained. Then, their trend can be monitored as depicted in Fig. 3.

In our example, the outcome of the assessment, after a positive response between week 0 and week 3, suggests that Mark's condition is less severe than the situation at the last measurement. Despite the fact that Mark follows doctor's pharmacological and non-pharmacological treatment plans, the disease condition does not get better in the next assessment, between week 3 and week 5, causing a worse response. This would trigger an alert in CPSMS, in order to perform proper updates in its treatment plans (Fig. 5).

5   Conclusion and Future Works

In this paper we have described the estimation model of our CPSMS for SpA [32]. The application is supposed to be used in the residential environment of SpA patients, but also to improve communication between patient and health personnel. Solutions of smart monitoring, sensor technologies, objective and subjective assessment, treatment plans, and guidelines are employed.

Our estimation model merges measures of signs and symptoms from heterogeneous sources of information. The disease status estimate of patients, that suffer from spondylarthropathy, is evaluated with different types of uncertainties, using a fuzzy rule-based evidential reasoning (FURBER) approach. A fuzzy rule-base, designed on the basis of a belief structure is exploited to capture uncertainty and non-linear relationships between these measures and the disease status. The approach is able to handle both vague information and ignorance or incompleteness which is not strong enough to make simple true or false judgments but with degrees of belief. Signs and symptoms measures are encoded as distributions on fuzzy linguistic variables, i.e., degrees of belief on assessment grades. Thus, information that come from precise data, random numbers and subjective judgments with uncertainty can be consistently merged under the unified model. Initial case study based on the Bath indices and the ASDAS index showed that such model is able to merge information that comes from disparate sources as other indices, sensor measures, as well as judgement of medical knowledge based on subjective scales. The inference of the rule-based system is implemented using the evidential reasoning algorithm.

The aggregation procedure of the rules preserves the original features of heterogeneous information sources and there is possibility for rule training and self-learning/updating in the rule-base. In fact, it is difficult to determine the elements of a rule-based system entirely subjectively. Also, a change in a rule weight or an attribute weight may lead to significant changes in the performance of a belief rule base. Moreover, the forms of fuzzy membership functions in the antecedent of a rule still remain an important factor for the system performance. For this reason, self-tuning optimal models for belief rule-based systems exist [48]. Based on these optimal models, revised models for self-tuning a FURBER for engineering system safety analysis has been investigated [24]. Supported by real data, in our next work, we'll perform the training of our rule base in an optimal way using expert judgments as well as statistical data. Input data, attribute weights, rule weights, and parameters of fuzzy membership functions will be combined to generate activation weights for rules, and all activated belief rules will be then combined to generate appropriate conclusions using the evidential reasoning approach. Such a combination process is formulated as nonlinear objective functions. The aim is to minimize the differences between observed outputs and the outputs of a BRB, whilst parameter specific limits and partial expert judgments can be formulated as constraints.

Another critical point to evaluate is that, in order to increase interpretability of the disease activity measures, as expressed through health scores, we need some criteria for identifying *improvement scores*. Improvement scores help to determine whether treatments really work, that is whether they actually produce clinically important improvement, allowing investigators, clinicians, regulators and patients to determine the efficacy (or lack thereof) of a given intervention and to communicate about response using the same metric [39]. The interpretation of change in health scores has been a topic of research for almost two decades [15, 17]. It is recognized that the statistical significance of a treatment effect, because of its partial dependency on sample size, does not always correspond to the clinical relevance of the effect. Statistically significant effects are those that occur beyond some level of chance. In contrast, clinical relevance refers to the benefits derived from that treatment, its impact upon the patient, and its implications for clinical management of the patient [16, 17]. For clinical relevance, one is interested in the *minimally important change* (MIC) of health status questionnaires. Changes in scores exceeding the MIC are clinically relevant by definition [8]. Different methods to determine the minimally important change on the scale of a measurement instrument have been proposed. These methods have been summarized by Crosby et al. [5], by distinguishing distribution-based and anchor-based methods. Criteria for disease activity improvement scores are therefore important for use in clinical practice, observational studies and clinical trials. Such criteria must be developed for our health score. In our real-data experiment we'll have to focus on the evaluation of clinically relevant cut-off values for such improvements scores.

This research has produced a novel tool for decisional support in treatment and self-management for spondylarthropathic patients where the related SpA disease status information is sparse and different kind of uncertainties are involved.

Acknowledgments This work has been supported by the Mobile Musculoskeletal User Self-management (MOSKUS) project funded by Research Council of Norway in the VERDIKT programme, grant number 227251. The authors would like to thank the MOSKUS project consortium for their valuable discussions and medical domain expertise.

References

1. Baturone, I., Barriga, A., Jimenez-Fernandez, C., Lopez, D.R., Sanchez-Solano, S.: Microelectronic design of fuzzy logic-based systems. CRC press (2000)
2. Binaghi, E., Madella, P.: Fuzzy dempster-shafer reasoning for rule-based classifiers. Int. J. Intell. Syst. **14**(6), 559–583 (1999)
3. Calin, A., Garrett, S., Whitelock, H., Kennedy, L., O'hea, J., Mallorie, P., Jenkinson, T.: A new approach to defining functional ability in ankylosing spondylitis: the development of the Bath Ankylosing spondylitis functional index. J. Rheumatol. **21**(12) 2281–2285 (1994)
4. Chen, H.-L., Yang, B., Wang, G., Liu, J., Chen, Y.-D., Liu, D.-Y.: A three-stage expert system based on support vector machines for thyroid disease diagnosis. J. Med. Syst. **36**(3), 1953–1963 (2012)
5. Crosby, R.D., Kolotkin, R.L., Williams, G.R.: Defining clinically meaningful change in health-related quality of life. J. Clin. Epidemiol. **56**(5), 395–407 (2003)
6. Dempster, A.P.: A generalization of Bayesian inference. J. R. Stat. Soc. Ser. B (Methodological) 205–247 (1968)
7. Dernis, E., Lavie, F., Pavy, S., Wendling, D., Flipo, R.-M., Saraux, A., Cantagrel, A., Claude-pierrre, P., Goupille, P., Le Loët, X., et al.: Clinical and laboratory follow-up for treating and monitoring patients with ankylosing spondylitis: development of recommendations for clinical practice based on published evidence and expert opinion. Joint Bone Spine **74**(4), 330–337 (2007)
8. de Vet, H.C., Terwee, C.B., Ostelo, R.W., Beckerman, H., Knol, D.L., Bouter, L.M.: Minimal changes in health status questionnaires: distinction between minimally detectable change and mini-mally important change. In: Health Qual. Life Outcomes **4**(1) 54 (2006)
9. Dogantekin, E., Dogantekin, A., Avci, D.: An expert system based on generalized discriminant analysis and wavelet support vector machine for diagnosis of thyroid diseases. Expert Syst. Appl. **38**(1), 146–150 (2011)

10. Dougados, M., Linden, S.V.D., Juhlin, R., Huitfeldt, B., Amor, B., Calin, A., Cats, A., Dijkmans, B., Olivieri, I., Pasero, G., et al.: The European spondylarthropathy study group preliminary criteria for the classification of spondylarthropathy. Arthritis Rheum. **34**(10), 1218–1227 (1991)
11. Garrett, S., Jenkinson, T., Kennedy, L.G., Whitelock, H., Gaisford, P., Calin, A.: A new approach to defining disease status in ankylosing spondylitis: the bath ankylosing spondylitis disease activity index. J. Rheumatol. **21**(12), 2286–2291 (1994)
12. Imriyas, K.: An expert system for strategic control of accidents and insurers risks in building construc-tion projects. Expert Syst. Appl. **36**(2), 4021–4034 (2009)
13. Irons, K., Jeffries, C.: The bath indices: outcome measures for use with ankylosing spondyli-tis patients: a synopsis of the creation of the bath indices and the research investigating the effectiveness of them. NASS (2004)
14. Issac Niwas, S., Palanisamy, P., Chibbar, R., Zhang, W.-J.: An expert support system for breast cancer diagnosis using color wavelet features. J. Med. Syst. **36**(5) 3091–3102 (2012)
15. Jacobson, N.S., Follette, W.C., Revenstorf, D.: Toward a standard definition of clinically sig-nificant change. Behav. Ther. **17**(3), 308–311 (1986)
16. Jacobson, N.S., Truax, P.: Clinical significance: a statistical approach to defining meaningful change in psychotherapy research. J. Consult. Clin. Psychol. **59**(1) 12 (1991)
17. Jaeschke, R., Singer, J., Guyatt, G.H.: Measurement of health status: ascertaining the minimal clinically important difference. Control. Clin. Trials **10**(4), 407–415 (1989)
18. Jenkinson, T.R., Mallorie, P.A., Whitelock, H., Kennedy, L.G., Garrett, S., Calin, A.: Defining spinal mobility in ankylosing spondylitis (AS). The bath AS metrology index. J. Rheumatol. **21**(9), 1694–1698 (1994)
19. Jones, S., Steiner, A., Garrett, S., Calin, A.: The bath ankylosing spondylitis patient global score (BAS-G). Rheumatology **35**(1), 66–71 (1996)
20. Kong, G., Xu, D.-L., Liu, X., Yang, J.-B.: Applying a belief rule-base inference methodology to a guideline-based clinical decision support system. Expert Syst. **26**(5), 391–408 (2009)
21. Kong, G., Xu, D.-L., Yang, J.-B.: Clinical decision support systems: a review on knowledge representation and inference under uncertainties. Int. J. Comput. Intell. Syst. **1**(2), 159–167 (2008)
22. Kumar, K.A., Singh, Y., Sanyal, S.: Hybrid approach using case-based reasoning and rule-based reasoning for domain independent clinical decision support in ICU. Expert Syst. Appl. **36**(1), 65–71 (2009)
23. Liu, J., Yang, J., Wang, J., Sii, H.: Review of uncertainty reasoning approaches as guidance for maritime and offshore safety-based assessment. J. UK Saf. Reliab. Soc. **23**(1), 63–80 (2003)
24. Liu, J., Yang, J.-B., Ruan, D., Martinez, L., Wang, J.: Self-tuning of fuzzy belief rule bases for engineering system safety analysis. Ann. Oper. Res. **163**(1), 143–168 (2008)
25. Liu, L., Yang, J.-B., Wang, J., SII, H.-S., Wang, Y.-M.: Fuzzy rule-based evidential reasoning approach for safety analysis. Int. J. Gen. Syst. **33**, 2–3 (2004). Taylor and Francis
26. Machado, P., Landewé, R., Lie, E., Kvien, T.K., Braun, J., Baker, D., van der Heijde, D., Assessment of Spondy-loArthritis international Society, et al.: Ankylosing spondylitis disease activity score (ASDAS): defining cut-off values for disease activity states and improvement scores. Ann. Rheum. Dis. **70**(1), 47–53 (2011)
27. Machado, P., Landewé, R., Lie, E., Kvien, T.K., Braun, J., Baker, D., van der Heijde, D., Assessment of Spondy-loArthritis international Society, et al.: Ankylosing spondylitis disease activity score (ASDAS): defining cut-off values for disease activity states and improvement scores. Ann. Rheum. Dis. **70**(1), 47–53 (2011)
28. Pantelopoulos, A., Bourbakis, N.G.: A survey on wearable sensor-based systems for health mon-itoring and prognosis. IEEE Trans. Syst. Man Cybern. Part C: Appl. Rev. **40**(1), 1–12 (2010)
29. Pearl, J.: Probabilistic Reasoning in Intelligent Systems: Networks of Plausible Reasoning. Morgan Kauf-Mann Publishers, San Francisco (1988)
30. Piury, J., Laita, L.M., Roanes-Lozano, E., Hernando, A., Piury-Alonso, F.-J., Gómez-Argüelles, J.M., Laita, L.: A Gröbner bases-based rule based expert system for fibromyalgia diagnosis. Revista de la Real Academia de Ciencias Exactas, Fisicas y Naturales. Serie A. Matematicas **106**(2) 443–456 (2012)

31. Poole, C., Singh, A., Freundlich, B., Koenig, A., Currie, C.: Estimating health related utility from clinically assessed disease activity severity in ankylosing spondylitis. Value Health **13**(3) (2010). ISSN: 1098-3015
32. Qi, J., Chen, L., Leister, W., Yang, S.: Towards knowledge driven decision support for personalized home- based self-management of chronic diseases. In: Proceedings of The 2015 Smart World Congress, pp. 1724–1729 (2015)
33. Rahaman, S., Hossain, M.S.: A belief rule based clinical decision support system to assess suspicion of heart failure from signs, symptoms and risk factors. In: 2013 International Conference on Informatics, Electronics & Vision (ICIEV), pp. 1–6. IEEE (2013)
34. Shafer, G., et al.: A mathematical theory of evidence, vol. 1. Princeton university press, Princeton (1976)
35. Shahadat, M., Khalid, M.S., Dey, S.: A belief rule-based expert system to diagnose influenza. In: The 9th International Forum on Strategic Technology (IFOST) (2014)
36. Sharaf-El-Deen, D.A., Moawad, I.F., Khalifa, M.: A new hybrid case-based reasoning approach for medical diagnosis systems. J. Med. Syst. **38**(2), 1–11 (2014)
37. Shortliffe, E.: Computer-Based Medical Consultations: MYCIN. Elsevier (2012)
38. Sii, H., Wang, J.: Safety assessment of FPSOs. The process of modelling system safety and case studies. In: Report of the Project. The Application of Approximate Reasoning Methodologies toOffshore Engineering Design EPSRC GR/R30624 and GR 32413 (2002)
39. Singh, J., Solomon, D., Dougados, M., Felson, D., Hawker, G., Katz, P., Paulus, H., Wallace, C.: Devel-opment of classification and response criteria for rheumatic diseases. Arthritis Rheum. **55**(3), 348–352 (2006)
40. Skalská, H., Freylich, V.: Web-bootstrap estimate of area under ROC curve. Aust. J. Stat. **35**, 325–330 (2006)
41. Smets, P.: Belief Functions Versus Probability Functions. Uncertainty and Intelligent Systems, pp. 17–24. Springer (1988)
42. Smolen, J.S., Braun, J., Dougados, M., Emery, P., FitzGerald, O., Helliwell, P., Kavanaugh, A., Kvien, T.K., Landewé, R., Luger, T., et al.: Treating spondyloarthritis, including ankylosing spondylitis and psoriatic arthritis, to target: recommendations of an international task force. Ann. Rheum. Dis. **73**(1), 6–16 (2014)
43. Šušteršič, O., Rajkovič, U., Dinevski, D., Jereb, E., Rajkovic, V.: Evaluating patients health using a hierarchical multi-attribute decision model. J. Int. Med. Res. **37**(5) 1646–1654 (2009)
44. Winston, W.L., Goldberg, J.B.: Operations Research: Applications and Algorithms, vol. 3. Duxbury Press Boston (2004)
45. Yang, J., Deng, M., Xu, D.: Nonlinear regression to estimate both weights and utilities via eviden-tial reasoning for MADM. In: Proceedings of 5th International Conference Optimization: Techniques and Applications, pp. 15–17 (2001)
46. Yang, J.-B.: Rule and utility based evidential reasoning approach for multiattribute decision analysis under uncertainties. Eur. J. Oper. Res. **131**(1), 31–61 (2001)
47. Yang, J.-B., Liu, J., Wang, J., Sii, H.-S., Wang, H.-W.: Belief rule-base inference methodology using the evidential reasoning approach-RIMER. IEEE Trans. Syst. Man Cybern. Part A: Syst. Hum. **36**(2), 266–285 (2006)
48. Yang, J.-B., Liu, J., Xu, D.-L., Wang, J., Wang, H.: Optimization models for training belief-rule-based systems. IEEE Trans. Syst. Man Cybern. Part A: Syst. Hum. **37**(4), 569–585 (2007)
49. Yang, J.-B., Sen, P.: A general multi-level evaluation process for hybrid MADM with uncertainty. IEEE Trans. Syst. Man Cybern. **24**(10), 1458–1473 (1994)
50. Yang, J.-B., Sen, P.: Multiple attribute design evaluation of complex engineering products using the evidential reasoning approach. J. Eng. Des. **8**(3), 211–230 (1997)
51. Yang, J.-B., Singh, M.G.: An evidential reasoning approach for multiple-attribute decision making with uncertainty. IEEE Trans. Syst. Man Cybern. **24**(1), 1–18 (1994)
52. Yang, J.-B., Wang, Y.-M., Xu, D.-L., Chin, K.-S.: The evidential reasoning approach for MADA under both probabilistic and fuzzy uncertainties. Eur. J. Oper. Res. **171**(1), 309–343 (2006)
53. Yang, J.-B., Xu, D.-L.: Nonlinear information aggregation via evidential reasoning in multiattribute decision analysis under uncertainty. IEEE Trans. Syst. Man Cybern. Part A: Syst. Hum. **32**(3), 376–393 (2002)

54. Yang, J.-B., Xu, D.-L.: On the evidential reasoning algorithm for multiple attribute decision analysis under uncertainty. IEEE Trans. Syst. Man Cybern. Part A: Syst. Hum. **32**(3), 289–304 (2002)
55. Zadeh, L.A.: Fuzzy sets. Inf. Control **8**(3), 338–353 (1965)
56. Zadeh, L.A.: Soft computing and fuzzy logic. IEEE Softw. **11** (1994)
57. Zadeh, L.A.: The Concept of a Linguistic Variable and its Application to Approximate Reasoning. Springer (1974)
58. Zimmermann, H.-J.: Fuzzy set theory. Wiley Interdisc. Rev.: Comput. Stat. **2**(3), 317–332 (2010)
59. Zimmermann, H.-J.: Fuzzy Set Theory and its Applications. Springer (2001)
60. Zochling, J.: Measures of symptoms and disease status in ankylosing spondylitis: ankylosing spondylitis disease activity score (ASDAS), ankylosing spondylitis quality of life scale (ASQoL), bath anky-losing spondylitis disease activity index (BASDAI), bath ankylosing spondylitis functional index (BASFI), bath ankylosing spondylitis global score (BAS-G), bath ankylosing spondylitis metrology index (BASMI), dougados functional index (DFI), and health assessment questionnaire for the spondylarthropathies (HAQ-S). Arthritis Care Res. **63**(S11), S47–S58 (2011)

Rescue System with Sensor Network for Vital Sign Monitoring and Rescue Simulations by Taking into Account Triage with Measured Vital Signs

Kohei Arai

Abstract Rescue system with sensor network for vital sign (Body Temperature, Heart Rate, Blood Pressure, Bless Rate and Consciousness) monitoring is proposed together with rescue simulations with consideration of triage by using the measured vital sign. Triage is a key for evacuation from disaster areas. Triage can be done with the gathered physical and psychological data with the sensor network for vital sign monitoring. Through a comparison between with and without consideration of triage, it is found that the time required for evacuation from disaster areas with consideration triage is 30 % less than that without triage.

Keywords Vital sign · Triage · Rescue simulation · Sensor network

1 Introduction

There are four major vital signs, body temperature pulse rate, blood pressure, and bless. In the emergency rescue point of view, fifth vital signal, JCS (Japan Coma Scale) or GCS (Glasgow Coma Scale) is added. The proposed physical and psychological health monitoring system is intended to monitor these five major vital signs for a triage in emergency rescue in particular for disable, elderly, paralyzed persons who need nursing, elderly care, and help for evacuation from disaster area when it occurs.

In emergency rescue processes, victims' location and attitude is important. Therefore, GPS receiver and WiFi beacon receiver as well as accelerometer are added to the aforementioned measuring sensors for body temperature pulse rate, blood pressure, bless rate, and Electroencephalogram: EEG, Electromyography: EMG for JCS and GCS status (psychological) monitoring. GPS receiver can be used for outdoor location determination while WiFi beacon can be utilized for indoor location estimation.

K. Arai (✉)
Graduate School of Science and Engineering, Saga University,
1 Honjo, Saga 840-8502, Japan
e-mail: arai@is.saga-u.ac.jp

© Springer International Publishing Switzerland 2016
L. Chen et al. (eds.), *Emerging Trends and Advanced Technologies for Computational Intelligence*, Studies in Computational Intelligence 647,
DOI 10.1007/978-3-319-33353-3_3

All sensors should be wearable and can be attached to ones' tall forehead. Attitude, the number of steps, calorie consumption etc. are also measured with smart phone, i-phone and the acquired sensor data are transmitted to smart phone and i-phone through Bluetooth like communication links. Through WiFi network or wireless LAN connection, acquired data can be collected in the designated information collection center. Then acquired data can be referred by the designated manager for triage (prioritize the victims in concern for rescue) and then the manager gives information of the specified victims who has to be rescue to the most appropriate rescue persons (take into account the distance between victims and rescue person, triage information road network topology, traffic condition etc.).

There are previously proposed methods and systems which allow physical health monitoring [1–5]. Most of previous methods and systems are not wearable and do not allow psychological status monitoring. The proposed physical and psychological health monitoring system is intended to monitor these five major vital signs. Instead of direct blood pressure measurement, indirect blood pressure measurement is proposed by using a created regressive equation with the measured body temperature, heart rate and the number of steps because it is hard to measure the blood pressure directly. Also, consciousness can be monitored by using acquired eye images and its surroundings on behalf of using EEG sensors, because EEG signals are used to be suffered from noises.

There are previously proposed evacuation and rescue methods and systems [6–8]. It may be possible to find that multi agent-based simulation makes it possible to simulate the human activities in rescue and evacuation process [9, 10]. A multi agent-based model is composed of individual units, situated in an explicit space, and provided with their own attributes and rules [11]. This model is particularly suitable for modeling human behaviors, as human characteristics can be presented as agent behaviors. Therefore, the multi agent-based model is widely used for rescue and evacuation simulation [9–13].

In this study, GIS map is used to model objects such as road, building, human, fire with various properties to describe the objects condition. With the help of GIS data, it enables the disaster space to be closer to a real situation [13–18]. Kisko et al. [17] employs a flow based model to simulate the physical environment as a network of nodes [17]. The physical structures, such as rooms, stairs, lobbies, and hallways are represented as nodes which are connected to comprise a evacuation space. This approach allows viewing the movement of evacuees as a continuous flow, not as an aggregate of persons varying in physical abilities, individual dispositions and direction of movement [19]. Gregor and Nagel [18] presents a large scale microscopic evacuation simulation [18]. Each evacuee is modeled as an individual agent that optimizes its personal evacuation route. The objective is a Nash equilibrium, where every agent attempts to find a route that is optimal for the agent [20]. Fahy [19, 21] proposes an agent based model for evacuation simulation [21]. This model allows taking in account the social interaction and emergent group response. The travel time is a function of density and speed within a constructed network of nodes and arcs [22, 23]. Gobelbecker and Dornhege [20] presents a method to acquire GIS data to design a large scale disaster simulation environment [20]. The GIS data is retrieved publicly through the website OpenStreetMap.org. The data is then converted to the Robocup Rescue Simulation system format, enabling a simulation on a real world scenario [24]. Sato and Takahashi [8] also proposed a method to create realistic maps using the open GIS data. The experiment shows the differences between two types of maps: the map generated from the program and the map created from the real data [25]. Ren et al. [9] presents an agent-based modeling and simulation using Repast software to construct crowd evacuation for emergency response for an area under a fire. Characteristics of the people are modeled and tested by iterative simulation. The simulation results demonstrate the effect of various parameters of agents [26]. Cole et al. [12] studied on GIS agent-based technology for emergency simulation. This research discusses about the simulation of crowding, panic and disaster management [27]. Quang et al. [15] proposes the approach of multi-agent-based simulation based on participatory design and interactive learning with experts' preferences for rescue simulation [15]. Hunsberger and Grosz [24], Beatriz et al. [25] and Chan and Leung [26] apply the auction mechanism to solve the task allocation problem in rescue decision making. Christensen and Sasaki [28] presents the BUMMPEE model, an agent-based simulation capable of simulating a heterogeneous population according to variation in individual criteria. This method allows simulating the behaviors of people with disabilities in emergency situation [28]. Based on the agent based method, a new rescue system is proposed by considering triage in rescue processes [29–33].

A rescue model for people with disabilities in large scale environment is proposed. The proposed rescue model provides some specific functions to help disabled people effectively when emergency situation occurs. Important components of an evacuation plan are the ability to receive critical information about an emergency, how to respond to an emergency, and where to go to receive assistance. Triage is a key for rescue procedure. Triage can be done with the gathered physical and psychological data which are measured with a sensor network for vital sign monitoring. Through a comparison between with and without consideration of triage, it may be possible to find that the time required for evacuation from disaster areas with consideration triage is less than that without triage.

The following section describes the proposed rescue system followed by rescue simulation model. Then simulation results with and without consideration of triage are shown. Finally, conclusion is described together with some discussions.

2 Proposed Rescue System and Decision Making Method

2.1 Rescue System

Figure 1 shows the entire system configuration of the proposed rescue system for disabled and elderly person together with physical and psychological health monitoring system.

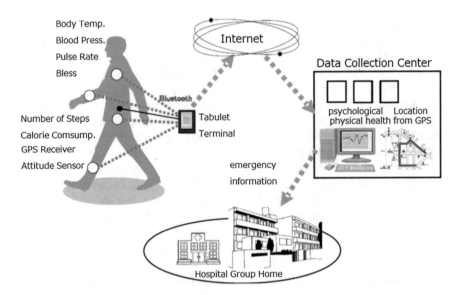

Fig. 1 Entire system configuration of the proposed wearable physical and psychological health monitoring system together with rescue system

Normal person who may rescue disabled and elderly person are situated somewhere around their district. The locations of disabled and elderly person as well as normal person are known with GPS receivers of which they have in their mobile devices. Also, their attitudes are known with their attitude sensors which are included in their mobile devices. On the other hand, wearable physical and psychological health sensors are attached to the proposed Arai's glass. These sensor data are transmitted to their mobile devices through Bluetooth communication links. The acquired sensor data together with location and attitude data are transmitted to the Health Data Collection Center: HDCC through WiFi communication links. Using these gathered data of each disabled and elderly person as well as normal person, data collection center makes decisions on who (normal person) rescues whom (disabled and elderly person) taking into account physical and psychological conditions and traffic condition which can be collected with the different method and system.

2.2   Rescue Model

Important components of an evacuation plan are the ability to receive critical information about an emergency, how to respond to an emergency, and where to go to receive assistance. I propose a wearable device which is attached to body of disabled people. This device measures the condition of the disabled persons such as their heart rate, body temperature and attitude; the device can also be used to trace the location of the disabled persons by GPS. The information will be sent to emergency center automatically. The emergency center will then collect the information together with information from rescue persons to assign which rescue person should help which disabled persons.

The centralized rescue model presented has three types of agents: rescue persons, disabled people and route network. The route network is also considered as an agent because the condition of traffic in a certain route can be changed when a disaster occurs.

Before starting the simulation, every agent has to be connected to the emergency center in order to send and receive information. The types of data exchanged between agents and emergency center are listed as below.

Message from agent

A1 To request for connection to the emergency center
A2 To acknowledge the connection
A3 Inform the movement to another position
A4 Inform the rescue action for victim
A5 Inform the load action for victim
A6 Inform the unload action for victim
A7 Inform the inactive status

Message from emergency center

K1 To confirm the success of the connection
K2 To confirm the failure of the connection
K3 To send decisive information

Before starting the simulation, every agent will send the command A1 to request for connection to the emergency center. The emergency center will return the response with a command K1 or K2 corresponding to the success or failure of their connection respectively. If the connection is established, the agent will send the command A2 to acknowledge the connection. After the initial process, all the connected agents will receive the decisive information such as the location of agents and health level via command K3; after that the rescue agents will make a decision of action and submit to the center using one of the commands from A3 to A7. At every cycle in the simulation, each rescue agent receives a command K3 as its own decisive information from the center, and then submits back an action command. The status of disaster space is sent to the viewer for visualization of simulation.

The disaster area is modeled as a collection of objects: Nodes, Buildings, Roads, and Humans. Each object has properties such as its positions, shape and is identified by a unique ID. Tables 1, 2, 3, 4, 5, 6 and 7 presents the properties of Nodes, Buildings, Roads and Humans object respectively. These properties are derived from RoboCup rescue platform with some modifications.

Table 1 Properties of node object

Property	Unit	Description
x, y		The x-y coordinate
Edges	ID	The connected roads and buildings

Table 2 Properties of building object

Property	Description
x, y	The x-y coordinate of the representative point
Entrances	Node connecting buildings and roads

Table 3 Properties of road object

Property	Unit	Description
StartPoint and EndPoint	[ID]	Point to enter the road. It must be the node or a building
Length and width	[mm]	Length and width of the road
Lane	[Line]	Number of traffic lanes
BlockedLane	[Line]	Number of blocked traffic lanes
ClearCost	[Cycle]	The cost required for clearing the block

Table 4 Properties of victim agent

Property	Unit	Description
Position	ID	An object that the victim is on
Position-In-Road	[mm]	A length from the Start-Point of road when the victim is on a road, otherwise it is zero
Health-Level	[health point]	Health level of victim. The victim dies when this becomes zero
Damage-Point	[health point]	Health level dwindles by Damage-Point in every cycle. Damage-Point becomes zero immediately after the victim arrives at a shelter
Disability-Type	Type [1...7]	Type of disability which is listed in Table 7
Disability-Level	[low/high]	Victim who has high disability level, will have higher Damage Point

Table 5 Properties of rescue person agent

Property	Unit	Description
Position	ID	An object that the rescue person is on
Position-In-Road	[mm]	A length from the Start-Point of road when the humanoid is on a road, otherwise it is zero
Current-Action	Type [1...3]	One of action listed in Table 7
Energy	Level [1...5]	Amount of gasoline in vehicle
Panic-Level	Level [0...9]	Shows the hesitance level of decision

Table 6 Action of rescue person agent

Action ID	Action	Description
1	Stationary	Rescue person stays still
2	Move-To-Victim	Rescue person go to location of victims
3	Move-To-Shelter	Rescue person carry victim to shelter

Table 7 Type of disability

Type	Description
1	Cognitive impairment
2	Dexterity impairment (Arms/hands/fingers)
3	Mobility impairment
4	Elderly
5	Hearing impairment
6	Speech and language impairment
7	Visual impairment

2.3   Decision Making Method

The decision making of rescue persons to help disabled persons can be treated as a task allocation problem [22–26]. The central agents carry out the task allocation for the rescue scenario. The task of rescue persons is to help disabled persons. A combinatorial auction mechanism is used to solve this task allocation problem. At this model, the rescue persons are the bidders; the disabled persons are the items; and the emergency center is the auctioneer. The distance and health level of each disabled persons are used as the costs for the bids. When the rescue process starts, the emergency center creates a list of victims, sets the initial distance for victims, and broadcasts the information to all the rescue person agents. Only the rescue person agents whose distance to victims is less than the initial distance will help these victims. Each rescue person agent will only help the victims within the initial distance instead of helping all the victims. The initial distance will help rescue persons in reducing the number of tasks that they have to do so that the decision making will be faster. The aim of this task allocation model is to minimize the evacuation time or the total cost to accomplish all tasks. In this case, the cost is the total rescue time.

2.3.1   The Criteria to Choose Disabled People

The rescue person's decision depends on the information of disabled people which receives from emergency center; therefore decisions must follow certain criteria to improve their relief activities. For example, the rescue persons must care about condition of disabled people; the more seriously injured people should have the

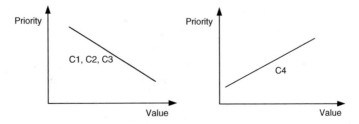

Fig. 2 Priority in the rescue person's decision

more priority even if they locate further than the others. There are several criteria that rescue persons should take in account before starting rescue process [15].

C1 Distance from rescue person to disabled people
C2 Distance from disabled people to nearest other disabled people
C3 Health level of disabled people
C4 Distance from disabled people to nearest other rescue person

Disabled people who have lesser values for criteria of C1, C2, C3 and greater values for criteria of C4 will have higher priority in the rescue person's decision process as shown in Fig. 2.

2.3.2   Determination Important light of Criteria

Referring to the decision making method which presents at [23], a programming with C# had been created with the following inputs: number of criteria (N=4); size of population (M=30); crossover probability (Pcross=90%), mutation probability (Pmut=10%), the number of reproduction (L=100); the pair-wise comparison among criteria is shown in Table 8. The solution obtained is w=(0.2882, 0.2219, 0.2738, 0.2161). These values are also considered as input parameters for rescue simulation. It can be changed by adjusting the pair-wise comparison in Table 8. For each rescue person, the cost to help certain victim is shown in Eq. (1).

$$C(v)_k = \sum_{i=1}^{4} w_i * v_i^k \tag{1}$$

where: w_i denotes the weight of the c_i criteria while v_i^k denotes the value of the ith criteria for the kth victim. The sign of value of criterion c_4 will be reversed when calculate the cost.

Table 8 Pair-wise comparison among criteria

Criterion	Linguistic preference	Fuzzy number	Criterion
C1	Good	(0.667, 0.833, 1)	C2
C1	Fair	(0.333, 0.5, 0.667)	C3
C1	Good	(0.667, 0.833, 1)	C4
C2	Poor	(0, 0.167, 0.333)	C3
C2	Fair	(0.333, 0.5, 0.667)	C4
C3	Good	(0.667, 0.833, 1)	C4

2.3.3 Forming Task Allocation Problem

Given the set of n rescue persons as bidders: $V = \{v_1, v_2, ..., v_n\}$ and set of m disabled persons considered as m tasks: $D = \{d_1, d_2, ..., d_m\}$. The distances from rescue persons to disabled persons; distances among disabled persons and health level of disabled persons and are formulated as follow:

$M[v_i, d_j]_t = \{mij|mij$: distances from rescue person v_i to victim d_j at time step t$\}$
$N[d_i, d_j] = \{n_{ij}|n_{ij}$: distances from victim d_i to another victim $d_j\}$
$H[d_i]_t = \{h_i|h_i$: health level of victim d_i at time step t: $h_{low} < h_i < h_{high}\}$

With the initial distance L. The normalization processes are shown in Eqs. (2)–(4).

Normalize $M[v_i, d_j]_t$: $M'[v_i, d_{j|t}] = \{m'_{ij}|m'_{ij} = [(1-0)(m_{ij} - 0)/(L - 0) + 0\}; 1 < i < n, 1 < j < m]\}$ (2)

Normalize $N[d_i, dj]_t = \{n_j^{'i}|n'_{ij} = (1-0)(m_{ij} - 0) + 1\}; 1 < i < n, 1 < j < m]\}$ (3)

Normarize $H[d_i]_t$; $H'[v_i, d_j]_t = \{h'_i|h'_{ij} = [(1-0)(h_i - h_{low}) + 0]; 1 < i < m\}$ (4)

The $B_id_{vi}(\{d_j, d_g, ...d_k, d_k\}, C)$ means that the rescue person will help victims $\{d_j, d_g, ... d_k, d_k\}$ with the total cost C. Total cost C is calculated by Eq. (1).

Let I is a collection of subsets of D. Let $x_{ij} = 1$ if the jth set in I is a winning bid and c_j is the cost of that bid. Also, let be $a_{ij} = 1$ if the jth set in I contains $i \in D$. The problem can then be stated in Eq. (5) [27].

$$\min \Sigma_{j \in I} c_j x_j, \text{ with constant } \Sigma_{j \in I} a_j x_j < 1, \forall 1 \in D \quad (5)$$

The constraint will make sure that each victim is helped by at most one rescue person at certain time step. For example, let's assume that rescue person A has the information of five victims, $\{d_1, d_2, ... d_4, d_5\}$. The initial distance is set to 200 m. The rescue person estimates the distance from him to each victims and selects only the victims who are not more than 200 m from his location. Assume that, the victim d_1 and victim d_2 are selected to help with the cost of 1.15. The bid submitted to the center agent is $B_i d_A = \{(d_1, d_2), 1.15\}$.

This optimization problem can be solved by Heuristic Search method of Branch-on-items [27]. This method is based on the question: "Which rescue person should this victim be assigned to?". The nodes in the search tree are the bids.

2.3.4 Example of Task Allocation Problem

To illustrate an example of a task allocation of rescue persons to help disabled persons, let's assume that there are five rescue persons and three disabled persons; The initial distance L is set to 200 m; $h_{low} = 100$; $h_{high} = 500$. The path in the search tree consists of a sequence of disjoint bids. Each node in the search tree expands the new node with the smallest index among the items that are still available, not including the items that have already been used on the path. The solution is a path, which has minimum

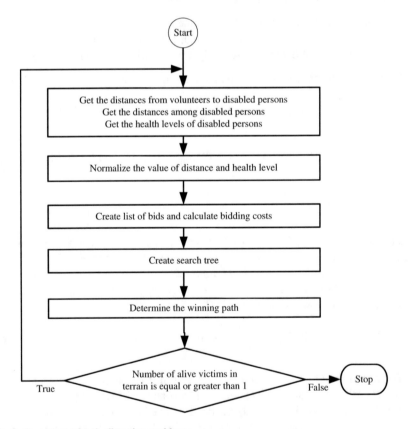

Fig. 3 Procedure of task allocation problem

cost in the search tree. Figure 3 shows the procedure of task allocation problem for helping disabled persons in emergency situation.

With initial distance 200, the rescue person v_i can help only victim d_g. The bid is formed as $B_{vi}(\{d_g\}, C)$. There are five criteria with important weigh: $w = (0.2882, 0.2219, 0.2738, 0.2161)$.

The cost C is calculated as below,

Distance from rescue person v_i to disabled people $d_g = 0.25$

Distance from disabled people d_g to nearest other disabled people $d_2 = 0.35$

Health condition of disabled people $d_g = 0.5$

The possible bids with costs are shown in Table 9.

The bid b2 and b7 have the same task {d1}; b5 and b8 have the same task {d2}; b1, b3 and b6 have the same task {b3}. The more expensive bids will be removed as shown in Table 10.

Then, the search tree is formed as shown in Fig. 4. The winner path is b2, b3, b1 which has the most minimum cost of 0.65. The task allocation solution: rescue person v2 will help disabled persons d1; rescue person v4 will help disabled person d2.rescue person v1 will help disabled person d3.

Table 9 Possible bids with costs

Bid	Rescue person	Disabled person	Cost
b_1	v_1	{d3}	0.18
b_2	v_2	{d3}	0.33
b_3	v_2	{d3}	0.30
b_4	v_2	{d1, d3}	0.74
b_5	v_3	{d2}	0.21
b_6	v_3	{d3}	0.38
b_7	v_4	{d1}	0.33
b_8	v_4	{d2}	0.14
b_9	v_4	{d1, d2}	0.55

Table 10 Tasks allocation and cost after removal of more expensive bids

Bid	Rescue person	Disabled person	Cost
b_1	v_1	{d3}	0.18
b_2	v_2	{d3}	0.33
b_4	v_2	{d1, d3}	0.74
b_8	v_4	{d2}	0.14
b_9	v_4	{d1, d3}	0.55

Fig. 4 Branch on items
based search tree

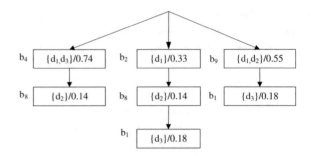

2.4 Path Finding in Gama Simulation Platform

After a rescue person makes the decision to help a certain victim, the path finding algorithm is used to find the route from rescue person agent to victim agent. The GIS data presents roads as a line network in graph type. The Dijkstra algorithm is implemented for the shortest path computation [8].

3 Experimental Results

In this section, I present experimental studies on different scenarios. I show the experimental results with traditional rescue model which not considering the updated information of victims and volunteers such as health conditions, locations, traffic conditions. The traditional rescue model provides fixed mission for which volunteers should help which victims. Whereas, my rescue model provides flexible mission for which volunteers should help which victims. The targets of volunteers can be changed dynamically according to current situation. The experimental results of my proposed rescue model are also presented to show the advantages comparing to traditional model.

The evacuation time is evaluated from the time at which the first volunteer started moving till the time at which all saved victims arrive at the shelters. The simulation model is tested using the Gama simulation platform [8, 10].

3.1 Preliminary Experimental Setting

The number of volunteers, the number of disabled persons, panic level of volunteer, disability level of victim and the complexity of traffic as parameters to examine the correlation between these parameters with rescue time is taken into account. The traffic complexity is function of the number of nodes and links in a road network.

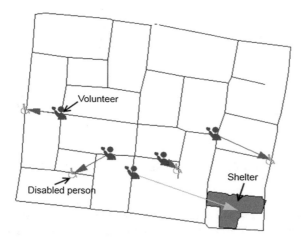

Fig. 5 Sample GIS map of disaster space

Figure 5 presents the sample GIS map consisting of four layers: road, volunteer, disabled person and shelter.

The initial health levels of disabled persons are generated randomly between 100 and 500 point. If the health level is equal or less than zero, the corresponding agent is considered as dead.

3.2 Experimental Results

3.2.1 Comparison with Traditional Model

The traditional model proposes the rescue process without knowing the updated information of victims and volunteers. In disaster space, the traffic condition is changed dynamically. Some road links can be inaccessible. My proposed method provides the updated traffic condition so that the path finding method can work effectively. The road map with 50 links is used to conduct the test with traditional model and my proposed model. The result is shown in Table 11.

Table 11 Comparison with traditional model

	Volunteer	Victim	Link	Rescue time	Dead victim
Proposed model	10	10	50	880	0
Traditional model	10	10	50	1300	1

Table 12 Rescue time and number of links

Volunteer	Victim	Link	Rescue time	Dead victim
10	10	50	880	0
		40	880	0
		30	1150	0
		20	1150	0

3.2.2 Simulation Result with Consideration of Complexity of Road Network

The correlation between the complexity of road network and the rescue time is concerned. The area of disaster space, the number of victims and volunteers, locations of victims and volunteer are not changed. The complexity of road network presents as the number of road links. The result is shown in Table 12.

The rescue time increase if number of link decrease at the same area of disaster space.

3.2.3 Simulation Result with Consideration of Panic Level of Volunteer

When emergency situation occurs, the volunteers are also getting panic. The panic probability of volunteers can be presented as the hesitance of volunteers in making decision to help disabled persons. In the simulation, There are assumed 10 levels of hesitance: 0, 0.1, 0.2, 0.3, 0.4, 0.5, 0.6, 0.7, 0.8, and 0.9. The hesitance level 0 means that there is no hesitance of making decision. These hesitance levels will be assigned to every volunteer agent. At each time step of simulation, a random value x $(0 < x < 1)$ will be generated. If x is equal or greater than hesitance level of volunteers, the corresponding volunteers will make decision to help disabled person; otherwise the volunteers will postpone the decision at this time step.

A sample GIS road map with 50 links is assumed. The correlation between panic probability of volunteer and rescue time is shown in Table 13.

3.2.4 Simulation Result with Consideration of Block Lane Percentage of Road Network

When emergency situation occurs, the road lane may block. The road is set as inaccessible condition if its number of block lanes is equal to the number of lanes. In the simulation, I assume that there are 10 levels of block road: 0, 0.1, 0.2, 0.3, 0.4, 0.5, 0.6, 0.7, 0.8, and 0.9. The correlation between the percentage of block road link, the rescue time and the number of dead victim is shown in Table 14.

Table 13 Simulation results with consideration of panic probability of volunteer

Volunteer	Victim	Link	Panic level	Rescue time	Dead victim
10	10	50	0	880	0
			0.1	900	0
			0.2	1150	0
			0.3	1150	0
			0.4	1500	1
			0.5	1750	1
			0.6	1760	1
			0.7	2200	1
			0.8	2600	2
			0.9	3250	3

Table 14 Simulation result with consideration of block lane percentage of road network

Volunteer	Victim	Link	Percentage of block road	Rescue time	Dead victim
10	10	50	0.1	950	0
			0.2	1000	0
			0.3	1050	0
			0.4	1200	0
			0.5	1325	1
			0.6	1700	1
			0.7	2300	1
			0.8	2700	3
			0.9	3450	4

3.2.5 Simulation Result with Consideration of Disability Level of Victim

The disability level of victim may affect to the rescue process. In order to facilitate the simulation, I assume that there are two level of disability: low and high. The victims, who have higher level of disability, will reduce the health level faster than victims, who have lower disability level. The correlation between the percentage of disability level of victim, the rescue time and number of dead victim is shown in Table 15.

3.2.6 Simulation Result with Consideration of Disconnectivity Between Agent and Emergency Center

In reality, when emergency situation occurs, the communication among objects may have problem. This problem of communication will affect to the rescue process. In

Table 15 Simulation result with consideration of percentage of high disability level

Volunteer	Victim	Link	Percentage of high disability level	Rescue time	Dead victim
10	10	50	0.1	880	0
			0.2	890	0
			0.3	1050	0
			0.4	1050	1
			0.5	1125	1
			0.6	1250	2
			0.7	1325	3
			0.8	1450	4
			0.9	1550	5

Table 16 Simulation result with consideration of disconnectivity between agent and emergency center

Volunteer	Victim	Link	Disconnectivity	Rescue time	Dead victim
10	10	50	10	960	0
			20	980	0
			30	1000	0
			40	1020	0
			50	1050	0
			60	1060	0
			70	1075	0
			80	1085	0
			90	1115	1

simulation, the disconnectivity by postponing the decision of volunteer for certain number of time steps is then taken into account. The result is shown in Table 16.

3.3 Realistic Experimental Setting

The realistic situation which is shown in Fig. 6 is simulated referring to the situation of Saga Prefecture in Japan. Wheel chairs denote victims while green colored facilities are shelters. Dark blue colored persons are rescue persons. After the decision making, rescue persons move to the assigned victims and to the nearest shelter along with the red arrows. The number of victims is 20 while the number of rescue persons is 10. The number of road links is 250. Panic level, Road block probability, Disability level are set at 0–1.0 with 0.1 step and Road disconnectivity is set at 0–90 with 10 step. The averaged rescue time for the conventional method is 13000 unit time steps while that for the proposed method with consideration of triage is 9000 unit time

Fig. 6 Realistic rescue simulations referring to the situation of Saga Prefecture in Japan

step. Therefore, the proposed rescue system is superior to the conventional rescue system without triage by around 30 %.

4   Conclusion

Vital sign can be monitored every sec. (Frequent measurements of body temp., heart rate, consciousness, no. of steps, etc.). Blood pressure can be monitored with the regressive equation of body temperature, heart rate and the number of steps, Consciousness can be monitored with eye camera (Psychological status: close, calm/irritation, etc.). Attitude (walk, lie down, stand up, sit down, etc.) and location can be monitored with smart phone together with vital sign.

The decisions to help victims are based on updated information from victims and volunteers therefore it can be change to adapt the current emergency situation. I also conduct the rescue simulation with considering the complexity of road network, the panic level of volunteers, the disability level of victims and the disconnectivity between agent and emergency center. The simulation results show that my model has less rescue time than traditional model which applies static decision making method. Triage can be done with the proposed system. With the triage, rescue time can be reduced by 30 % for typical realistic situations.

Acknowledgments The author would like to thank all the patients who are contributed to the experiments conducted. The author also would like to thank Professor Dr. Takao Hotokebuchi, President of Saga University for his support this research works. Also, author would like to thank Dr. Tran Sang Xuan of Vinh University, Vietnam for his effort to this research.

References

1. Arai, K.: WeIarable healthy monitoring sensor network and its application to evacuation and rescue information server system for disabled and elderly person. Int. J. Res. Rev. Comput. Sci. **3**(3), 1633–1639 (2012)
2. Arai, Kohei: Wearable computing system with input output devices based on eye-based Human Computer Interaction: HCI allowing location based Ib services. Int. J. Adv. Res. Artif. Intell. **2**(8), 34–39 (2013)
3. Arai, Kohei: Vital sign and location/attitude monitoring with sensor networks for the proposed rescue system for disabled and elderly persons who need a help in evacuation from disaster areas. Int. J. Adv. Res. Artif. Intell. **3**(1), 24–33 (2014)
4. Arai, Kohei: Method and system for human action detection with acceleration sensors for the proposed rescue system for disabled and elderly persons who need a help in evacuation from disaster areas. Int. J. Adv. Res. Artif. Intell. **3**(1), 34–40 (2014)
5. Arai, Kohei: Frequent physical health monitoring as vital sign with psychological status monitoring for search and rescue of handicapped, disabled and elderly persons. Int. J. Adv. Res. Artif. Intell. **2**(11), 25–31 (2013)
6. Kaprzy, Edt., Arai, K.: Rescue System for Elderly and Disabled Persons Using Iarable Physical and Psychological Monitoring System, Studies in Computer Intelligence, vol. 542, pp. 45–64. Springer (2014)
7. Obelbecker, G., Dornhege, M.: Realistic cities in simulated environments—an open street map to Robocup rescue converter. In: Online-Proceedings of the Fmyth International Workshop on Synthetic Simulation and Robotics to Mitigate Earthquake Disaster (2009)
8. Sato, K., Takahashi, T.: A Study of Map Data Influence on Disaster and Rescue Simulation's Results. Computational Intelligence Series, vol. 325. Springer, Berlin (2011)
9. Ren, C., Yang, C., Jin, S.: Agent-Based Modeling and Simulation on Emergency, Complex 2009. Part II, LNICST **5**, 1451–1461 (2009)
10. Zaharia, M.H., Leon, F., Pal, C., Pagu, G.: Agent-based simulation of crowd evacuation behavior. Int. Conf. Autom. Control Model. Simul. **529–533** (2011)
11. Quang, C.T., Drogoul, A.: Agent-based simulation: definition, applications and perspectives. In: Invited Talk for the biannual Conference of the Faculty of Computer Science, Mathematics and Mechanics (2008)
12. Cole, J.W., Sabel, C.E., Blumenthal, E., Finnis, K., Dantas, A., Barnard, S., Johnston, D.M.: GIS-based emergency and evacuation planning for volcanic hazards in New Zealand. Bull. NZ Soc. Earthq. Eng. **38**(3) (2005)
13. Batty, M.: Agent-Based Technologies and GIS: Simulating Crowding, Panic, and Disaster Management. Frontiers of Geographic Information Technology, vol. 4, pp. 81–101 (2005)
14. Patrick, T., Drogoul, A.: From GIS Data to GIS Agents Modeling with the GAMA Simulation Platform. TF SIM (2010)
15. Quang, C.T., Drogoul, A., Boucher, A.: Interactive learning of independent experts. Criteria for rescue simulations. J. Univers. Comput. Sci. **15**(13), 2701–2725 (2009)
16. Taillandier, T., Vo, D.A., Ammyoux, E., Drogoul, A.: GAMA: a simulation platform that integrates geographical information data, agentbased modeling and multi-scale control. In: Proceedings of Principles and practice of multi-agent systems, India (2012)
17. Kisko, T.M., Francis, R.L., Nobel, C.R.: EVACNET4 User's Guide. University of Florida (1998)
18. Gregor, M.R., Nagel, K.: Large scale microscopic evacuation simulation. Pedestrian Evacuation Dyn. 547–553 (2008)
19. Fahy, R.F.: EXIT89—high-rise evacuation model—recent enhancements and example applications. In: Interflam '96, International Interflam Conference—7th Proceedings, pp. 1001–1005. Cambridge, England (1996)
20. Gobelbecker, M., Dornhege, C.: Realistic cities in simulated environments- an open street map to RoboCup Rescue converter. In: 4th International Workshop on Synthetic Simulation and Robotics to Mitigate Earthquake Disaster (SRMED 2009), Graz, Austria (2009)

21. Fahy, R.F.: User's Manual, EXIT89 v 1.01, An Evacuation Model for High-Rise Buildings, National Fire Protection Association. Quincy, Mass (1999)
22. Nair, R., Ito, T., Tambe, M., Marsella, S.: Task allocation in the rescue simulation domain: a short note, Volume 2377 of Lecture Notes in Computer Science, vol. 751–754. Springer, Berlin (2002)
23. Boffo, F., Ferreira, P.R., Bazzan, A.L.: A comparison of algorithms for task allocation in robocup rescue. In: Proceedings of the 5th European Workshop on Multiagent Systems, pp. 537–548 (2007)
24. Hunsberger, L., Grosz, B.: A combinatorial auction for collaborative planning. In: Proceedings of the Fourth International Conference on Multiagent Systems (2000)
25. Beatriz, L., Silvia, S., Josep, L.: Allocation in rescue operations using combinatorial auctions. Artif. Intell. Res. Dev. **100**, 233–243 (2003)
26. Chan, C.K., Leung, H.F.: Multi-auction approach for solving task allocation problem. Lect. Notes Comput. Sci. **4078**, 240–254 (2005)
27. Sandholm, T.: Algorithm for optimal winner determination in combinatorial auctions. Artif. Intell. **135**, 1–54 (2002)
28. Christensen, K.M., Sasaki, Y.: Agent-based emergency evacuation simulation with individuals with disabilities in the population. J. Artif. Soc. Soc. Simul. **11**(3), 9 (2008)
29. Arai, K., Sang, T.X.: Fuzzy genetic algorithm for prioritization determination with technique for order preference by similarity to ideal solution. Int. J. Comput. Sci. Netw. Secur. **11**(5), 229–235 (2011). May
30. Arai, K., Sang, T.X.: Multi agent-based rescue simulation for disable persons with the help from rescue persons in emergency situations. In: Int. J. Res. Rev. Comput. Sci. **3**(2) (2012)
31. Arai, K., Sang, T.X.: Fuzzy genetic algorithm for prioritization determination with technique for order preference by similarity to ideal solution. Int. J. Comput. Sci. Netw. Secur. **11**(5), 229–235 (2011). May
32. Arai, K., Sang, T.X., Uyen, N.T.: Task allocation model for rescue disable persons in disaster area with help of rescue persons. Int. J. Adv. Comput. Sci. Appl. **3**(7), 96–101 (2012)
33. Arai, K., Sang, T.X.: Decision making and emergency communication system in rescue simulation for people with disabilities. Int. J. Adv. Res. Artif. Intell. **2**(3), 77–85 (2013)

An Approach for Detecting Traffic Events Using Social Media

Carlos Gutiérrez, Paulo Figueiras, Pedro Oliveira, Ruben Costa
and Ricardo Jardim-Goncalves

Abstract Nowadays almost everyone has access to mobile devices that offer better processing capabilities and access to new information and services, the Web is undoubtedly the best tool for sharing content, especially through social networks. Web content enhanced by mobile capabilities, enable the gathering and aggregation of information that can be useful for our everyday lives as, for example, in urban mobility where personalized real-time traffic information, can heavily influence users' travel habits, thus contributing for a better way of living. Current navigation systems fall short in several ways in order to satisfy the need to process and reason upon such volumes of data, namely, to accurately provide information about urban traffic in real-time and the possibility to personalize the information presented to users. The work presented here describes an approach to integrate, fuse and process tweet messages from traffic agencies, with the objective of detecting the geographical span of traffic events, such as accidents or road works. Tweet messages are considered in this work given their uniqueness, their real time nature, which may be used to quickly detect a traffic event, and their simplicity. We also address some imprecisions ranging from lack of geographical information, imprecise and ambiguous toponyms, overlaps and repetitions as well as visualization to our data set in the UK, and a qualitative study on the use of the approach using tweets in other languages, such as Greek. Finally, we present an application scenario, where traffic information is processed from tweets massages, triggering personalized notifications to users through Google

C. Gutiérrez · P. Figueiras · P. Oliveira · R. Costa (✉) · R. Jardim-Goncalves
Centre of Technology and Systems, Faculdade de Ciências e Tecnologia,
Universidade Nova de Lisboa, UNINOVA, Lisboa, Portugal
e-mail: rddc@uninova.pt

C. Gutiérrez
e-mail: csg@uninova.pt

P. Figueiras
e-mail: paf@uninova.pt

P. Oliveira
e-mail: pgo@uninova.pt

R. Jardim-Goncalves
e-mail: rg@uninova.pt

© Springer International Publishing Switzerland 2016
L. Chen et al. (eds.), *Emerging Trends and Advanced Technologies
for Computational Intelligence*, Studies in Computational Intelligence 647,
DOI 10.1007/978-3-319-33353-3_4

Cloud Messaging on Android smartphones. The work presented here is still part of on-going work. Results achieved so far do not address the final conclusions but form the basis for the formalization of a domain knowledge along with the urban mobility services.

Keywords Machine learning · Geo-parsing · Information retrieval · Social networks · Classification · Traffic events

1 Introduction

A few years ago information about events of any kind came almost exclusively from specialized information. With the increase of social networks' usage, information has become more decentralized; as each user become a consumer/provider of information. This paradigm becomes relevant when it is crucial to disseminate information in real-time about an event. A good example where social networks had a great impact in disseminating events in real-time and namely for disaster relief, was by the time of the Haiti earthquake in January 2010, when posts to Facebook, Flickr, YouTube and Twitter were numerous [1]. Research revealed that those messages sent during the Haiti earthquake that were considered "actionable" and more useful than the others were generally the ones that included locations [2]. Moreover, Twitterers themselves are more likely to pass along, or re-tweet, messages that include geo-location and situational updates [3], indicating that Twitterers find such messages important.

Location estimation using the content in social networks has its own challenges, namely the uncontrolled and the limited content, are two important drawbacks. In fact, one of the most important limitations in extracting useful information from a tweet is its 140-character limit. Additionally, users can make spelling mistakes, and use varying conventions and abbreviations; the quality of content is not as good as in news articles. In addition to the credibility of the content, the credibility of user profiles and their geo-references are questionable as well. Not all users have to provide their GPS locations, or their city of residence in their profiles. As a result, the most obvious problem is uncertainty and the lack of rich and reliable data. Moreover, tweet density depends heavily on the population and Twitter usage in a region. That means, if an event occurs in a small town, the number of Twitterers posting tweets about that event would be considerably lower than the number of users if the event would have occurred in a huge capital. Especially, once an event is broadcasted on the media or shared among people via phone calls or Internet, people from anywhere can post a tweet about it, making the location center more obscure to detect. Therefore, a practical problem to solve is how to normalize these tweets according to the user density of the area, and obtain a fair result to compare and evaluate tweets from different locations.

In the scope of this work, we focus on the analysis and detection of traffic events, where information disseminated by social networks can have an important role. Social networks can be seen as a mechanism which allows the detection of very small events, such as a damaged car in a side street. In addition to that, the interval between the occurrence of a traffic incident and the publication of a tweet about such incident usually tends to be much less, when compared to the time required for a news agency to broadcast about such particular event.

Extracting useful information from tweets presents some challenges, as described by the authors in [4, 5]. Taking into account that the information is completely unstructured, tweets can contain grammatical errors and abbreviations. Each user has its own style of writing, the information can be incomplete (e.g. a street name with no more information about city, etc.), false or not credible. More specifically, and to what this work is concerned, determining geographic interpretations for place names, or toponyms, involves resolving multiple types of ambiguity. From our point of view, and considering traffic related tweets, the relevance of a tweet message is directly proportional with the number of similar tweets posted by others in a short period of time. Moreover, in order to automatically recognize the relevance of tweet message for the purpose of detecting traffic events is a challenging task (e.g. the tweet "Excessive speed is the main cause of car accidents in Liverpool" is not relevant for the purpose of the work to be addressed here, but it can be considered as a traffic related tweet).

In the scope of this work, our analysis takes into account tweets posted by regional traffic agencies, where the problem concerned with the credibility of tweet messages is delimited and is not part of the presented work. Nevertheless, it is important to state that, traffic agencies act in the scope of this work as a provider and not as consumer of information. Into what this work is concerned, the main objective is not trying to solve all issues related with geo-parsing tweet messages, rather than, to present a framework which is capable to process tweet messages from regional traffic agencies and notify drivers about the status of the mobility network in real-time. The idea is to increase the geo-referencing performance of tweets, by fusing and integrating tweets from several traffic agencies. This means that spatial scope and density of tweets will be enlarged by aggregating sources from several traffic agencies, the expectation is to increase the precision of geo-referencing when analysing tweets from an isolated traffic agency perspective. The main particularity regarding geolocation is that the tweets used are not already geolocated, meaning that our approach does not use the geolocation field of tweets; rather it tries to infer traffic events' locations through NLP techniques and mechanisms, which per se is a challenge that is far to be solved, being an hot topic within social-media mining works. This also means that there is a need for the right way to disambiguate between places with similar names.

The computational framework is able to extract a set of contextual information (named entities), such as the location of event. Such named entities are extracted from tweets, by integrating an NER (entity named recognizer) engine, able to parse tweet messages from regional agencies, with the objective to notify promptly users, about traffic status. The approach presented here, implements a computational framework able to: (i) classify a set of traffic-related tweets adopting machine learning techniques, (ii) extract a set of contextual information such as: the location and type of event, (iii) geo locate the event on a map and (iv) the follow up of the incident i.e. monitoring incidents' evolution. Furthermore, the authors tried to use the framework with tweets written in other languages, namely Greek, in order to see if a simple translation/transliteration step, before performing the Name Entity Recognition process, would sustain the same level of performances.

The work presented here, builds upon the work previously developed in the FP7 European research project MobiS-318452: "Personalized Mobility Services for Energy Efficiency and Security through Advanced Artificial Intelligence Techniques" [6]. The main goal was to create a new concept and solution of a federated, customized and intelligent mobility platform, applying novel Future Internet technologies and Artificial Intelligence methods to monitor, model and manage the urban mobility complex network of people, objects, natural, social and business environment in real-time.

The paper is organized as follows: Sect. 2 presents the related work in the area of event detection through crowdsourcing and social networks. The section is concluded by describing the main differences between our approach and other relevant works found the in scientific literature. The major functionalities that comprise our approach are explained in Sect. 3. Section 4 presents the overall process, and how the functionalities were implemented, the tools used and the interdependencies between components that implement the functionalities. The results are presented and analysed in Sect. 5. Finally, Sect. 6 presents our main conclusions and additional consideration on future work.

2   Related Work

Event detection on social networks may be considered as a subdomain of webmining, so the publications on this subject are of various natures. Nevertheless, it will be described here some recently relevant research works, which have been done in the field of event detection (not only about traffic) using Twitter. Furthermore, crowdsource-based apps, such as Waze or Trapster, can be seen as an evolution of social networking towards traffic control and information providing. The main difference here is that such applications are not only specific for traffic-related matters, so the user has to install and use it with the single purpose of providing and retrieving traffic information, but they also have the issue of distracting the users, since these applications imply using them while driving.

The work presented here can be seen as complementary to these traffic-related applications in the sense that, on one side, it is more general and can be used with different sources that do not have necessarily to be linked to traffic information providers per se, such as news web sites, Facebook and Twitter posts or any other social network or information provider. The use of Twitter, or of Facebook for that matter, is a good showcase for real-time, faster-than-news information mining retrieval, since usually news travel faster through social networks than they do through news providers. On the other hand, traffic-related applications can use the information coming from this kind of approach, so as to not bother the users with questions about traffic while they are driving, which can present strong security issues.

The downside of this approach, when comparing to traffic-related applications, is that the information is not explicit: these applications already have specialized forms to gather information about accidents and other traffic events, such as their location and severity; this does not happen with the presented approach, in which all information is retrieved via Natural Language Processing techniques that can present faults and be imprecise.

Concerning traffic events, the authors in [7–9] present different approaches to detect events from Twitter. In [7], authors use an approach to extract parts of tweet messages as verbs and prepositions in order to parse tweets. Enabling in this way the separation of tweets that refer to a traffic incident localized to a certain point and those located between two points. Unlike the presented approach, input tweets must contain traffic keywords such "accident" or "traffic congestion", and at least one verb and one location. In [8] several types of features are used to train a machine learning classifier: character n-grams, syntactic features, spatial and temporal unigram features, term frequency—inverse document frequency (tf-idf) [10] and linked open data features [11]. In our approach, classifier uses only tf-idf features and we obtain similar performances. Twitraffic [9] is a system to filter tweets about traffic and extract the sentiment of the tweet. For instance, "Really bad traffic—M6, M42" has a negative sentiment score and "The M25 is so calm with no traffic" a very positive score. Our framework does not take into account the sentiment analysis component. Finally, none of previous authors addresses the monitoring of incidents evolution.

The authors in [12] implement event detection method by first identifying tweets that are geographically and temporally close to each other. Generated tweet clusters are then evaluated based on several features including the tweet content, in order to decide whether the cluster is actually a new event or not. Li and co-workers present a solution for searching, ranking, and analysing crime and disaster related tweets in Twitter [13]. While plotting a tweet on a map, they first check its GPS tag. If it is not GPS annotated, they look for a place name in the tweet content. If it does not exist either, they use the location in the profile of its creator.

Concerning event detection and management in a more general way, the bibliography is broader. In [14] the authors propose an approach to detect any kind of incident from an emergency broadcaster. The main difference with our approach is that the data is extracted from an incident database and not from social networks. This has the advantage of the homogeneous structure of text messages, which is not the case of incidents contained in the tweets. Social media serve only to enrich the information on events, while our input is directly tweets. A system for managing crisis situations through tweets' analysis is presented in [15]. Tweets analysed were previously filtered by the user of the system. It does not include algorithms to classify the content of a single tweet. Other systems like [16], completely crowdsource-based, are specialized in the analysis of mobile data and not just social networks. Finally, authors in [17] present a crisis management system that incorporates tweet clustering and burst detection algorithms [18].

More recently, Anantharam et al. developed a somewhat similar procedure for tweet event extraction, not only focused on traffic but on generic city events [19]. This approach uses a hybrid NER procedure that has an annotation step in order to annotate the locations and events in the tweet text. Giridhar et al. identify possible events, by clustering apparent locations, via POS tagging and grammar-based rules [20]. They also analyse the quality of the extracted locations. In this case, the work does not find the events by evaluating the event types, but only through locations. Nevertheless, the aim is to extract a single address from a cluster of tweets.

Finally, Wang et al. present a method to analyse if tweets' topics are related to traffic events, using a Latent Dirichlet Allocation-based algorithm, and they compare it with a SVM-based algorithm [21]. In this case, the authors do not refer any kind of location extraction from tweets; rather, they use the geolocation tag of the tweet itself which could lead to false event locations. For a more exhaustive analysis on spatiotemporal analyses of Twitter data, namely for event detection refer to the work of Steiger et al. [22].

3 Framework

The framework presented here adopts a step-wise approach, composed by several functionalities which are performed sequentially in order to detect traffic events from tweets. Extracting traffic related patterns from tweets, require several analysis components. The first one is related to the system input classification, where data is obtained from Twitter and needs to be filtered, by separating relevant traffic related tweets from non-traffic related tweets. Once a relevant tweet about traffic is detected, the process for feature extraction is executed, enabling to detect: (i) the event type, (ii) toponyms, and (iii) temporal information. Finally, a clustering algorithm is used in order to trace the traffic event evolution, by capturing the most up-to-date related tweets.

3.1 Classification

As mention previously, the objective of the classification module is to classify the content of the tweets into two main classes: traffic related events and non-traffic related events. The purpose of the presented work is to be able to process large quantities of tweet messages within a 'big data' context, for that reason it was analysed a considerable amount of tweets and without any constraint about the content or structure of a tweet. To classify a tweet into a positive or a negative class, we prepared the training data and devised a classifier using a support vector machine (SVM) based on features such as keywords in a tweet, the number of words, and the context of target-event words. By preparing positive and negative examples as a training set, we can produce a model to classify tweets automatically into positive and negative categories, where positive tweets are the ones which are traffic related and negative tweets are non-traffic related tweets. Relevant literature [8, 17], has shown that the adoption of SVMs in solving problems of this nature, has brought significant improvements. SVM has shown to be a suitable model for this type of problem. The classification is performed in two different phases (learning and testing phase). The SVM separates the input data into two classes through a decision function that is learned during a learning process.

3.2 Event Type Classification

After the filtering in the previous step, and eliminating non-relevant tweets for our purpose, the "event type classification", refers to classifying a tweet according to the type of traffic event. For this work, it was considered eight different classes of events: (i) traffic jam, (ii) roadwork, (iii) freight traffic, (iv) road closure, (v) ice, (vi) wind and snow, (vii) traffic accident and (vii) others. The objective here, is that each relevant tweet can be labelled with the correct event type.

3.3 Translation/Transliteration

In the case of tweets that are not written in English, a intermediate translation and transliteration step is needed. The authors tried to use Greek tweets for testing purposes. This proved to be difficult, because most translation engines, such as Google Translator and Bing Translator, have still a long way to go in terms of translating full sentences, and understanding if a certain word is qualified to be translated, or should just be transliterated. One example of this could be found in a Greek street name, such as "Παρασκευή Οδός" that can be translated as "Friday Street". When applying the translation, the NER process will not recognize the resulting translation as a location.

Transliteration works better, but it also presents some issues. Almost all map providers, such as Google Maps or OpenStreetMaps, have location names in English, and it is difficult to geolocate the appropriate transliterated equivalent of a street name written in Greek. One way to solve this problem was to find certain words that mean or relate to roads or avenues, in order to infer if Friday is a street or just the day of the week, for instance.

3.4 Named Entity Recognition

The Named Entity Recognition (NER) functionality involves the processing of tweet and identifying occurrences of certain words or expressions which belong to a particular set of categories referred as Named Entities (NE). Examples of name entities could be: locations, products, name of persons, organisms.

Geo-parsing is a subset of named entity recognition, which identifies entities in text and classifies them into categories such as agent, organization and location. Geo-parsing is performed automatically based on language rules. A model based on language rules is then applied to unseen instances to find location words in unseen text. The text of a tweet is not only unstructured, it is also brief. If not supplemented with metadata, it is potentially difficult to parse due to lack of context. It has been found that a data set too small or with insufficient context for the location word as in the geo-parsing of word or phrase search queries is liable to yield low recall and precision results [23].

For the purpose of the work being tackled here, we are only interested in detecting locations which appear on a tweet. The adoption of a NER functionality within our framework will support the extraction of the location feature from the tweet message. All the extracted location entities from a tweet, will provide the basis for detecting the location of a particular traffic event. We have tested several NER implementations that will be further discussed in Sect. 4.

3.5 Temporal Information

When analysing events in social networks and in particular tweet messages, one important aspect besides detecting "where" the event took place, is also "when" such event has occurred. One possible approach to detect the novelty of a tweet message could take into account, the date and time which the tweet message was posted. Within the scope of this work, we are particularly interested in analysing the tweet message, and detect whether the event is occurring, if it has already occurred or if it is something that will happen in a near future. To do so, it is necessary to extract and analyse temporal expressions and verbs used in the tweet message, supported by (NLP) natural language processing techniques such as POS (part of speech) tagging.

3.6 Geolocation

The Geolocation module is capable to geo-reference the locations detected by NER using a geo-reference engine (e.g. Google Geocoding). The authors chose not to use the actual geolocation field within the tweets because, when posting about traffic events, users do not necessarily do it on the events' locations. This happens because most of the times users are driving by the event, and they cannot post a tweet while driving, which in turn makes the geolocation field of the tweet inaccurate.

A particular challenge to take into account in this step relates to the accuracy of the geo-referencing, which sometimes is affected by ambiguity on several toponyms or lack of precision form the NER engine in identifying locations on tweets. A specific traffic event can be confined to a particular place (a point in a map), or between two different places. For the last case, the presentation of the event in a map is confined into a region, which refers to road segments where the traffic event relates to. Disambiguation between places is made by analysing the proximity of all the locations found by the geolocation step. This means that, after locating all the regions that relate to locations within the tweet's text, the framework performs the geographical intersection between all regions in order to pinpoint the right locations and to disambiguate between them.

3.7 Real-Time Clustering

One particular behavior of social networks and especially on twitter is its real-time nature; relevant events have a tendency to be spread out by several users in a short period of time. Therefore real-time clustering can be assumed as a way to identify tweets which refer to the same subject, in our case, to the same traffic event even if posted by different users. In this step, we are particularly interested in analyzing the reliability of a tweet messages. The idea is not to analyze a tweet message separately, but to analyze a set of similar tweets. When considering a traffic event which was referred in a single tweet, during a limited period of time, very little can be concluded about the credibility of the tweet message itself. But, if during the same period of time, similar tweets are posted by different and are not simply re-tweets, the probability that such traffic event is actually happening becomes more credible, meaning that more concrete conclusions can be drawn from that particular event. For this step, we propose the adoption of a clustering algorithm which is intended to be performed in real time within a short time frame, simultaneously with crawling (i.e. obtaining tweets from Twitter API). The result will be clusters of tweets, which relate to the same event. This will not only increase the credibility of traffic events, but on the other hand, it will enrich the knowledge about a particular event with several tweets, enabling on this way to track the evolution of such event.

4 Implementation

In order to implement the approach described, several technologies, tools and APIs were used in every module of the framework. In this section we present details about it. Figure 1 shows the tools used in each module.

4.1 Classification

For implementing the classification module, we relied on a well-known open source data mining tool called RapidMiner [24]. This tool was used to implement the SVM learning model and the classification of new tweets. SVM was trained with a dataset

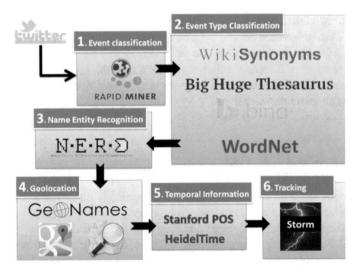

Fig. 1 Overall view of the system, with technologies, API and tools used in every step

of 10,000 tweets, which 5,000 tweets are considered "positive" i.e. they contain relevant information about traffic events and 5,000 tweets considered "negative" i.e. they contain general information. Similarly, the test is performed with 1,000 tweets of each type. For the sake of clarity, to mention that we are dealing here with binary classification, meaning that one tweet can be classified as "positive" or "negative".

Tweet messages about traffic events exclusively for the UK region were extracted through Twitraffic, and used in order to create a dataset of positive tweets. The dataset containing negative tweets was built by a crawling twitter API with predefined keywords related to sporting events, TV series, politics, technology, etc. The idea was to randomly create a dataset of general scope tweets, which were not related with traffic events. Table 1 shows some examples of tweets in the dataset.

In order to build a feature vector and train the classifier, we use the classical vector-space model [25] applied into the tweet corpus. First, most common words are selected. Then n-grams (n = 1, 2 or 3) are calculated and the most common are selected too. Feature vector of tweets are represented as a vector containing the tf-idf score of these selected words and n-grams. The assessment regarding the performance of the classifier is further analysed in Sect. 5.

4.2 Event Type Classification

Each type of a traffic event is classified taking into account a set of relevant words. For example, traffic jam can be associated to words like "crash" or "snarl up", and road closure to words like "obstruction" or "barrier". To create a list of similar words,

Table 1 Examples of tweets in training dataset

Tweet	Label
Traffic is very slow on the M1 North out of Parktown towards the Buccleuch Interchange	Pos
Traffic a nightmare near odsal m606 total nightmare everywhere around it then towards Leeds	Pos
My god the traffic on A13 is sooo bad	Pos
Need to be at uni in half an hour and I m stuck in hideous tailbacks on the M53 badmorning	Pos
I dont think there s one close by last time I checked the closest one was 70 miles away	Neg
Its 6:20 pm early days in a looooooong night at work	Neg

several thesaurus have used: WikiSynonyms (Wikipedia), Big Huge Thesaurus, Bing Synonyms and WordNet. The presented work took into account 8 classes of traffic events, where for each class we constructed a list of corresponding synonyms. The event type classification mechanism relies on matching synonyms with terms available on the tweet message. For instance, the tweet "N2—one lane closed due to snow" contains "closed" and "snow" it will be classified as "Road closure" and "Wind and Snow" event type, meaning that, for this particular classification, we are considering a multiclass classification.

4.3 Named Entity Recognition

In order to extract locations from tweet messages, several NER engines were tested: Alchemy, OpenCalais, Stanford NER, and NERD. A novelty of our system is that we include different parsers to identify location words. Informal results of our experiment showed that, in many cases, the four engines disagree on what is and what is not a location. The results of the comparison are shown in Sect. 5, nevertheless NERD was chosen because it showed the best performance. As an example the location contained in the tweet "Severe M62 West Yorkshire—One lane closed on M62 westbound between J26, M606 (Chain Bar) and J25, A644 (Brighouse)" were only able to be detected by NERD.

4.4 Temporal Information

The temporal information extraction from tweets is still in development phase. The approach to be followed here, aims at extracting temporal information from tweets, using a Part-of-Speech (POS) tagger for analysing the verbal form, combined with the approach presented in [8] used for analysing the time expressions. The idea is combining POS rules like "is + gerund implies present tense" and time mentions like "yesterday" to decide about the event timeline.

4.5 Geolocation

For the task of extracting the GPS coordinates from the entities detected by NER engine, three distinct geolocation external applications were used: Google Geocoding, GeoNames and Nominatim. The goal is to have the most accurate information about where the event has occurred. Several levels of accuracy are being considered here, namely: road, city, region and country level. It is possible that an event involves several places at the same time, such a serious accident which produces a traffic jam on a highway linking two cities. As mentioned before, the more similar tweets about a particular event are analysed, the more accurate can be the geolocation of the event itself.

Adopting three different geolocation services sometimes lead to different results, which required a disambiguation approach to tackle possible inconsistencies. If the first result of at least two geolocation engines are in agreement on a location then we will use them. Otherwise, if results are ambiguous, we take the first result from Google Geocoding since this engine return generally most numerous results from a query. For instance, for the query "A2" we can get the A2 highway at Ziesar in Germany, or A2 freeway in UK. This approach is obviously very naïve and has to be improved and complemented with the tracking component.

It is worth to mention that some terms labelled by the NER engine as locations might not be correct. For example the term "vehicle" could have been labelled as a location. In this case, a simple search on one of geocoding systems is able to detect these false positives.

4.6 Real-Time Clustering

The trace and tracking of tweet messages using a clustering algorithm, is still in development. Nevertheless, we intend to follow a similar approach as described in [26], where authors presented a system for "first story detection" namely detecting the first tweet posted about a certain event. Others approaches like the clustering method described in [17] will be tested, and now our efforts are focused on modelling a progressive IDF and adapting a distributed real-time computation system like Storm to our framework. Progressive IDF is needed to best describe events with *tf-idf* in a real-time context.

5 Results

In this section we will present and analyse the results concerning the assessment of the classification and NER modules. As previously explained in Sect. 4, temporal information and real-time clustering components are still in development phase.

5.1 Preliminary Analysis

The analyses were conducted in order to discover patterns on the data set, which was created using tweets from Twitraffic, an API which provides traffic related tweets for UK from regional traffic agencies. We focused on tweets' temporal and geographic distribution. Most of these tweets come from traffic agencies that post tweets either about generally UK or a region/motorway.

Figure 2 shows the temporal distribution of tweets per agency. From this analysis and according to the chart, it can conclude that a very important event happened between May 27th and 28th. In fact, with a request on our dataset we obtained similar tweets at least 50 times: *"M1 northbound between J14 and J15 |Northbound|Multiple Vehicle Accident"*. With a more exhausting analysis, we discover that this accident happened near to St. Albans region, and that's why TrafficStAlbans agency had an important amount of posted tweets on that day.

Figure 3 depicts the number of tweet messages posted, which explicitly refer to the most important highways in the UK. For such, the "UkSocialTraffic" agency was analyzed and it is possible to detect that the event on May 27th have the same impact.

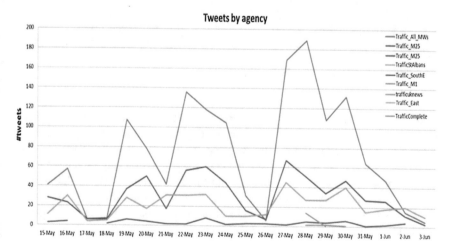

Fig. 2 Number of tweets by agency and day

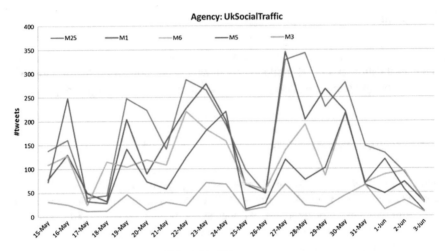

Fig. 3 Detail of tweets posted by UkSocialTraffic agency

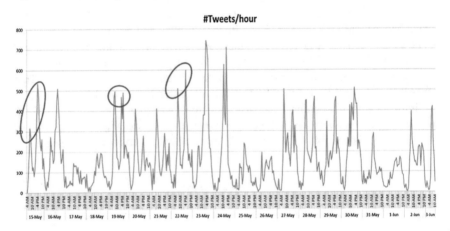

Fig. 4 Total number of tweets by hour

We also conducted our analysis in order to detect the daily behavior of traffic related tweet messages, and we found interesting patterns related to specific hours of the day, as depicted in Fig. 4. Almost every day we see that around 8 am and 5 pm are moments when people post more tweets, which possible may be related to the fact that people are going from home to work and vice-versa. The circular areas in Fig. 4 indicate these two distinct periods for different days of the week for a given highway. We may assume that, such phenomena could be heavily related with the rush hours in the morning and evening periods.

Table 2 Train and test performances in event classification

Train	Accuracy: 0.893		
	Negative	Positive	Precision
Prediction negative	4326	390	0.917
Prediction positive	674	4610	0.872
Recall	0.865	0.922	Macro F1: 0.894
Test	Accuracy: 0.955		
	Negative	Positive	Precision
Prediction negative	993	83	0.922
Prediction positive	7	917	0.992
Recall	0.993	0.917	Macro F1: 0.956

5.2   Classification

This section describes the analysis of the results provided by classification module. For assessing this work, we used the classical performance metrics, precision and recall. In our case, precision is the ratio of the number of relevant tweets retrieved to the total number of irrelevant and relevant tweets retrieved. Recall is the percent of all relevant tweets that is returned by the classifier. We compute also accuracy and macro F1 score. Accuracy is calculated as the sum of correct classified tweets by the total number of classified tweets, and macro F1 is the average of precision and recall. Table 2 shows the testing and training performances, where we get very promising results both regarding training and testing phases, similar to performances shown in [7, 8].

5.3   Named Entity Recognition

The comparison of the different NER engines was performed using 50 tweet messages that were selected randomly from our training dataset. For each tweet, we manually extracted all the locations, which make a total of 142 locations across all tweets. The 50 tweets that were selected, do not follow any particular structure, but contain a lot of information corresponding to one or more locations. Table 3 shows the percentage of entities detected for each system. The differences in performances led the authors to conclude that NERD engine presents the best performances because it uses Open-Calais and Alchemy in addition to other NER systems. Although the percentage of detection is high (over 80 %), there are also entities that were tagged incorrectly as locations. For instance, common words like "road", "junction" or "eastern" can be detected as locations by NERD. However, the precision will be improved in our case as in next step (geolocation) we perform a disambiguation of incorrect entities.

Table 3 Location extraction performances

	OpenCalais	Alchemy	Stanford	NERD
% Detection	50.7	39.43	35.21	80.98

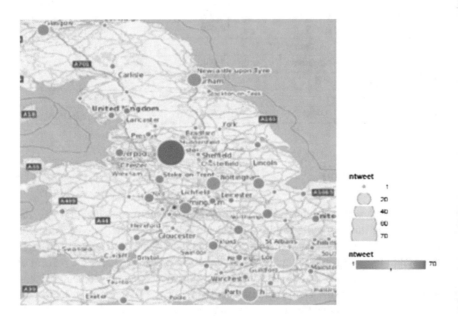

Fig. 5 Visual interface of geolocation and clustering shows some relevants *points* with traffic events, especially around Manchester and London

5.4 Geolocation

The visualization of results is an important functionality in our framework, because users will interact directly with the interface to get more information about events. In Fig. 5 we show a traffic event clustering. In red and yellow, we can see the most important locations with traffic events.

The authors also developed a Web interface in which users can check the information extracted from tweets. Figure 6 shows the overall view of the web interface.

By clicking on one of the tweet, users can check the type of traffic event the tweet is referring to (in this case, a road closure) and all the location entities extracted with the NER process, as well as their geolocated counterparts (Fig. 7). Also, in the map, the bounding box resulting from the geometrical intersection of all the geolocated regions can be seen (the red square). Moreover, the road segment in which the event occurred is also highlighted in black, as a result of finding all the road segments belonging to A47 (the road mentioned in the tweet) within the highlighted intersection area.

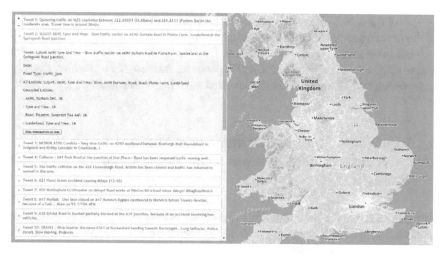

Fig. 6 Twitter event viewer

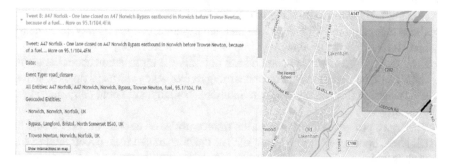

Fig. 7 View of the intersection of locations caught within the tweet's text, and corresponding road

5.5 The Case for Greek Tweets

The authors tested the procedure with Greek tweets, extracted from MyRouteGr, a Greek traffic information Twitter account. To enable Greek tweets to be processed in the same way, the process can be described as follows: (i) First, each tweet is translated to English, in order to apply the tweet classification and event classification procedures; (ii) a NER step to catch location-related words within the tweet, such as "road", "highway" or "avenue". This step is very important to disambiguate between entities that may be locations, such as the "Friday Street" example. The idea is that, when the NER step catches a location-related word linking to a road or avenue, it also looks for words starting with capital letters before the that location-related word. This of course has the inconvenient of just finding locations that start with capital letters; (iii) finally the words are put together, as in "Friday Street", and transliterated, in order to serve as inputs for the geolocation procedure.

One of the drawbacks of this step is the undesired effect of lowering the precision of the overall procedure, due to the number of errors it can produce. Furthermore, because tweets don't follow any syntactical rules, users often do not write location names starting with capital letters, which poses an issue to this procedure. Even so, some Greek tweets were positively classified as traffic events and well geolocated.

6 Application Scenario: Twitter Event Mining with Personalized Notification via Google Cloud Messaging

The methodology for mining traffic events from tweets was used in order to send notifications to users that could be affected by such events. This application scenario can be summarized as follows:

- Processed tweets that corresponded to a positively geolocated traffic event are considered relevant to be disseminated to urban commuters;
- Users that plan their itinerary using their smartphone, can in this way, be immediately inform if a traffic event will impact on their travel plan;
- Each individual travel plan, would have not only the information about the actual route itself, but also information concerning its frequency and usual times of arrival and departure (e.g. once a week, at Saturdays beginning at noon; daily, starting at 9:00 am);
- Then, two processes would run in the framework's server-side, the first would be triggered upon a new traffic event, such as the ones mined from tweets, and would look into the existing routes to check if the time and location of the event would affect any of the existing routes; the second would be triggered upon a new user route, and would look for traffic events that could affect the new route;
- A prototype Android application would work as interface in case a traffic event affects a route, then notification is sent to the user which was affected by the event. This notification is sent through the Google Cloud Messaging (GCM) service (Fig. 8), and it personalized to the affected user(s) only. In the case of a new event affecting multiple existing routes, the notification would be sent to all the affected users;
- Finally, another situation is triggered if user is already on its way (current route), the application would check if the user's actual location is before or after the location of the event. If the user's location is before the event, then the notification is sent (Fig. 8); otherwise, the system would not notify the user, since he/she will not be affected by the event.

Fig. 8 Sample GCM notification of traffic events on route

7 Conclusions and Future Work

The objective of the work described in this paper is to establish a computational framework, able to detect traffic-related events in real-time, using social networks. The framework aims to be flexible enough not only to be applied to Twitter, but also to other social networks. As described in this paper, we investigated the real-time nature of Twitter, in particular for traffic event detection. Semantic analyses were applied to tweets to classify them into a positive and a negative class. We consider each Twitter user as a sensor, and set a problem to detect an event based on sensory observations. Name-entity recognition was used to extract location entities in a tweet message, and pinpoint those locations on a map. A temporal analysis was also performed, using NLP techniques in order to detect the novelty of a tweet message. From an implementation perspective, we developed a novel approach to notify people promptly of a traffic event. Furthermore, the approach was slightly modified in order to test it with tweets written in other languages than English, by adding a translation/transliteration procedure, in order to not only classify tweets in Greek as possible traffic events, but also to aid in the geolocation process of the tweets.

An application scenario using tweet messaging processing for notify users about traffic events is presented. Such application scenario makes use of Google Cloud Messaging service, in order to notify affected users in real-time about traffic events.

Despite the challenges associated with the real-time nature and length limitation that distinguish Twitter messages from other social networks, as future work, some improvements and additional functionalities need to be taken into account, namely the aspects concerning the credibility of a tweet message. We propose to adopt a real-time clustering approach which clusters messages from a stream of tweets. A tweet message tends to be more credible, if several users post similar messages in a very short period of time. Other factors may also influence the credibility of tweets, like the profile of each user and historical usage of Twitter by each user. Such capability is per se a challenge.

Other improvements need to take into account the elimination, as far as possible, of the ambiguities generated by the name-entity recognizer. Although the development of another NER algorithm is out of the scope of this work, the ambiguities can be minimized, if processing a set of similar tweets, instead of a single tweet. Such approach will provide more accurate information about the location in which the event took place. The computational framework presented here is still on an early development stage, but some preliminary results have been achieved and lead the authors to conclude that results are promising as discussed in the results section.

References

1. Gao, H., Barbier, G., Goolsby, R.: Harnessing the crowdsourcing power of social media for disaster relief. IEEE Intell. Syst. **26**(3), 10–14 (2011)
2. Munro, R.: Subword and spatiotemporal models for identifying actionable information in Haitian Kreyol. In: Proceedings of the Fifteenth Conference on Computational Natural Language Learning, Portland (2011)
3. Vieweg, S., Hughes, A., Starbird, K., Palen, L.: Microblogging during two natural hazards events: what twitter may contribute to situational awareness. In: Proceedings of the SIGCHI Conference on Human Factors in Computing Systems, Atlanta (2010)
4. Bifet, A., Frank, E.: Sentiment knowledge discovery in twitter streaming data. Discov. Sci. 1–15
5. Gimpel, K., Schneider, N., O'Connor, B., Das, D., Mills, D., Eisenstein, J., Heilman, M., Yogatama, D., Flanigan, J., Smith, N.A.: Part-of-speech tagging for Twitter: annotation, features, and experiments. In: Proceedings of the 49th Annual Meeting of the Association for Computational Linguistics: Human Language Technologies: short papers (2011)
6. Costa, R., Figueiras, P., Maló, P., Jermol, M., Kalaboukas, K.: MobiS—personalized mobility services for energy efficiency and security through advanced artificial intelligence techniques. In: 5th KES International Conference on Intelligent Decision Technologies, Sesimbra (2013)
7. Wanichayapong, N., Pruthipunyaskul, W., Pattara-Atikom, W., Chaovalit, P.: Social-based traffic information extraction and classification. In: International Conference on Telecommunications (2011)
8. Schulz, A., Ristoski, P., Paulheim, H.: I See a car crash: real-time detection of small scale incidents in microblogs. In: ESWC 2013 Satellite Events: Revised Selected Papers (2013)
9. http://twitraffic.co.uk/
10. Singhal, A., Choi, J., Hindle, D., Lewis, D., Pereira, F.: AT&T at TREC-7. In: Proceedings of the Seventh Text REtrieval Conference (1999)
11. Paulheim, H., Fümkranz, J.: Unsupervised generation of data mining features from linked open data. In: Proceedings of the 2nd International Conference on Web Intelligence, Mining and Semantics (2012)
12. Walther, M., Kaisser, M.: Geo-spatial event detection in the twitter stream. In: Proceedings of the 35th European conference on Advances in Information Retrieval, Moscow (2013)
13. Li, R., Lei, K., Khadiwala, R., Chang, K.: TEDAS: a twitter-based event detection and analysis system. In: IEEE 28th International Conference on Data Engineering, Washington (2012)

14. Abel, F., Hauff, C., Houben, G.J., Stronkman, R., Tao, K.: Semantics + filtering + search = twitcident. Exploring information in social web streams. In: Proceedings of the 23rd ACM Conference on Hypertext and Social Media (2012)
15. Rogstadius, J., Vukovic, M., Teixeira, C.A., Kostakos, V., Karapanos, E., Laredo, J.A.: Crisis-Tracker: crowdsourced social media curation for disaster awareness. IBM J. Res. Dev. 57(5), 1–4 (2013)
16. Okolloh, O.: Ushahidi, or'testimony': web 2.0 tools for crowdsourcing crisis information. Participatory Learn. Action 59(1), 65–70 (2009)
17. Yin, J., Lampert, A., Cameron, M., Robinson, B., Power, R., Using social media to enhance emergency situation awareness. IEEE Intell. Syst. 52–59 (2012)
18. Fung, G.P.C., Yu, J.X., Yu, P.S., Lu, H.: Parameter free bursty events detection in text streams. In: Proceedings of the 31st international conference on Very large data bases (2005)
19. Anantharam, P., Barnaghi, P., Thirunarayan, K., Sheth, A.: Extracting city traffic events from social streams. ACM Trans. Intell. Syst. Technol. 9(4), p. Article 39 (2014)
20. Giridhar, P., Abdelzaher, T., George, J., Kaplan, L.: On quality of event localization from social network feeds. In: The Seventh International Workshop on Information Quality and Quality of Service for Pervasive Computing. St. Louis, MO (2015)
21. Wang, D., Al-Rubaie, A., Davies, J., Clarke, S.S.: Real time road traffic monitoring alert based on incremental learning from tweets. In: IEEE Symposium on Evolving and Autonomous Learning Systems. Orlando, FL (2014)
22. Steiger, E., de Albuquerque, J.P., Zipf, A.: An advanced systematic literature review on spatiotemporal analyses of twitter data. Trans. GIS (2015)
23. Guillén, R.: GeoParsing web queries. In: Advances in Multilingual and Multimodal Information Retrieval, pp. 781–785. Springer, Berlin (2008)
24. Rapid-I GmBH, RapidMiner (2012). http://rapid-i.com. Accessed 3 Sept 2012
25. Salton, G., Wong, A., Yang, C.: A vector space model for automatic indexing. Commun. ACM 613–620 (1975)
26. Petrović, S., Osborne, M., Lavrenko, V.: Streaming first story detection with application to twitter. In: Human Language Technologies: The 2010 Annual Conference of the North American Chapter of the Association for Computational Linguistics (2010)

Applying Supervised and Unsupervised Learning Techniques on Dental Patients' Records

Syed Mohtashim Abbas Bokhari and Shoab Ahmad Khan

Abstract The research presents a process for applying data mining techniques on dental medical records comprised of oral conditions and different dental procedures that are performed on various patients. The dental expert decides to pursue a set of procedures based on the examination and diagnostics. Digital dentistry is becoming more and more active now, hence this research addresses the issues in exploiting the digital data at its potential like heterogeneous data gathering, access restrictions or inadequate patient data and lack of expert systems to utilize the data. It proposes a way to deal with the dental medical records and apply data mining. Having gathered the dental data and prepared it through pre-processing techniques, unsupervised learning techniques were applied to perform clustering in order to discover interesting patterns and assigning these a label class. Mostly the patients lie in the mild and moderate dental patient's class. The most common problem that is being noticed in patients is tooth cavity with a treatment named "resin-based composite—one surface, posterior". Using this labelled data set, supervised learning algorithms were applied to train and test the data for predicting the targeted class accurately. A comparison between classification algorithms based on their accuracy was made to filter out the best outcome. An expert system has also been developed to support the idea, ease up the decision making process and automate the manual practices that are being used. It provides quick recommendations to the medical expert in examining the patient depending upon the diagnosis. Research reveals that decision tree runs better than others on our data set with highest accuracy in predicting the Patients' targeted classes.

S.M.A. Bokhari (✉) · S.A. Khan
Department of Computer Engineering, College of EME,
National University of Sciences and Technology (NUST), H-12, Islamabad, Pakistan
e-mail: mohtashim_abbas@yahoo.com

S.A. Khan
e-mail: shoabak@ceme.nust.edu.pk

© Springer International Publishing Switzerland 2016
L. Chen et al. (eds.), *Emerging Trends and Advanced Technologies
for Computational Intelligence*, Studies in Computational Intelligence 647,
DOI 10.1007/978-3-319-33353-3_5

1 Introduction

Data mining is a multidisciplinary sub-field of computer science. It is a complete process to discover worthy patterns in large datasets that involves various methods at the same time such as; artificial intelligence, machine learning, statistics and database systems [1, 2]. Summarizing, data mining is the extraction or "mining" of knowledge from large datasets [3]. The dataset used in this research is of dental domain.

In real time patient comes to a dentist for treatment and after examination he has to perform certain procedures on a patient. This paper aims to focus on the applying mining algorithms on processed data. Data is assembled through different resources, but major part is from the Armed Forces Institute of Dentistry (AFID), Pakistan. The dental community is now endeavoring to collect the data in such a way that helps pulling out worthy knowledge. The data mining can be applied for the decision making in the best interest of the public. Currently, decision tree analysis, neural network, logistic regression and so on are being applied to different medical fields, like in the prediction of diseases, treatment plan, trends or pattern recognition. And the outcomes deliver a clinical structure and provide a model to medical problems. Different dental forecasting models are provided to measure and anticipate the condition of the patients having dental problems [4].

The research offers a framework for handling dental medical records from Pakistan. Our emphasis is on mining the processes, particularly unsupervised as well as supervised learning techniques to extract knowledge from different logs and predicting the nature of the records. Classification and clustering are other data mining techniques are derived from the main concept of machine learning [5]. Till now, process mining has been utilized in diverse situations, such as manufacturing, clinical procedures and hospitals. Mining involves obtaining a quantitative understanding of these processes.

Literature review shows that previous work of mining in dental domain only focuses on a few problems and procedures but the data set used in the handles more than 220 dental procedures at the same time. In Pakistan knowledge/pattern extraction, clustering and classification is being done on cardiac data [6]. In dental domain, association rules based mining is used to discover different rules helpful for the medical experts treating the patients [7]. Some unsupervised learning techniques have also been used to extract different hidden patterns in the dentistry domain [8]. In extension to this work, the goal of this research is to apply classification and develop an expert system to use data for decision making. As the development of classifier precision by training is one of the major research area in machine learning, so for this purpose we used unsupervised learning methods to analyze various clusters and label the clusters and then applied different classifiers to predict the targeted class accurately. An expert system has been hence developed to automate AFID use cases and for medical experts making decisions. To date, there are very limited data mining based applications in dentistry. However, many applications of process mining in the healthcare sector are well-known [4, 9–15].

2 Review of Literature

It's difficult to get to the relevant information since the use of computers has been constantly increasing the storage of data in the field of dentistry. As the data is in raw form that needs to be processed to analyse different interesting patterns and trends. Usually, the knowledge is mined from the professional's manual way. Certainly, the manual processes don't work anymore if the amount of data grows and scope increase. To manage such a large volume of data and explore it, various computing technologies are necessary [16, 17].

Data mining has now appeared to be an essential area that helps in the extraction of useful knowledge from large data sets of unstructured and seemingly unusable data. Data mining plays a vital part in several applications like business or health care organizations, e-commerce, science and engineering. To support and help the medical practitioners, information and knowledge based systems are also being developed. Healthcare institutions keep large data sets of patients as medical reports, patient's history and also in electronic form etc. [18]. The data is generally unstructured, complex, noisy and discrete [19]. In such data the valuable knowledge is hidden that need to be explored and used for the betterment of human beings. In order to mine or make an effective use of the data, the unstructured form of data is transformed into useful information to help the domain experts for better clinical judgments.

The data mining has played part to contribute in the medical field. As compared to other data mining domain applications, medical data mining plays a vital role and it has some unique characteristics. In the healthcare sector, the data mining is basically used to predict diseases in a timely manner. There are several data mining techniques such as clustering, classification and rule based mining etc. used to predict diseases [20]. Many diseases like, heart problems, diabetes, hepatitis and cancer can be identified through data mining [21]. Benko and Wilson [22] refer different healthcare organizations which use data mining to meet their long-term necessities.

In the past, for patients, mostly three methods used in the field of data mining, artificial neural networks (ANN), decision tree and logistic regression analysis. ANN comprises of many artificial neurons and neuron links each other in the form of a network to replicate mankind's neural structure. The decision tree method is very effective in medical science to predict, analyse and explore the outcomes. The decision tree is an influential classification algorithm that is becoming more widespread with growing data and data mining as a field. Decision tree algo dissects the records recursively into various groups; each of these is having a common set of characteristics. The key purpose of the decision tree is to arrange a tree structure by dividing raw data recursively into numerous subsets depending upon relativeness. Having established the decision tree, each leaf node indicates expert's knowledge, forming the rules from root to leaf. Well-known decision tree algorithms are Quinlan's ID3, C4.5 and C5 [23]. While Logistic regression is a type of technique in the statistical analysis that is mostly used in forecasting binary or multi-class dependent variables.

Balasubramanian and Umarani [24] identified the risk factors associated with the high level of fluoride content in water through clustering algorithms and finds

meaningful hidden patterns which give meaningful decision making to this socioe-
conomic real world health hazard. Artificial neural networks (ANNs) was used to
construct a mathematical model [25] with input factors as age, smoker status, type of
toothbrush, brushing and consumption of pickled food, fizzy drinks, orange, apple,
lemon and dried seeds. The outcome was the sum of tooth surface loss scores for
selected teeth. This paper [16] adopted the cancer data to perform data mining for
effectively controlling the treatment course of cancer and improving the quality of
the life. The knowledge being excavated to probe the causes of diseases and looking
for the key indicators relative to the oral cancer. Mans et al. [26] used data mining
to provide the workflow for placement of the implant process in dental management
to enhance performance. While Polášková et al. [27] emphases on the same implant
service and develops a decision support system to aid the medical practitioners to be
able to decide where to put the implant exactly depending upon some attributes like
width of the bone horizontally vertically. A combination of supervised and unsupervised learning techniques used to predict liver diseases with accuracy 71–91 % [28].
Literature review shows that there is little work done in dentistry. The work focuses
only few problems and dental procedures, but overlooks other important dental procedures. However this research gathers, analyses and labels more than 220 dental
procedures at the same time.

The aim is to process the unstructured dental data from Pakistani hospital and
apply mining to analyse different clusters of processed data to extract useful mining
patterns along with labelling of clusters by assigning class variables. After balancing
this labelled dataset we intend to apply then classification techniques and use the
dental dataset to make timely decisions. And finally heading towards developing an
expert system to use the data to make intelligent recommendations for the medical
experts. Similar work has been done in other medical fields in Pakistan. But little
research has been carried out in the domain of dentistry as already discussed in
introduction.

3 Methodology

The problem identification and understanding is an important part of the data mining
in order to solve that problem. Data relevant to clinical history, statistical patient data,
procedures and impressions of a patient are preserved in the AFID but the format of
the stored data is in unstructured format that cannot be used for analysis, processing
and decision making unless made into a structured form. Clinical decisions are based
only on perception and experience of a doctor that may lead to undesirable biased
mistakes or unnecessary medical expenses. This ultimately has negative effects on
the quality of medical service. This large number of knowledge enriched data has to
be processed to get useful information because this info is very helpful in making
worthy medical decisions. Yet, there is a need of an effective framework for medical
records particularly in the domain of dentistry in Pakistan to discover and predict
hidden patterns along with the patient's condition.

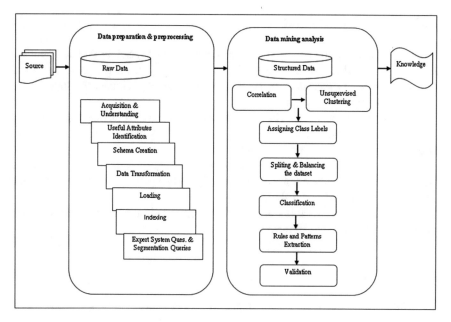

Fig. 1 Process workflow

The Fig. 1 shows a methodology that was adopted to carry out this research on dental medical records. The model is represented and explained below by its flexible phases. The figure displays all the phases of the model in order. Each phase has some inputs, actions and outcomes. The contribution in this research is to explore the data of dental domain in a way to achieve worthwhile results. Data was in an unstructured form which was then refined to unleash meaningful patterns.

3.1 Process

3.1.1 Data Acquisition Phase

First the acquisition and understanding of the raw data was done to identify the usefulness of data. Understand, analyse and model the data, heading towards filtering suitable attributes to support decision making. Data analysis comprises several facts and approaches, covering different techniques under a variety of names, in various domains. This research was carried out in close association with the Armed Forces Institute of Dentistry (AFID), Pakistan. Hence the real data from AFID is being used. This collected data includes unstructured statistical records of 750 dental patients.

3.1.2 Attributes Identification

After understanding the data with the collaboration of experts in AFID, important attributes and entities from the raw data were identified. At this stage it's done manually while at other step we shall filter out valued attributes through feature selection and related techniques.

Entity—An entity is basically an object that may have a physical existence like particular person, car, house, or an employee—or it may have a conceptual one like a company, a job, or a university course.

Attribute—An attribute is something that describes the entity. Considering the same above example, the employee is the entity and employee's name, age, address, salary and job etc. are the attribute.

The entities listed in Table 1 are created in the database after understanding the nature of the dental medical records.

3.1.3 Schema Creation

A schema was then created for the defined medical entities and relationships were identified to make it in an organized way. So that database could be used for proper application and mining purposes as well. A database schema is its structure formed and presented in a formal language, backed through the database management system (DBMS). It is basically a set of formulas known as integrity constraints that are enforced on a database to make sure the compatibility between different parts of the schema. All constraints are described in the alike language. Having passed through different stages, conceptual schema is then converted into an explicit mapping; the database schema. This shows that how real world entities are molded in the database.

Table 1 Entities with their details

Entity	Description
R_Address	Contains patient address details
R_Disease	Contains patient diseases other than dental that may affect medication
R_Medication	Contains medication info of dental patients
R_Procedure	Contains procedures that are performed on patients
R_Categories	Procedures are divided into various categories
R_Patient	Contains all patients' related data
R_Treatment	Contains a treatment plan for dental patients
R_Visits	Contains patients' visit details
R_Charges	Contains fee schedules for different treatments

3.1.4 Transform Data and Loading

The data, collected from different reports, is noisy and discrete that needed transformation. The data were then transformed through scripts and put into the MySQL database. It includes following pre-processing steps.

Data was transformed through scripts. These scripts transform a set of data values from the format of source system into a format that is required and suitable for the destination system.

Data transformation is basically pre-processing of unstructured data and we dissect it into the following steps:

(a) Data Cleansing

In data cleansing the missing and inconsistent data were dealt. The noisy data was made smooth and outliers were identified and removed. Because the data needs to be made consistent and clean before extracting the knowledge. The scripts as well as rapid miner tool was used to do it iteratively.

(b) Nominal to numeric conversion

Some attributes in the data require nominal to numeric conversion because K-Means algorithm can't handle nominal data. So some attributes had to be converted into numeric format under data transformation.

(c) Data Normalization

Attribute values were then scaled to fall within a specified range using Rapid-Miner5, the Java based tool.

(d) Data Reduction

In this final step of data transformation volume of data was reduced by eliminating the redundant rows. Now the data is ready to be loaded.

3.1.5 Loading into the DB

After data transformation, transformed data was loaded the into a destination repository. There are many methods in ETL to load the data. Some involve build-in SQL insert statement and physically insert every record as a fresh row into the table of the targeted warehouse. There is also a concept of bulk loading to load the data in chunks. We chose normal loading instead of bulk to insert all the data in one go. Loading portion is generally a bottleneck of the entire process.

3.1.6 Indexing

The indexing technique was used in MySQL to access the DB faster. It enhances the pace of data search operations on a table, but at the expense of further writes and storage space to sustain the index data structure. Indexes are basically used to figure out data in a quick way without a need of searching each row of database table every time when the table is accessed. Indexes can be formed by utilizing at least one or more attributes of a table to effectively access ordered tuples.

3.1.7   Expert System Questions

The next step is determination of the questions that can be answered through the data. We discussed with AFID domain experts to identify the questions that can be answered and expected patterns that can be extracted through mining operations.

3.1.8   Data Mining Operations

Data Mining, is knowledge discovery and in fact a process to dig up and analyze huge data sets for the sake of extracting the sense and meaning of the data. Data mining operations and tools are used to predict trends that eventually help in making proactive knowledge-driven decisions.

Data mining operations were then performed on the dataset mainly supervised learning algorithms and unsupervised clustering on processed data. Unsupervised clustering help us labeling while supervised algorithm aid predicting the targeted class. The will be helpful to medical practitioners to treat their patients wisely.

3.1.9   Knowledge Extraction

Having clustered the data into various clusters, we analyzed those clusters to find valuable outcomes and make the dataset labeled to use it for supervised learning. Eventually we utilized the data and automated the processes to develop an expert system. We shall discuss this in detail in results and discussion section.

3.1.10   Validation Through Domain Experts

Outcomes, patterns and the results identified through the mining are then validated through the domain experts. Domain experts validate the results because rules are made by these experts.

3.2   Tools Used in the Research

In order to apply data mining operations, there are various data mining tools accessible in the marketplace, each has its own advantages and disadvantages. These tools assist and help the analysts and developers in the process of data mining. Generally, these tools can be classified into three categories: text-mining tools, traditional data mining tools and dashboards [29].

Machine Intelligence is believed to be one of the basic units of data mining [30, 31]. It includes the study and examination of different computational methods that are probable to develop an automatic way with practice and intelligent training.

Table 2 Tools

Tool	Usage
Rapid miner 5.3	Rapid miner® by rapid is used as a key analytical tool with different extensions for data preprocessing, mining and implementation of different algorithms
MySQL	MySQL was used as a database for storage and restructuring after getting raw data from AFID
Microsoft excel 2010	Microsoft excel 2010 along with MySQL is used for results visualization and analysis

List of some well-established, updated and new tools of data mining are presented on Kdnuggets web [32] and an article [33].

Table 2 shows the tools and their purpose that were chosen to carry out and aid our research.

3.3 Unsupervised Clustering

Unsupervised learning techniques use unlabelled dataset therefore, named as unsupervised. The data doesn't have any target attribute, but it is then explored to find some intrinsic structures in it. We use mainly K-Means algorithm, an unsupervised learning technique to make the random clusters and analyse them. K-Means are widely used in medical, scientific and industrial applications to solve the clustering problem. Using K-means approach for clustering, some important features are extracted and based on those features different patterns in the dental patient's database are analysed. K-Means helped the domain experts labelling the data accordingly. The coming section will show the experimental results and analysis.

3.4 Labelling the Data Set for Clustering

After making unsupervised clusters of real time processed data set, domain experts were accessed to label those clusters and have their worthy opinion. Depending upon medical practitioner's recommendation while classifying the data, the degree of symptoms lies in different compartments as follows:

- None or Normal
- Mild Dental Patients
- Moderate Dental Patients
- Severe Dental Patients.

3.5 Supervised Learning Techniques

Supervised learning is the machine learning technique to infer a function from labelled training data. The training data consist of a set of training examples. In supervised learning, each example is a pair consisting of an input object (typically a vector) and a desired output value (also called the supervisory signal). Having labelled the data, supervised learning algorithms were applied mainly decision tree, KNN, decision stump and artificial neural networks. These algorithms were compared on the basis of their accuracy in predicting the label class.

3.6 Dental Expert System (DES)

Expert System is a computer application program to gather and present information in such a way that could be helpful in making timely decisions. It helps the medical experts not only in decision making but also eases up the procedures by automating the manual on going processes. According to the uses cases of AFID a DES has also been developed to categorize and classify the dental medical records. It provides suggestions and recommendations to the medical experts while examining the patient.

4 Results and Discussion

Having discussed with many physicians and domain experts, we have prepared to get important attributes out of raw data. Patients of various ages with different ailments were taken. Total data collected from AFID is around 750 patients; Men 332 (44 %) and Women 400 (56 %).

Age of the patient is a vital attribute and we distribute age into five different categories as follows.

- Very Young (1–20 years)
- Young (21–40 years)
- Mature (41–50 years)
- Old (51–60 years)
- Very Old (Age > 61).

4.1 Attributes Selection

Initially correlation matrix was applied to find important attributes on the basis of their hidden behaviour towards each other. The rapid minor supports many built in

Table 3 Selected attributes for clustering

Table	Type
R_Procedure	VARCHAR
R_Diagnosis	VARCHAR
R_Teeth_No	VARCHAR
Pr_Time consuming	Integer
Pr_Critical	Integer
Pr_Time to heal	Integer
Pr_Complexity	Integer
P_ Gender	Binary
P_Age	Integer
P_Marital status	Binary
P_City	VARCHAR (50)

learning algorithms for correlational attributes. We selected correlation matrix and applied on 750 patients' records that were created in MySQL and then exported to Excel 2013.

The records of the database contain 30 attributes, out of which 16 attributes were chosen through correlation matrix. And eventually we have taken 11 attributes on the basis of expert's opinion as shown in Table 3.

4.2 K-Means Method

This k-Means algorithm proceeds with an input parameter k and dissects the set of n objects into k number of clusters in a way where resulting intra-cluster resemblance is high while the inter-cluster resemblance is low. Cluster likeness is determined in regard to the mean value of the objects in a cluster, which can be observed as the cluster centroid.

After making the data structured, an unsupervised learning technique; K-Mean clustering was used with k value 4 to analyse data as shown in the Fig. 2. There are four clusters separated by any specific attribute as per requirement and analysis.

On the basis of diagnosis and total of the ranks as shown in Table 4, the panel of doctors determines a patient to be in one of the following classes:

- None or Normal
- Mild Dental Patients
- Moderate Dental Patients
- Severe Dental Patients

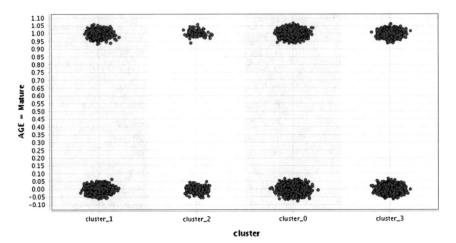

Fig. 2 K-Means clustering

Table 4 Class labels

Procedure	Diagnosis	Time consuming	Critical	Time to heal	Complexty	Sum	Class
Periodic oral evaluation	Oral hygiene	1	1	0	1	4	None
Crown- porcelain/ceramic substrate	Cavity	5	5	2	5	17	Moderate
Sealant—per tooth	Cavity	2	2	2	2	8	Mild
Complete denture maxillary	Missing teeth	5	3	3	3	14	Moderate
Comprehensive orthodontic treatment of the adolescent dentition	Restoration	10	9	7	10	36	Severe

To determine the state of the dental patient, we need to determine the following attributes based on the diagnosis:

Time to heal: A patient's time to heal after a particular procedure is performed

Complex: Complexity of a procedure

Critical: Criticality of a procedure, possibility that the patient may get infected by other infections as well.

Time Consuming: Rank on the basis of time required to perform a particular procedure

Hence, there are total four clusters, clustered in such a way that they belong to 4 different classes; normal, mild, moderate, severe dental patients. Most of the patients belong to mild and moderate dental patient's class and very few to severe class. Based on the deep analysis of results of the clusters, some patterns significant to dental problems are extracted. These are stated below:

- The greater no. of male patients under the age of 20 in dataset indicate that they are more prone to have treatments; "sealant—per tooth" and "Unerupted tooth" respectively.
- The results show that young people whose age generally lie in 21–30 years range, tend towards "resin-based composite—one surface, posterior".
- Trend shows, in mature age, male patients most often need to take treatment named "amalgam—one surface, primary or permanent".
- Old and very aged men tend to have "surgical removal of erupted tooth requiring removal of bone and/or sectioning of tooth and including elevation of mucope-riosteal flap if indicated" and "complete denture—maxillary" respectively.
- Females same as males with age less than 20, are more likely to have "sealant—per tooth".
- In females from age 21–30, trend shows an incline towards this treatment; "resin-based composite—one surface, posterior".
- Females of mature and old age, both are likely to have "amalgam—one surface, primary or permanent".
- For females older than 60, there are greater chances to have treatment; "crown—porcelain fused to high noble metal".

4.3 Supervised Learning Techniques

Data has been made structured by pre-processing techniques and labelled through unsupervised clustering. Now supervised learning techniques mainly classification algorithms are applied on the labelled data to predict the class. But data needs to be further tweaked in order to apply classification algorithms. So following two steps, balancing and splitting were performed on the dataset before using the classifiers.

4.3.1 Balancing the Dataset

Data set should be stratified that means balanced and evenly distributed in the classes to remove the suspicion of bias. But this may cause decline in the accuracy at the same time.

4.3.2 Splitting the Dataset

Data is then split in to two equal parts, one is used to train the data and the other one is for the testing purposes. So that a model has developed for the new records for the testing.

4.3.3   Classifiers

The research includes following different classifiers to predict the class;

- Decision Tree
  Decision tree learning uses a decision tree as a predictive model that plots observations about an item to deduce its targeted value. It's one of the predictive modelling methods used in data mining and machine learning. In the tree structures, leaves represent class labels and branches represent conjunctions of features that lead to those class labels.
- K-Nearest Neighbours Algorithm (KNN)
  KNN is one of the top ten data mining algorithms used for the classification and regression. It is known as lazy learner that makes predictions based on KNN labels assigned to test sample [34]. KNN is a type of instance-based learning, or lazy learning, where the function is only approximated locally and all computation is deferred until classification. The k-NN algorithm is among the simplest of all machine learning algorithms. KNN works as follows [35]:

  – It assembles a training set of data D
  – It selects the initial value of K, As there is no standard way followed to set the value of K, so it is selected randomly based on experimental results. The value of K is fixed according to the required results of the sample data. This is why selection of K-value is still an issue.
  – It measures the distance between the sample point X and to its K neighbours by using the Euclidian distance formula. The distance between these samples is defined as:

$$d(X, Y) = \text{sqrt}(\sum_{i=1}^{n}(xi - yi) \tag{1}$$

- Decision Stump
  A decision stump is a machine learning model consists of a level-one decision tree. It is a decision tree with one internal node (the root) that is immediately connected to the terminal nodes (its leaves). A decision stump makes a prediction based on the value of just a single input feature. Sometimes also called 1-rules. It's a tree graph where the leaves denote the classification results while nodes show some predicate. It's a tree traverse process where the start is taken from root and then we descend further [36].
- Artificial Neural Networks (ANNs)
  In machine learning and cognitive science, artificial neural networks are a family of statistical learning models motivated by biological neural networks (the central nervous systems of animals, in particular the brain) and are used to approximate the functions that depend on a large number of inputs and are usually not known. Artificial neural networks are generally presented as systems of interconnected "neurons" that exchange messages between each other. These connections have numeric weights that can be adjusted based on experience, making neural nets adaptive to inputs and capable of learning.

Table 5 Classification algorithms, accuracy results

S. No	Classifier	Accuracy (%)
1	Decision tree	95.65
2	KNN	93.43
3	Decision stump	75.20
4	Neural network	60.01

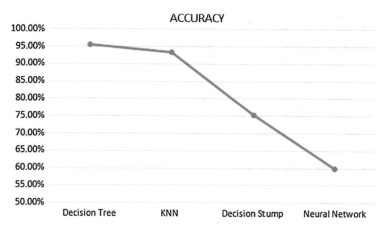

Fig. 3 Accuracy comparison

Experiments have been performed on the dataset and results have also been analysed respectively. The research consists of multiple classes of dental patients as mentioned before. The classifiers applied to predict the targeted patient class and the results evaluated, support decision tree classifier on the basis of its accuracy. Table 5 shows the classifiers and their accuracy level.

Figure 3 shows the graphical representation of the attained results. It can be seen through the comparison of applied classifiers that decision tree has better results over the others with an accuracy of 95.65 %. The next closest is KNN but the other two results have no match with the formers.

4.4 Automation of the Manual Processes

Automation of the manual processes that were being used in AFID was done by developing a dental expert system in JAVA using wave-maker tool. DES takes the user information from the patient including his name, age, gender, location etc. Having examined the patient, medical expert selects the diagnosis from the provided options (covers a wide variety of dental problems) as shown in Fig. 4. And the DES decides accordingly the main category or department of the patient to refer to. There

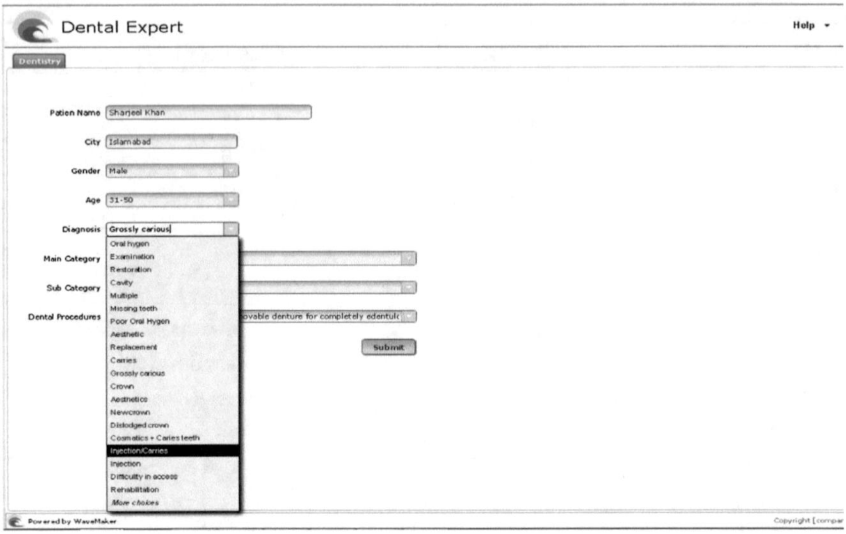

Fig. 4 Dental expert system, diagnosis

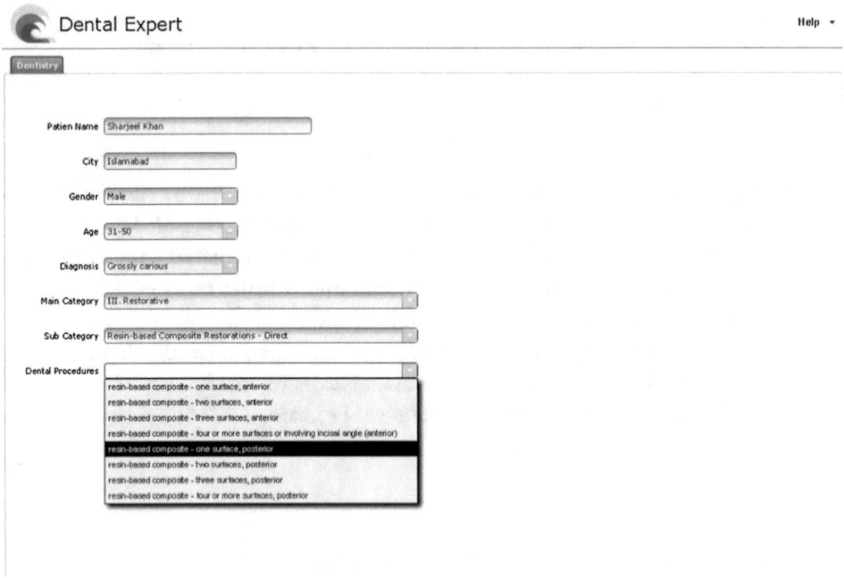

Fig. 5 Dental expert system, procedure categories

are twelve main categories according to the departments in AFID. The DES further divides these main categories into 46 subcategories to be more decisive.

As shown in Fig. 5, On the basis of main category, the DES filters the subcategories related to the patient. Then lastly the DES recommends the dental procedures to the

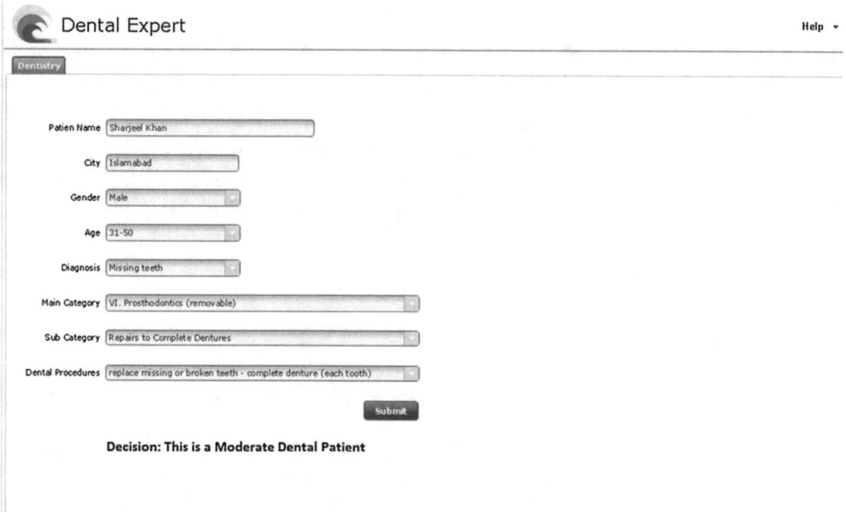

Fig. 6 Dental expert system, patient's condition

medical expert to perform on that patient. Having taken all the inputs, the systems depicts the patient's condition.

As shown in the Fig. 6, In this case DES reveals the patient belongs to a moderate class or cluster and hence needs moderate care while being cured.

5   Conclusion

The K-Means was applied using Rapid Minor 5.3. It clustered data into various chief cluster units along with their class labels. The clusters were varied in four divisions. Not only demographic analysis through clustering of the dental medical records was done but labelling of the dataset was also accomplished. "After balancing the labeled dataset" different classifiers were used to train, test and then predict the class accurately. We found decision tree runs better than others on our dataset in terms of its accuracy in predicting the targeted patient class. Most of the records belong to mild and moderate dental patient's class and very few of them belong to severe dental class.

An expert system has also been developed to automate the manual use cases and to ease up the processes and decision making for medical experts. The DES provides possible treatment options according to the inputs and categorizes the patient as soon as he comes to the medical expert. In case, the patient is of severe condition, the system identifies the dental patient as severe and hence he is provided extra care while being cured.

Healthcare organizations have huge datasets which come from assorted sources. That data is not always suitable in structure or quality. Medical analysis is a significant but difficult task that ought to be accomplished precisely and proficiently and its computerization would be very expedient and beneficial. As we made a lot of effort to make it a well-structured data so we further intend to apply the association rule based mining. As a future work other important attributes and procedures like medication etc. may also be included that could play important role in decision making.

References

1. Fayyad, U., Shapiro, P.G., Smyth, P.: From data mining to knowledge discovery in databases. AI Mag. **17**(3), 37 (1996)
2. Nordhausen, K., Hastie, T., Tibshirani. R., Friedman, J.: The elements of statistical learning: data mining, inference and prediction. Int. Statist. Rev. **77**(3), 482–482 (2009)
3. Han, J., Kamber, M., Pei, J.: Data Mining: Concepts and Techniques. The Morgan Kaufmann Series in Data Management Systems. Morgan Kaufmann Publishers (2011)
4. Shital, C.S., Andrew, K., Michael, A., Donnell, O.: Patient recognition data mining model for BCG-plus interferon immunotherapy bladder cancer treatment. Comput. Biol. Med. **36**, 634–655 (2006)
5. Depeursinge, A., Iavindrasana, J., Hidki, A., Cohen, G., Geissbuhler, A., Platon, A., Müller, H.: Comparative performance analysis of state-of-the-art classification (2010)
6. Fatima, M., Basharat, I., Khan, S.A., Anjum, A.R.: Biomedical (cardiac) data mining: Extraction of significant patterns for predicting heart condition. In: 2014 IEEE Conference on Computational Intelligence in Bioinformatics and Computational Biology, pp.1–7, 21–24 May 2014
7. Abbas, S.M., Basharat, I., Khan, S.A., Qureshi, A.W., Ahmed, B.: A framework for clustering dental patients' records using unsupervised learning techniques. In: IEEE Science and Information Conference 2015, London, UK
8. Abbas, S.M., Basharat, I., Khan, S.A., Qureshi, A.W., Ahmed, B.: Discovering patterns in dental medical records using association rule based mining. In: The 2015 International Conference on Soft Computing and Software Engineering (SCSE 2015), University of California, Berkeley, USA
9. Ghattas, J., Peleg, M., So_er, P., Denekamp, Y.: Learning the context of a clinical process. In: Rinderle-Ma, S., Sadiq, S., Leymann, F. (eds.) BPM 2009 International Workshops, Ulm, Germany, September 7, 2009. Revised Papers, volume 43 of Lecture Notes in Business Information Processing, pp. 545–556. Springer, Berlin (2010)
10. Mans, R.S., Schonenberg, M.H., Leonardi, G., Panzarasa, S., Quaglini, s., Van der Aalst, W.M.P.: Process mining techniques: an application to stroke care. In: Andersen, S.K. (ed.) Health Beyond the Horizon—Get IT There, Proceedings 21st International Congress of the European Federation for Medical Informatics, MIE 2008, volume 136 of Studies in Health Technology and Informatics, pp. 573–578. IOS Press (2008)
11. Mans, R.S., Schonenberg, M.H., Song, M.S., Van der Aalst, W.M.P., Bakker, P.J.M.: Process mining in healthcare: a case study. In: Azevedo, L., Londral, A.R. (eds.) Proceedings of the First International Conference on Health Informatics (HEALTHINF 2008), vol. 1, pp. 118–125. INSTICC Press (2008)
12. Poelmans, J., Dedene, G., Verheyden, G., van der Mussele, H., Viaene, S., Peters, E.: Combining business process and data discovery techniques for analyzing and improving integrated care pathways. In: Perner, P. (ed.) Advances in Data Mining. Applications and Theoretical Aspects 10th Industrial Conference, ICDM 2010, volume 6171 of Lecture Notes in Computer Science, pp. 505–517. Springer, Berlin (2010)

13. Rebuge, A., Ferreira, D.R.: Business process analysis in healthcare environments: a methodolgy based on process mining. Inf. Syst. **37**(2), 99–116 (2012)
14. Shan, Y., Jeacocke, D., Murray, D.W., Sutinen, A.: Mining medical specialist billing patterns for health service management. In: Roddick, J.F., Li, J., Christen, P., Kennedy, P.J. (eds.) Conferences in Research and Practice in Information Technology, vol. 87, pp. 105–110 (2008)
15. Yang, W.S., Hwang, S.Y.: A process-mining framework for the detection of health care fraud and abuse. Expert Syst. Appl. **31**, 56–68 (2006)
16. Li, H.H., Wang, S-L., Chiang, Y.Y., Fu, S.K.: Using data mining to investigate dental cancer claiming data. National Chung-Cheng University (1), Central Taiwan University of Science and Technology (2), PRZEGLĄD ELEKTROTECHNICZNY (2013)
17. Chou, S.M., Lee, T.S., Shao, Y.E., Chen, I.F.: Mining the breast cancer pattern using artificial neural networks and multivariate adaptive regression splines. Expert Syst. Appl. **27**, 133–142 (2004)
18. AbuKhousa, E., Campbell, P.: Predictive data mining to support clinical decisions: an overview of heart disease prediction systems. In: IEEE 2012 International Conference on Innovations in Information Technology (IIT), pp. 267–272
19. Hartigan, J.A., Wong, M.A.: Algorithm AS 136: a k-means clustering algorithm. Appl. Stat. **100–108** (1979)
20. Vijayarani, S., Sudha, S., Comparative analysis of classification function techniques for heart disease prediction. Int. J. Innovative Res. Comput. Commun. Eng. **1**(3) (2013)
21. Parpinelli, R.S., Lopes, H.S., Freitas, A.A.: An ant colony based system for data mining: applications to medical data. In: Proceedings of the Genetic and Evolutionary Computation Conference (GECCO-2001), pp. 791–797
22. Benko, A., Wilson, B.: Online decision support gives plans an edge. Managed Healthc. Executive **13**(5), 20 (2003)
23. Han, J., Kamber, M.: Data Mining: Concepts and Techniques. Morgan Kaufmann Publishers, USA (2011)
24. Balasubramanian, T., Umarani, R.: An analysis on the impact of fluoride in human health (dental) using clustering data mining technique. Pattern Recogn. Informat. Med. Eng. (2012)
25. Al Haidan, A., Abu-Hammad, O., Dar-Odeh, N.: Predicting tooth surface loss using genetic algorithms-optimized artificial neural networks. Comput. Math. Methods Med. **2014**, 7 Article ID 106236
26. Mans, R., Reijers, H.A., Wismeijer, D., van Michiel, J.I.M.: Processes, Mining, in Dentistry, IHI'12, January 28–30. Miami, Florida, USA (2012)
27. Polášková, A., Feberová, J., Dostálová, T., Kříž, P., Seydlová, M.: Clinical decision support system in dental implantology. Mefanet J **1**(1), 11–14 (2013)
28. Nanyue, W., Dawei, H., Tongda, L., Zengyu, S., Yanping, C.: Comparative study of pulse-diagnosis signals between 2 kinds of liver disease patients based on the combination of unsupervised learning and supervised learning. IEEE Int. Conf. Bioinform. Biomed. (2013)
29. http://www.theiia.org/intAuditor/itaudit/archives/2006/august/data-mining-101-tools-and-techniques/. Accessed 24 March 2014
30. Michalski, R.S., Carbonell, J.G., Mitchell, T.M.: Machine Learning: An Artificial Intelligence Approach. Tioga Publishing Company (1983). ISBN 0-935382-05-4
31. Kotsiantis, S.B.: Supervised machine learning: a review of classification techniques. Informatica (03505596), **31**(3) (2007)
32. http://www.kdnuggets.com/2013/09/mikut-data-mining-tools-big-list-update.html. Accessed 24 March 2014
33. Mikut, R., Reischl, M.: Data mining tools. Wiley Interdisc. Rev. Data Min. Knowl. Discov. **1**(5), 431–443 (2011)
34. Yan, Z., Xu, C.: Combining, KNN algorithm and other classifiers. In: Sun, F., Wang, Y., Lu, J., Zhang, B., Kinsner, W., Zadeh, L.A. (eds.) Proceedings of 9th IEEE International Conference on Cognitive Informatics (ICCI'10) . 978-1-4244-8040-1/10/$26.00 2010 IEEE

35. Yan, S., Shipin, L., Yiukun, W.: KNN classification algorithm based on the structure of learning [J]. Comput. Sci. **34**(12) (2007)
36. Yang, G., Zhou, Z., Yu, X.: Hyperspectral imagery classification based on gentle adaboost and decision stumps. J. IEEE **978** (2009). 1-4244-4994-1/09

Technology in Primary Schools: Teachers' Perspective Towards the Use of Mobile Technology in Children Education

Rabail Tahir and Fahim Arif

Abstract Today technology is progressively being recognized as a significant learning tool for helping young children in developing their cognitive, social and learning skills. Now a day's even young children are exposed to the latest technology such smartphones, tablets and e-readers as observed by many teachers and parents. The new mode of technology is considered to present some potential as an educational tool. Many new platforms are available for the educational media content. Undoubtedly technology is an important element in the lives of most children now days. Although many schools have also incorporated the use of technology as a learning tool in their curriculum still some researchers and teachers have lots of concerns regarding the use of technology in schools and specifically the use of mobile technology by young children. In this paper we have conducted a survey with the teachers of 12 different primary schools in Pakistan ($N = 104$). This paper is an attempt to investigate the use of technology in primary schools for children and teachers. The paper also explores the attitude of teachers towards the use of mobile technology for primary school age children and specifically in the context of education by using educational or learning applications (apps) for children both in homes and in school environment. This paper also sheds light on the use of technology in primary schools and also aspires to provide the guidance in order to overcome the concerns of teachers regarding technology usage in schools and to increase the accessibility of mobile technology in education for young children.

Keywords Mobile technology · Technology in school · Mobile learning · Teachers perspective · Educational applications

R. Tahir (✉) · F. Arif
Department of Computer Software Engineering, Military College of Signals,
National University of Sciences and Technology (NUST), Islamabad, Pakistan
e-mail: rabailtahir.mscs19@students.mcs.edu.pk; rabailtahir_4@yahoo.com

F. Arif
e-mail: Fahim@mcs.edu.pk

© Springer International Publishing Switzerland 2016
L. Chen et al. (eds.), *Emerging Trends and Advanced Technologies
for Computational Intelligence*, Studies in Computational Intelligence 647,
DOI 10.1007/978-3-319-33353-3_6

1   Introduction

Today technology is progressively being recognized as a significant learning tool for helping young children in developing their cognitive, social and learning skills. Although some researchers are against the use of technology for the learning of children, in spite of this the effects of technology have been extensively recognized in the area of education and have proved to be very positive for the development of young children. Children who make use of technology show better language skills, intelligence, structural knowledge and problem solving skills as compared to children who do not use technology for their learning [1]. Today's children are more exposed to the advance technology even at an early age. The experiences with the latest technology can surely pave the way for extraordinary learning opportunities. On the other hand without an educational element, technology cannot reach to its full potential for supporting the learning and development of children. The educational component in early childhood programs often means that an adult is nearby, interacts with the children and also provide the opportunities for peer-to-peer learning in order to encourage and help children gain the skills they require for success [2].

Undoubtedly technology is an important element in the lives of most children now days. Many schools have also incorporated the use of technology as a learning tool in their curriculum in order to provide opportunities for teachers and children to become comfortable with the use of modern technology as an effective learning tool. While simultaneously, some other researchers in the field are unsure about use of educational technology in classrooms for young children [3]. Teachers and parents have lots of concerns regarding the use of computers and mobile technology by young children. They also find it difficult to understand if it is for the best because these technology products were not there in their childhood. Some think that technology is valuable for the future lives of children whereas others are of the thought that children should be playing outside or reading a book rather than using technology [4]. Teachers fear that if computer technology penetrates into teaching than by the use of programmed instruction and electronic worksheets children will fail to benefit from the key experiences that aid in their development. These concerns however are not baseless, as there are many popular software programs available for young children that emphasize on learning at the cost of developmentally inappropriate methods. Numerous advances in the field of educational technology, however, may lead the childhood educators to rethink about the potential of technology usage in classrooms [3]. Conventionally, the teachers of preschool send much time in planning and preparing the learning materials, which usually leads to teachers feeling fatigued. Moreover such materials require a lot of storage. On the contrary, interactive technology is much more attractive, flexible, and sustainable. The interactive devices are considered to make the preschool children learn their study materials much easier. Whereas the interactive technology is also effective for primary school children as the interactive whiteboards are more popular in the primary schools than in preschools. An introduction to iPad for preschoolers is a possible solution. More over it is evident that Apple iPad has been widely used in the education activities of preschoolers since 2011 [5].

The national standards of educational technology developed by International Society for Technology in Education (ISTE 1998) provides a clear idea of skills and knowledge that young children should be acquiring in the early years. These standards can serve as a guideline for selecting the applications and programs for children usage. As children reach the primary school age, they usually start to use technology tools as a part of their academic studies in school. Children should be exposed to developmentally appropriate, collaborative and creative uses of technology [3]. Mobile devices are increasingly becoming popular and their capability and quality is increasing because of the technological breakthrough in data networks and advancements in the wireless bandwidth [6]. Mobile learning (m-learning) is possibly the largest growth area in the entire field of ICT in education. It comprises of any kind of learning that is intervened through a mobile device. These devices include smartphones, digital media players, tablets and personal digital assistants. They are different from portable devices like laptops which can also be moved to different locations but they still lack the flexibility and convenience of smaller handheld devices [7]. Mobile learning can complement the other learning and teaching methods or even replace them. Mobile technology is important for education because it offers: (i) wireless access to internet resources, other students and educational materials, (ii) a number of communications channels on a single device, such as, email, text and voice messaging and (iii) cheaper as compared to technology such as desktops or laptops. According to a Student Monitor survey in United States, the usage of student's mobile phones is 40 % in junior high schools and 75 % in the high schools and 90 % colleges. In some countries such as United Kingdom, Sweden, Italy and Czech Republic the usage of mobile devices is as high as 100 %. Moreover, students in Germany, China, and Philippines are using the mobile devices to study math, learn English; spellings and also to access university lectures [6].

According to a recent research the mobile phone usage by younger students will shift to smart phones in the next 3 years whereas PDAs no longer be popular apart from where they assist in specialist courses. In spite of the significantly growing market, still there is a lack of research work in this field in many countries, including Pakistan market. Previous researches focused on older children's use of technology and tend to focus on the screen-based media mainly computers. Up to now very less attention has been paid towards the young children use of technology and role of technologies for education of young children. Therefore research in this area in still immature [8].

Our previous research [9] examined the attitude of parents towards the use of mobile technology specifically educational applications for children aged six to ten years in Pakistan. This research aims to investigate the use of technology in primary schools and teacher's usage of mobile technology to aid teaching and for improving children education. This research is carried out to explore the perspective of teachers towards the use of mobile technology and educational apps for primary school age children. This age group is considered because the primary/elementary school years are considered as the important period of child mental, physical and emotional development and child starts to develop logical and concrete operational thinking. It is the first step and foundation for a child. Therefore the analysis is performed on the basis of a survey conducted with the teachers of primary schools in Pakistan. The main aim of this research is to highlight the use of technology in Pakistani schools and the attitude of teachers towards the use of mobile technology and educational apps for young children addressing the key concerns in use of technology: both by them to support their teaching practices and by children for their education and learning. It also presents some key findings and suggestions in order to overcome the issues and key concerns of teachers and to make mobile technology more effective and acceptable to be employed by schools for children learning. This is the first study in Pakistan to the best of our knowledge to attempt to quantify, on the national level, the attitude of teachers towards the use of mobile technology for children. The research investigates the followings aspects: use of technology in primary schools of Pakistan (both public and private schools), primary school education, favorite activities of children in school and technology usage by teachers to support their teaching practices. It also presents a measure of teachers' experiences with educational applications used for children: investigating the subjects or topics teachers think could be learnt most by children with the use mobile technology, technology resources they feel are most effective and the considered benefits of mobile technology and educational apps. This research study also explores the key concerns or obstacles in the use of mobile technology for children: finding out the reasons why teachers and school administration restrict the use of mobile devices with children. Finally, the demographic information such as gender, age, school description and classes they teach is also presented.

This paper includes following sections: (ii) preliminary research, (iii) research methodology, (iv) results and discussions, (v) conclusion and future work.

2 Preliminary Research

2.1 Mobile Technology for Children Learning

Mobile technologies are connecting with nearly all aspects of our lives including different tools, devices, applications, virtual environments and online environments for educational opportunities, providing 24/7 learning at the learner's preference. Mobile devices are supporting educators to develop a new community of learning for and by the students of today making use of, tablets, iPads, smartphones, iPod and other mobile devices to stay connected. Mobile learning is described as the ability to provide or obtain the educational content on mobile devices such as smartphones, PDAs, tablets and mobile phones. The educational content is the digital learning assets that includes any form of media or content that is made available on a mobile device. Mobile learning using the handheld devices is in its early stages in terms of both pedagogies and technologies. Consequently still there is some dispute amongst researchers regarding the way mobile learning should be defined: in terms of technology and devices; mobility of learners and learning, experience of learner while learning with mobile devices and also the advantages and disadvantages of learning [10]. The new generation of these mobile technologies is changing the ways of children learning. As children are always excited to use mobile devices, therefore mobile learning is very engaging and it provides children with new ways of relating their experiences with abstract knowledge. This development in the educational technology has presented an important shift in the way new technology can be used to stretch the minds of young children. However on the other hand it is also observed that when children are given a mobile device to be used, in support of some other activity they usually gets distracted and concentrates only on the device. This can isolate children from others around them, reading and listening to what is on the mobile device only. Hence it is very important to ensure that children do not focus on technology too much and avoid over excessive use [11].

In [12] researcher aimed to propose a theory of learning intended for a mobile society. It includes both learning that is supported by the mobile devices such as mobile phones, personal audio players, portable computers and also the learning in an era described by the mobility of people as well as knowledge. The convergence between the new learning and the new technology can be seen as presented in the Table 1.

In [13] researcher surveyed the university students to determine the patterns of mobile devices usage, functions of mobile phone they use and types of the educational activities they think are useful for mobile phones. Two types of materials were developed and introduced for studying English as a foreign language for learners in Japan on mobile devices and presented the reactions of students to such type of learning activities on mobile phones. Similarly in [14] researcher aimed to develop and uphold a realistic understanding of digital technology and young people with a view to support that researchers play a meaningful and useful role in order to support young people technology usage. The paper reflects the actual uses of digital technology and digital information among young people.

Chiong et al. [15] focused on the new types of digital media and their influence on young children and their families in United States (US). The study explored how we can deploy the mobile devices and applications (apps) particularly to help support children learning. It has three parts: the first part discussed the new trends in mobile devices, particularly pass-back effect, that is when the adult passes their own personal device to the child. The second part discussed the results of studies undertaken to

Table 1 Convergence between new technology and learning	New technology	New learning
	Personal	Personalized
	User-centered	Learner- centered
	Mobile	Situated
	Networked	Collaborative
	Ubiquitous	Ubiquitous
	Durable	Lifelong

find out the effectiveness and feasibility of using mobile apps to encourage learning amongst preschool and the early elementary school aged children. Finally, the third part presented the implications of these findings for education and research. Each part of the study explored mobile learning from a different angle.

2.2 Concerns Regarding Children Technology Usage

Teachers and parents have lots of concerns regarding the use of computers and mobile technology by young children. Some think that technology is valuable for the future lives of children whereas others are of the thought that children should be playing outside or reading a book rather than using technology [4]. Some of the researches in this area include following.

In [16] researcher explored the current studies on computers and children. Initial research regarding technology focused on the issues such as access and time children were spending while using technology. As the use of technology became more common, research topics shifted to the issues related to the content and its consequences on children. Now current research on children's technology usage is again also following this pattern. But the increased level of interactivity and communication possibilities these days with mobile technology brings both the chances of enhanced learning and the increased risk of harm as well. As a result, the research regarding the outcomes of exposure to different types of content is being considered very important.

Shields et al. [17] reviewed the concerns related to children use of computers and this research focused on the factors which should be kept in mind when making choices regarding the use of technology for children. As technology is becoming important in everyday lives it is important to assure that the access to technology leads to positive learning experiences for children and also for their development and growth.

A number of myths have emerged about the experience of children with technologies. Plowman et al. [4] in their research selected seven statements both in favor and against the use of technology by children in order to represent the perception from parents, teachers and media. This research work focused on the detailed case studies of 3–4 years old children living in Scotland in order to investigate these myths. The key finding of this paper illustrated a description of technologies encountered by children at home, the influence of family practices on children's use of technology, and why it is important for teachers to know about the experience of children with technology used at home.

This research [3] also highlighted some of the main issues encountered in technology usage with young children. The author described some practices where students significantly incorporated the technology into science curriculum. Coauthors Rosie and Erin also illustrated their experiences regarding integrating technology in the science units in classrooms. The case studies discussed in this research work relate to science curriculum of early childhood, but the applications can also be used in other curriculum areas.

Chou et al. [18] presented the findings from a pilot project conducted in four 9th grade classrooms of geography in a large K-12 school in United States for one to one learning with iPads. The research findings highlighted many technical challenges and possible opportunities for both students and teachers. The positive effects of iPad integration on the learning of student include increased engagement, improved digital literacy, and increased time for the projects. The negative effects mainly include distraction by the numerous irrelevant websites and apps. Considering instructional activities, the positive effects of mobile technology includes the improved teaching practices with updated information and implementation of student centered activities. Whereas challenges faced are the lack of teacher selected mobile apps and also the need for additional time for preparation and conducting training.

2.3 Design of Educational Technology

Mobile technologies now days have a vast potential to transform the education. Therefore these technologies should be designed and implemented in a way that they are appropriate to the cultural and social context of learning. The use of mobile technology in global educational context brings forth the technological as well as sociocultural challenges. Particularly the technology tools and the applications recognized in developed nations may not be well accepted and pose challenges in the developing countries. Therefore 'one technology for all contexts' or 'one size fits all' does not practically work in this case as different people may have different concerns regarding technology which should be overcome to make technology effective and helpful [19]. Each era of technology has shaped education to some extent in its own image. The technological acceptance of education is not to argue, but there is a mutually constructive convergence between the basic technological impacts on culture and the modern educational practices and theories. In this new era of global digital communication, there is an increasing interest in the relationship between mobile technology and learning. However we need an appropriate concept of education for this mobile age. A lot of theories of learning have been evolved but nearly all were based on the assumption that learning or education occurs in a school, by some trained teacher. A few of researchers have developed a theory-based concept of learning outside classroom and schools but none have focused on the mobility of learners and the learning [20].

Lee et al. [5] in their research focused on two main areas: the design of Child Computer Interaction (CCI), and the effectiveness of CCI. Mike et al. [12] paper presented a framework for the design of educational technology that would aid learning from any location for a lifetime. A theory of lifelong learning by technology is defined in this paper and also indicated how it presents the requirements for hardware, software, interface design and communication of a handheld learning device. The researcher concluded with a formative evaluation of a tutor system for 7–11 years old children.

2.4 Integrating Technology in Education

A lot of money is being spent in both private and public areas to give children access to technology in homes and in schools. It is important to notice what part technology is playing in children growth physically, intellectually, physiologically and socially. It is also important to see whether technology is increasing or decreasing the gap between rich and poor [17].

Information and communication technologies (ICT) are broadly considered as improving learning in both schools and homes due to which there is an increase in their rapid adoption in the developed societies [21]. ICTs are progressively becoming important in education. Many educators accentuate student centered and creative pedagogical approaches provided by the technology tools, while the others only focus on the role of these technologies in creating well connected and well-informed global citizens. Despite a difference in the goals of educators and those of governments, ICTs are becoming popular in teaching and learning [7]. In 2008–2009, schools in United Kingdom spent round £880 million on ICT. Almost half of the primary school children use digital resources of one kind or the other at least weekly (46 % in math's, 43 % in English, and almost 30 % in science) however only less than one in ten secondary school children (10 % science, 8 % English and 7 % math's). So, with the support of government policies to give internet access for every school and every child, with the support of industry in a variety of digital education initiatives, and with families getting internet access even at home it is evident that digital technology would be as significant in twenty-first century as the book in nineteenth century [21]. In spite of the assumption that incorporation of ICT impacts the whole school system, research directed on ICT in schools is usually limited to the analysis of variables at class level only. In contrast to this research, the present study [22] investigated the integration of ICT from a school improvement approach. More specifically, it explored the local policy of school with respect to integration of ICT from both the perceptions of teachers and the principal's perspective. A sample size of 53 primary school principals was selected in order to answer the research questions. Additionally, the interview data were also supplemented with the survey data obtained from 574 teachers from same schools. According to the results the school related policies have a considerable effect on use of ICT in class. The results also showed that the school policies are mostly underdeveloped or underutilized [22].

This paper [23] discussed the strategies and challenges to access and utilizes the learning arenas in schools which can be alternative to formal learning arena. Lilla et al. in [2] discussed the criteria and offered a practical plan and tool for using, evaluating and integrating the educational technology that is developmentally appropriate for children in early childhood programs, involve tools to support teachers in successfully implementing the technology, and to make the educational technology integrated into the curriculum and classrooms.

The popularity of mobile technology has dramatically increased the in recent years. The penetration of mobile phones is over 100 % in the most developed countries indicating duel device ownership According to the "Horizon Report" mobile devices would be extensively adopted in education in the coming future. According to a latest report from Ericsson, research has indicated that by 2015, 80% of people will be accessing internet from mobile devices. In terms of importance in education the internet capable mobile devices will possibly outnumber the computers within a year. Schools around the world have started to explore the affordances of mobile technologies for teaching a range of skills across different curriculum areas. As the mobile technologies are new for the classroom environment, there is very little published research is available on the type of devices currently being employed at different levels and how they are included in different areas of curriculum, what benefits and challenges are involved, and how these are handled. This study tried to fill this gap with a thorough research on current practices in the independent schools in western Australian [7]. Mark et al. [7] in their paper presented the adoption of mobile technology in 10 independent schools in Australia, based on the interviews conducted with staff in 2011. According to the results iPads were considered as the most popular device, followed by iPhone and iPods. In most schools mobile learning (m-learning) was still at an experimental stage only, but the subject matter related to the need to incorporate mobile devices into learning ecology were already emerging. With focus on their role in collaboration, promoting production or consumption, personalization, and creating untarnished learning spaces. Used for both pedagogical and organizational purposes, mobile technology was considered as increasing student motivation, with empirical facts showing improved student learning according to a small scale research conducted by 2 schools. Some of the challenges faced were the need to wisely manage technology, ethical issues involved in its use, and the possible roles of staff in its consumption. Professional development (PD) was considered important for staff and almost all schools selected in this study had planned to expand their use of mobile technologies in classrooms in future.

2.5 Teachers and Technology for Education

The educators have accepted the use of mobile technology as an effective learning tool for children and have also integrated them into distance education. This research work [24] compared mobile learning with electronic learning and ubiquitous learning. The researcher also highlighted the technological and pedagogical aspects of mobile learning presented in the prior research studies. Hence this paper provided useful information regarding the concepts associated with mobile learning and the way different technologies can be incorporated into learning and teaching effectively. According to this research mobile technology exhibit a unique attribute to provide an efficient face to face communication whilst students are making use of mobile devices in the classroom. The ownership of mobile devices increases the involvement of user in the learning process. According to many researchers the lower cost of mobile devices as compared to other technology is also an added advantage [24].

Norazah et al. [6] performed a descriptive statistics in order to examine the use of mobile technology by the lecturers for teaching purpose. A sample size of 20 lecturers from University (UNISEL), Malaysia (Faculty of Industrial Art and Design Technology) was selected as the participants. The results indicated that the lecturers acknowledge the notion of mobility in learning and the game play is considered as an interesting concept to support the classroom learning.

Shazia Mumtaz [25] presented the literature related to effective uptake of information and communication technology (ICT) by teachers. Research work showed several factors that influence the decision of teachers to use ICT in classroom. These factors included, quality of hardware and software, ease of use, access to resources, incentives to change, support in their school and national and polices, background in formal computer training and commitment to professional learning. The paper highlighted the role of pedagogy and also suggested that the viewpoint of teachers regarding learning and teaching with ICT are fundamental to integration. It is recommended that the successful implementation of ICT require focusing on three interconnected frameworks for transformation: school, teacher and the policy makers.

Sugar et al. [26] investigated the beliefs of teachers about adoption of technology as a deliberate, intentional, reasoned, decision-making process, as described in Ajzen's Theory of Planned Behavior (1985). Both quantitative and qualitative data were obtained from teachers in 4 schools located in United States. Overall results showed that the decisions of technology adoption were inclined to the individual attitudes of teachers towards the technology adoption, which were based on some particular basic personal beliefs regarding the outcomes of adoption. External support and relative resources (such as funding) were trivial factors affecting the technology adoption decisions of teachers. From the results obtained, author recommended that the school administrators should closely work with teachers in order to address the concerns regarding technology adoption and also provide support and resources. Moreover the educational software designers should also focus on designing effective future technology for teachers.

For years, researchers have been interested in finding the factors affecting the use of mobile technology in classroom. This research [27] focused on the educational beliefs of teachers as precursor of computer use, and at the same time controlling the influence of demographical variables (age, gender etc.) and the technology related variables (general computer attitudes, computer experience etc.). Multilevel modeling was used in order to find out the variation in determinants of computer usage in classroom ($N = 525$). The research supported the assumption that teachers' beliefs are important factors in explaining the technology adaptation by teachers in the classroom. The results showed that traditional beliefs have a negative impact while the constructivist beliefs have a positive effect on the classroom use of computers.

Henryk et al. [28] investigated the analytics of a number of personal variables of elementary school teachers regarding their use of computers available for teaching. The sample size was 170 teachers from elementary schools ($N = 170$). Teachers from 4 different schools were asked to fill the questionnaires that encompassed the following categories: levels of computer use, teacher focus of control, relevance of computers to teaching, innovativeness and self-confidence in computers usage.

Moreover, data on gender, age and years of experience with computer were also collected. Logistic regression method was used to examine the relationships between computer use and teacher characteristics. About half of the sample of teachers in this study reported not using computers for teaching. Innovativeness and self confidence were closely related to teachers' computer use according to the results obtained from questionnaire. The research results suggested that these two variables need to be taken into account when scheduling training intervention or for the differential staffing for unifying teachers and their computer usage.

3   Research Methodology

The methodology used in this research is survey. For this research study, 104 teachers from 12 primary schools in Pakistan have been approached; only teachers teaching primary classes i.e. grade 1–5 were approached for the survey. We requested the participants to fill the questionnaire according to their perspective about the use of mobile technology for children and what they actually practice. In Pakistan the age of elementary/primary school children range approximately from 6 to 10 years therefore teachers were asked to keep this age in mind when answering questions.

The teacher's survey was conducted in Islamabad, Pakistan. We approached 12 primary schools located within Islamabad, Pakistan and distributed the survey questionnaires among teachers and asked them to fill the questionnaires. The next day survey questionnaires were collected from them. Additionally multiple Facebook groups and websites for teachers and different elementary/primary schools in Pakistan also proved to be very helpful in distributing questionnaire and collecting survey data. A total of 120 teachers were approached but only 104 provided us with their feedback. From Islamabad we got response from 56 teachers, 27 were form Lahore and rest of the 21 respondents were from different cities Faisalabad, Karachi, Hyderabad and Sargodha.

All participants were told about the purpose of this survey by adding an introductory paragraph in the start of the survey explaining the purpose of this study. The participants were also ensured that their personal data will be kept confidential and will not be disclosed without permission. The survey questions covered several topic like education in primary schools, technology usage in schools for children, favorite activities of children in school, attitude of teachers towards children use of mobile technology in schools and at home, technology resources beneficial for teaching, reasons to restrict the use of mobile technology in school, benefits of using educational apps for children, effectiveness of educational applications in children learning and the content they choose for children's educational technology. The participants were asked a total of 20 questions. The survey questions encompass the following categories presenting the summary of the survey findings.

- Demographic Information
- Experience with Mobile and Educational Applications

- Technology Usage in School
- Elementary School Learning
- Favorite Activities of Children
- Teacher's Attitude towards Technology Usage for Children
- Reasons to Restrict the Use of Mobile Technology in School
- Technology Resources Used to Support Teaching Efforts
- Mobile technology Usage Considered Beneficial for Teaching
- Benefits of Mobile Educational Applications for Children
- Selection of Educational Content

It should be noted that the findings in this study are based on the teachers' responses to the questions regarding use of mobile technology in schools. Teachers from both public and private sector schools participated in the survey however the ratio of private school teachers was higher so that might influence the results. This research specifically explores the use of mobile technology with focus on educational applications.

4   Results and Discussions

This section covers the results obtained through the survey conducted with teachers. In this study the term "children" refers to the children of primary school age and "teachers" refer to the teachers of primary school children.

4.1   Demographic Information

4.1.1   Gender

Out of 104 participants, 35 were male and 69 were female. The distribution of survey participants with respect to gender is shown in Table 2.

Table 2 Gender distribution

	Male	Female	Standard deviation	Total responses (N)
Data	35	69	0.47	104
Percentage (%)	33.65	66.35		

4.1.2   Age

Out of 104 participants, 44 were in the age range of 20–29 years, 41 were in the age range of 30–39 years and 19 were above 40. The distribution of survey participants with respect to age is shown in Table 3.

4.1.3   School Description

Teachers from both public and private sector schools participated in the survey as there is a lot of difference in the public and private schools in Pakistan in terms of technology and other facilities. Among the survey participants, 62 were from private schools (59.62 %) and 42 were teaching in public school (40.38 %). Table 4 depicts the results of school description.

4.1.4   Financial Background of Children Families

According to teachers who participated in the survey, the financial background of majority of the children studying in these schools were from middle income families (66.02 %), whereas 46.6 % were from high income and 40.78 % from low income families. Table 5 presents the results of financial background of families.

Table 3   Age distribution

	20–29	30–39	40 above	Total responses (N)
Data	44	41	19	104
Percentage (%)	42.30	39.42	18.27	

Table 4   School description

	Public/government school	Private school	Standard deviation	Total responses (N)
Data	42	62	0.49	104
Percentage (%)	40.38	59.62		

Table 5   Financial background of families

	Low income	Middle income	High income	Total responses
Data	42	68	48	103
Percentage (%)	40.78	66.02	46.6	

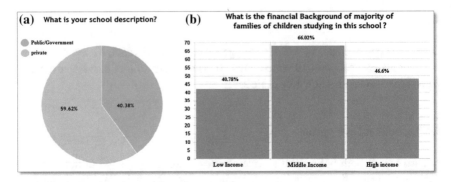

Fig. 1 School description and financial background

Figure 1a diagrammatically represents the results of school description of teachers who participated in the survey and (b) diagrammatically represents the results regarding the financial background of families of children studying in different schools according to the teachers who participated in the survey.

4.1.5 Grades Taught

All the teachers who participated in the survey were teaching in primary school (grade 1 to grade 5), 60 % were teaching grade 2, 53.3 % were teaching grade 1, 51.7 % grade 3, 20.0 % grade 4 and 18.3 % were teaching grade 5. All the teachers who participated were teaching primary/elementary school age children i.e. 6–10 years old except three who did not respond to the age range of children they were teaching. The results are shown in Table 6.

Figure 2 presents the results of participant distribution with respect to grades they teach in the form of bar graph

4.2 Experience with Mobile and Educational Applications

All teachers were asked about their experience regarding mobile technology. They were asked about their knowledge about mobile applications in general and educational apps in particular. 85.6 % of the teachers said that they know about mobile apps, 5.8 % said that they don't know and 8.6 % said that they somewhat know but not used much. Conversely they when asked about educational apps, 71.2 % teach-

Table 6 Grades taught

	Grade 1	Grade 2	Grade 3	Grade 4	Grade 5	Other	Responses (N)
Data	59	67	70	33	27	0	104
Percentage (%)	56.73	64.42	67.31	31.73	25.96	0	

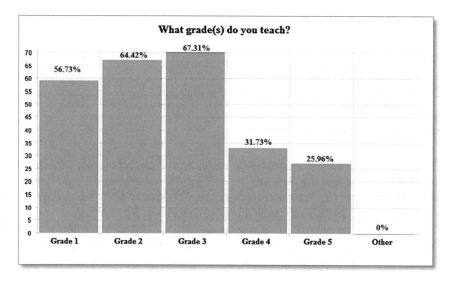

Fig. 2 Participant distribution with respect to the grades they teach

Table 7 Experience with mobile and educational applications

	Mobile applications			Educational applications			Total responses
	Yes i know	Somewhat know	Don't know	Yes i know	Somewhat know	Don't know	
Data	89	9	6	74	20	10	104
Percentage (%)	85.6	8.6	5.8	71.2	19.23	9.6	

ers knew about educational apps, 9.6 % said they don't know and 19.23 % said they somewhat know about these. Table 7 presents the results.

4.3 Technology Usage in School

4.3.1 Technology Usage

The surveys responded were asked about the use of technology in the primary schools for children. According to the results, 63 participants said that there is use of some kind of technology for young children in schools they teach. While according to 41 participants there was no use of technology for children in the primary schools where they were teaching. Table 8 summarizes the results for technology usage in schools.

The analysis of results showed that participants teaching in private schools indicated technology usage in schools while the teachers of public schools replied with no technology usage in schools for children. Hence according to the survey results the public schools in Pakistan do not provide technology facilities for primary school children whereas on the contrary private schools do provide some kind of technology in schools for children. Whereas further examination of results indicates that the types of technology available in private schools also vary from school to school.

Table 8 Technology usage in schools

	Yes	No	Standard deviation	Total responses (N)
Data	63	41	0.49	104
Percentage (%)	60.58	39.42		

4.3.2 Type of technology Used

The participants of survey were asked about the type of technology available in the schools they teach for young children. According to the survey results, 54 participants responded desktop PC, 47 responded projectors, 42 said none of the technology was used in school, 12 participants reported interactive tables, 7 participants said game devices were used and only 2 participants responded laptops. The following was the order (which respect to frequency of use) of devices used in primary schools in Pakistan. Table 9 presents the results of type of technology used in schools showing the percentage of use.

1. Desktop PC
2. Projectors
3. Interactive tables
4. Game devices
5. Laptops

The analysis of results obtained from teachers' survey showed that none of the public schools provided any kind of technology for primary school age children in Pakistan. The results also indicated that technology usage in Pakistani schools for primary/elementary school age children is limited to mainly desktop PC and projectors only while a few schools use interactive tables or game devices. On the other hand in developed countries, the modern mobile technology such as smartphones and tablets are increasingly being used in schools for young children as an effective medium of children education and learning. Figure 3 provides the diagrammatical representation of results regarding technology usage and type of technology.

Table 9 Types of technology usage in primary schools

	Desktop PC	Laptop	Smartphone	Tablet/ pads	Projector	Game devices	Interactive tables	None	Total responses
Data	54	2	0	0	47	7	12	42	104
Percentage (%)	51.92	1.92	0	0	45.19	6.73	11.54	40.38	

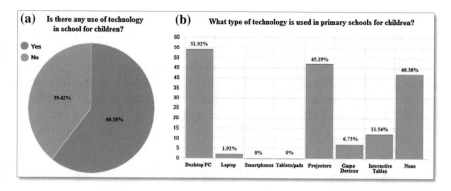

Fig. 3 Technology usage and type of technology

4.4 Primary/Elementary School Learning

4.4.1 Subjects

he participants were asked about the subjects taught in elementary/primary classes (i.e. 1–5 grades). According to the results the following were the main subjects taught in primary classes in Pakistan: Urdu, English, Math, Science, Social Studies, Islamiyat, Drawing, Computers, However the results showed that in most public schools computer was not taught as a subject to the primary school children. This indicated a gap in curriculum design of private and public schools in Pakistan.

4.4.2 Concepts

The important concepts which primary school age children learn in school (specifically related to English, maths, science, drawing and others) according to the survey results (responses that were most frequently repeated) are as follows:

Math: Addition and subtraction, multiplication and division, ascending and descending order, tables, greater than, lesser than, fractions, expanded form, odd and even, numbers in words etc.

English: Reading, writing, vocabulary, words, compound words, scramble words, syllable, grammar such as nouns, pro-nouns, verbs, adjectives etc., punctuation and sentence formation.

Science: Living and non living things, human body parts, organ system, animals, parts of plant and animals, natural and manmade things, sounds, types of soil, seasons.

Drawing, Geometry etc.: Names of shapes, different angles, different colors and their names, painting, drawing fruits, vegetables, objects, scenery, flower and mathematical shapes.

Other Concepts: Health, hygiene, moral values, and social habits.

This information is important in order to design and develop educational apps for children that are relevant to their age and matches the curriculum and concepts taught in schools.

Participants were asked about different subjects and concepts taught in primary schools in order to understand the curriculum and find ways in which technology integration can aid in effective learning of these concepts and pinpoint the subjects in which technology incorporation will be most helpful.

4.5   Favorite Activities of Children in School

The teachers were questioned about the favorite activities of children in primary school, the survey results showed that following activities were found most pleasing by children: using computer or other available technology, talking and playing with their friends, drawing, extracurricular activates, art work, spelling competition, reciting rhymes, acting on poems, listening stories, math questions, matching shapes/pictures, solving puzzles.

4.6   Teacher's Perspective Towards Technology

4.6.1   Technology Usage for Children Education

Teachers were asked whether they find technology useful for educational purpose for children at this young age. According to the survey results, 57.69 % found technology very useful, 25 % said that it is somewhat useful and 17.31 % think that it is not useful for education purpose for children at this young age of 6–10 years. Table 10 presents the results.

4.6.2   Mobile Technology Usage for Children Education

When survey participants were asked "Do you think mobile technology such as smartphone applications can help in children education at this age?", 70 participants responded very helpful whereas 34 participants though smartphones are not helpful in children education. Table 11 presents the results for mobile technology usage.

Table 10 Technology usefulness for children education

	Very useful	Somewhat useful	Not useful	Total responses (N)
Data	60	26	18	104
Percentage (%)	57.69	25	17.31	

Table 11 Attitude towards mobile technology usage

	Yes	No	Standard deviation	Total responses (N)
Data	70	34	0.47	104
Percentage (%)	67.31	32.69		

4.6.3 Use of Mobile Educational Applications in Schools for Children

The teachers were asked about the suggested use of mobile educational applications in school for children. The results indicated that 54.18 % were in favor of using mobile educational apps in schools whereas 45.19 % of teachers were against the use of mobile educational apps in school for children. Table 12 illustrates the results for the use of mobile educational apps in schools.

4.6.4 Use of Mobile Educational Applications in Homes for Children

The participants were asked "Do you suggest the use of mobile educational applications at home by parents for improving the learning skills of young children?". Majority of the participants 86 out of 104 suggested the use of educational applications at home by parent while only 17 were against use at home also, whereas 1 participant did not respond to this question (Table 13 presents the results).

4.7 Reasons to Restrict the Use of Mobile Technology in School

When teachers were asked about the reason of restricting the use of mobile apps in schools even for educational purpose, following were the major reasons according to the survey results:

1. If mobile technologies are allowed in schools students will get distracted and will not focus on studies.

Table 12 Use of mobile educational apps in schools

	Yes	No	Standard deviation	Total responses (N)
Data	57	47	0.50	104
Percentage (%)	54.18	45.19		

Table 13 Use of mobile educational apps in homes

	Yes	No	Standard deviation	Total responses (N)
Data	86	17	0.37	103
Percentage (%)	83.5	16.5		

2. It is an expensive device and school cannot afford to provide mobile technologies for every student.
3. Students can misuse mobile technology.
4. Students can cheat using mobiles in class tests.

4.8 Technology Resources Used to Support Teaching Efforts

Participants were asked about the technology resources they currently use to support or supplement their teaching practice. According to the survey results following were the order of resources most commonly used by teachers to support teaching efforts. Table 14 presents the results in detail.

1. Educational websites
2. Mobile phone educational apps
3. Web based interactive games
4. Social Media
5. Google Maps/Google Earth

Figure 4 represents the above results in the form of bar graph.

4.9 Mobile Technology Usage Considered Beneficial for Teaching

The uses of mobile technology considered most beneficial for teaching according to the survey results are in the following order:

1. Educational apps
2. Educational websites
3. E books
4. Photos/videos
5. Games
6. Other

Table 14 Technology resources used to support teaching efforts

	Educational apps	Google maps	Web based interactive games	Social media	Educational websites	None	others	Total responses (N)
Data	54	17	25	23	60	28	7	104
Percentage (%)	51.92	16.35	24.04	22.12	57.69	26.92	6.73	

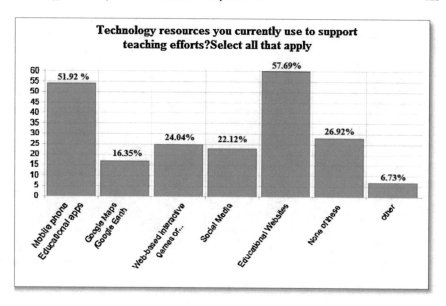

Fig. 4 Technology recourses to support teaching efforts

Table 15 Mobile technology usage considered beneficial for teaching

	Educational apps	Games	Educational websites	E-books	Photos/ videos	Others	Total responses (N)
Data	86	13	82	46	43	13	100
Percentage (%)	86	13	82	46	43	13	

Four participants did not answer this question. Table 15 represents the detailed results for the uses of mobile technology considered beneficial for teaching. The results are diagrammatically presented in Fig. 5.

4.10 Benefits of Mobile Educational Applications for Children

The survey participants were asked about their perspective towards the usefulness or benefits of mobile technology specifically mobile educational applications in order to determine the role of mobile technology in children education. Table 16 presents the results regarding teacher's perspective towards the role of mobile technology (educational apps) for children education and Fig. 6 shows the bar graph for the obtained results.

The participants were also asked about the viewpoint regarding the benefits of using mobile technology (educational applications) for children; the results obtained are depicted in Table 17. Figure 7 shows the bar graph of the results.

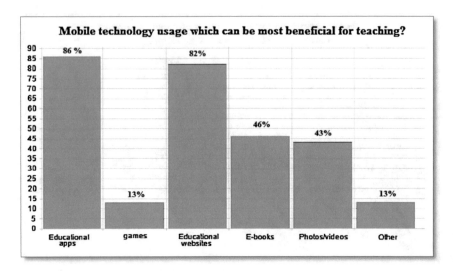

Fig. 5 Mobile technology usage considered beneficial for teaching

Table 16 Role of mobile educational apps in children education

	Useful teaching tool used by teachers for children	Useful self-learning tool used by children	Useful teaching tool used by parents for children	Not useful	Total responses (N)
Data	57	53	90	13	104
Percentage (%)	54.81	50.96	86.54	12.5	

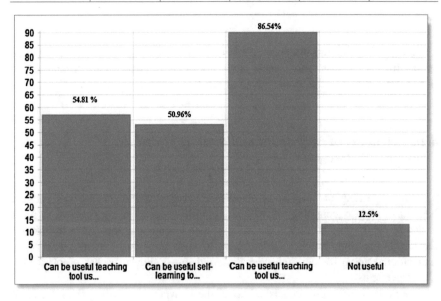

Fig. 6 Teachers perspective towards the role of mobile technology

Table 17 Benefits of using mobile technology (educational applications) for children (N = 104)

Benefits of mobile educational applications	Data	Percentage (%)
To support and emphasize on content being taught	62	59.62
To increase student motivation to learn	59	56.73
To make students more technology-literate	60	57.69
To respond to a variety of learning styles	13	12.5
To provide additional practice to learners/students	65	62.5
None of these	9	8.65

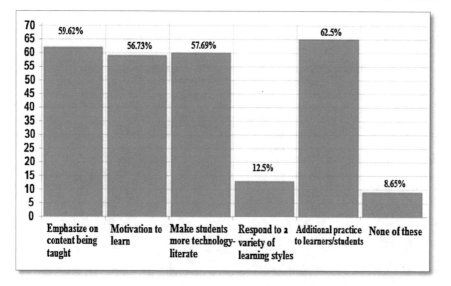

Fig. 7 Benefits of using smartphone educational apps for children

4.11 Educational Content Selection

The teachers who participated in the survey were asked about the content they prefer or suggest can be best learned with educational apps for primary school age children (6–10 years). Table 18 illustrate the results for educational content selection.

According to the survey results majority of the teachers were of the thought that educational apps for math and vocabulary are rather more useful for primary school age children. Additionally educational apps for quiz, memory games and drawing were also recommended for children in this age range. Figure 8 presents the results obtained from the survey participants in the form of a bar graph.

Table 18 Educational content selection (N = 104)

Educational content	Data	Percentage (%)
Math	86	82.69
Science	27	25.96
English	36	34.62
Vocabulary/alphabets	89	85.58
Poems/stories	23	22.12
Drawing	48	46.15
Quiz games	67	64.42
Memory games	49	47.12
Other topics	3	2.88

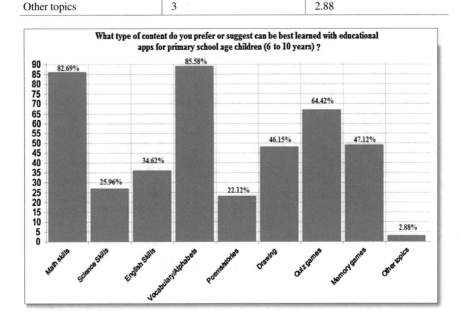

Fig. 8 Teacher's preference regarding selection of content for educational apps for children

5   Conclusion and Future Work

On the basis of results obtained from teacher's survey presented in this research work, a few preliminary thoughts are presented about the implications. These implications translate this research into practice and can help the software engineers, developers, designers, and educators to define as well as improve the role mobile technology can play in children education and make positive interactions in order to improve children learning skills by the use of mobile technology.

- Government should incorporate the use of technology for children education in the public schools in order to improve the access of mobile educational apps for children's from low-income families. As according to the survey results public schools in Pakistan do not provide technology resources for children in primary schools.

- Most of the young children from low income families have very limited access and knowledge of technology. Therefore children need access and time in order to develop the skills, knowledge and comfort for the use of technology before they actually start using technology independently for any defined purpose such as education. So all primary schools must define policies to incorporate the use of technology at this level to make children technology literate which has become a significant element of this increasingly digital world.

- For integration of technology in the classrooms, teachers should understand the impact it will have on the learning or educational outcomes of young students.

- Instead of making use of technology for only practice purpose, teachers should make use of technology as an important instructional tool to improve the learning skills of students.

- The successful integration of technology in schools highly depends on the school administrators. They should provide an individualized and characterized process of training and implementation.

- The teachers should understand how technology executes inside the restricted settings of a classroom.

- For the successful adaptation of technology, teachers must be open to transform their role in the classroom environment because when technology is used as a tool, the teacher take role of a facilitator and the students take on a practical role in education and learning.

- Teachers should determine what young students need to learn in conformity with the outlined curriculum standards and then choose from a variety of different applications to help children accomplish these goals. Teachers should incorporate the use of mobile technology specifically mobile educational applications and consider it as a tool available to support teaching efforts in order to provide effective learning opportunities for children and emphasize on the content being taught.

- It is very important for teachers to be well-versed about a range of available technology suitable for children. Since as educators it is their responsibility to assist children to utilize technology in safe and educational ways.

- Schools should provide professional development programs for teachers because in order to integrate technology in education, teachers should be well aware of the pros and cons of different available technologies and the ways in which technology might support their teaching practices and the curricular goals. They should know not only know how to use technology but also the criterion to use technology in order to make it useful for children education.

- There are also many issues of support from teachers and the school environment in incorporating technology (specifically mobile technology) in schools for young children. Therefore developers should provide administrative feature for teachers and school administration to control the settings of mobile devices according to their requirements.

- Provide password protection so that children can only access the educational content allowed for them on mobile technology and should not waste their time.
- Find ways to avoid children exposure to inappropriate content through mobile technologies.
- Children must learn to use technology same as they learn all other things in their developmental stage of life, within their own time and pace.
- It is important to make use of technology as it motivates the young children and also offer a different learning style.
- The adults (specifically parents) must keep a check on the use of technology by children to be in support of learning experiences and school curriculum that children are engaged in.
- Encourage the use of mobile educational apps as a great teaching tool for both parents and teachers providing additional learning for children.
- Consider the need to improve and provide more educational apps related to science and poems/stories for children learning as these apps had the lowest ratings in the survey.

In future work researchers should encourage research in this area and conduct more research studies with even larger samples, focusing on different types of applications and control groups, in order to provide generalizability and reliability. Moreover future research should investigate the long-term effects of mobile apps on learning, usage and interest. Also examine the particular issues of interface design and learning related to the cognitive functioning, age and time. As noted that throughout this study, the survey focuses on teachers' attitude about the use of mobile technology for children education, focusing on the educational apps. Certainly, there are many other different aspects and actors involved in the children's technology usage that would definitely bring their own perspectives to the data and would bring together different lessons and implications from it. Therefore future research should focus on other aspects such as how to incorporate mobile technology in classrooms and the role of government and school administration.

Acknowledgments The authors acknowledge and appreciate the cooperation and support provided by the groups and people who participated in this study. We appreciate and thank all the teachers for sharing invaluable opinion regarding their perspective towards technology usage for children education. We are also thankful to all the schools that helped us in getting the surveys filled by teachers.

References

1. Couse, L.J., Chen, Dora, W.: A tablet computer for young children? Exploring its viability for early childhood education. J. Res. Technol. Educ. **43**(1), 75–98 (2010)
2. McManis, L.D., Gunnewig, S.B.: Finding the Education in Educational Technology with Early Learners, Technology and Young Children (2012)
3. Murphy, K., DePasquale, R., McNamara, E.: Meaningful connections: using technology in primary classrooms. Young Child. **58**(6), 12–18 (2003)

4. Plowman, L., McPake, J.: Seven myths about young children and technology. Childhood Educ. **89**(1), 27–33 (2013)

5. Lee, L.-C., Wei, W.-J.: Child-computer Interaction Design and Its Effectiveness Research and Practice in Technology Enhanced Learning, vol. 8.1, pp. 5–19 (2013)

6. Suki, N.M., Mohd Suki, N.: Are lecturers' ready for usage of mobile technology for teaching. World Acad. Sci. Eng. Technol. **54** (2009)

7. Pegrum, M., Oakley, G., Faulkner, R.: Schools going mobile: a study of the adoption of mobile handheld technologies in Western Australian independent schools. Australas. J. Educ. Technol. **29.1** (2013)

8. Plowman, L., McPake, J., Stephen, C.: Just picking it up? Young children learning with technology at home. Camb. J. Educ. **38**(3), 303–319 (2008)

9. Tahir, R., Arif, F.: Mobile technology in children education: analyzing parents' attitude towards mobile technology for children. In: Proceedings of the Science and Information Conference (SAI) 2015, July 28–30. ISBN: 978-1-4799-8546-3 (2015)

10. Franklin, T.: Mobile learning: at the tipping point. Turk. Online J. Educ. Technol.-TOJET **10**(4), 261–275 (2011)

11. Druin, A.: Mobile Technology for Children: Designing for Interaction and Learning. Morgan Kaufmann (2009)

12. Sharples, M., Taylor, J., Vavoula, G.: A theory of learning for the mobile age. In: Andrews, R., Haythornthwaite, C. The Sage Handbook of E learning Research, pp. 221–247. Sage Publications (2007)

13. Thornton, P., Houser, C.: Using mobile phones in education. In: Proceedings. The 2nd IEEE International Workshop on Wireless and Mobile Technologies in Education (2004)

14. Selwyn, N.: The digital native—myth and reality. Aslib Proceedings, vol. 61. No. 4. Emerald Group Publishing Limited (2009)

15. Chiong, C., Shuler, C.: Learning: Is there an app for that. Investigations of young (2010)

16. Wartella, E.A., Jennings, N.: Children and computers: new technology. Old concerns. The Future of Children, pp. 31–43 (2000)

17. Shields, M.K., Behrman, R.E.: Children and computer technology: analysis and recommendations. The Future of Children, pp. 4–30 (2000)

18. Chou, C.C., Block, L., Jesness, R.: A case study of mobile learning pilot project in K-12 schools. J. Educ. Technol. Dev. Exch. **5.2**, 11–26 (2012)

19. Keengwe, J., Bhargava, M.: Mobile learning and integration of mobile technologies in education. Educ. Inform. Technol. 1–10 (2013)

20. Sharples, M., Taylor, J., Vavoula, G.: A Theory of Learning for the Mobile Age. In: Andrews, R., Haythornthwaite, C. (eds.) The Sage Handbook of E learning Research, pp. 221–47. London, Sage (2007)

21. Livingstone, S.: Critical reflections on the benefits of ICT in education. Oxford Rev. Educ. **38**(1), 9–24 (2012)

22. Tondeur, J., Van Keer, H., van Braak, J., Valcke, M.: ICT integration in the classroom: challenging the potential of a school policy. Comput. Educ. **51**(1), 212–223 (2008)

23. Mifsud, L.: Alternative learning arenas-pedagogical challenges to mobile learning technology in education. In: Proceedings IEEE International Workshop on Wireless and Mobile Technologies in Education, 2002 (2002)

24. Park, Y.: A pedagogical framework for mobile learning: categorizing educational applications of mobile technologies into four types. The International Review of Research in Open and Distributed Learning, vol. 12.2, pp. 78–102 (2011)

25. Mumtaz, S.: Factors affecting teachers' use of information and communications technology: a review of the literature. J. Inf. Technol. Teach. Educ. **9**(3), 319–342 (2000)

26. Sugar, William, Crawley, Frank, Fine, Bethann: Examining teachers' decisions to adopt new technology. J. Educ. Technol. Soc. **7**(4), 201–213 (2004)

27. Hermans, R., Tondeur, J., van Braak, J., Valcke, M.: The impact of primary school teachers' educational beliefs on the classroom use of computers. Comput. Educ. **51.4**, 1499–1509 (2008)

28. Marcinkiewicz, H.R. Computers and teachers: factors influencing computer use in the classroom. J. Res. Comput. Educ. **26.2**, 220–237 (1993)

Designing, Implementing and Testing an Automated Trading Strategy Based on Dynamic Bayesian Networks, the Limit Order Book Information, and the Random Entry Protocol

Javier Sandoval and Germán Hernández

Abstract This paper evaluates, using the Random Entry Protocol technique, a high-frequency trading strategy based on a Dynamic Bayesian Network (DBN) that can identify predictive trend patterns in foreign exchange orden-driven markets. The proposed DNB allows simultaneously to represent expert knowledge of skilled traders in a model structure and to learn computationally from data information that reflects relevant market sentiment dynamics. The DBN is derived from a Hierarchical Hidden Markov Model (HHMM) that incorporates expert knowledge in its design and learns the trend patterns present in the market data. The wavelet representation is used to produce compact representations of the LOB liquidity dynamics that simultaneously reduces the time complexity of the computational learning and improves its precision. In previous works, this trading strategy has been shown to be competitive when compared with conventional techniques. However, these works failed to control for unwanted dependencies in the return series used for training and testing that may have skewed performance results to the positive side. This paper constructs key trading strategy estimators based on the Random Entry Protocol over the USD-COP data. This technique eliminates unwanted dependencies on returns and order flow while keeps the natural autocorrelation structure of the Limit Order Book (LOB). It is still concluded that the HHMM-based model results are competitive with a positive, statistically significant P/L and a well-understood risk profile. Buy-and-Hold results calculated over the testing period are provided for comparison reasons.

J. Sandoval (✉)
UNAL UExternado, Bogotá, Colombia
e-mail: jhsandovala@unal.edu.co; javier.sandoval@uexternado.edu.co

G. Hernández
UNAL, Bogotá, Colombia
e-mail: gjhernandezp@unal.edu.co

© Springer International Publishing Switzerland 2016
L. Chen et al. (eds.), *Emerging Trends and Advanced Technologies for Computational Intelligence*, Studies in Computational Intelligence 647,
DOI 10.1007/978-3-319-33353-3_7

1 Introduction

Learning profitable trading strategies requires the combination of expert knowledge and information extracted from data. In this context, a novel approach to tackle the financial price prediction problem is to express expert knowledge in a Dynamic Bayesian Networks (DBN) structure to model the temporal relationships of complex non-deterministic financial market variables. This new approach has been successfully applied in other domains and can be also implemented to predict price movements and create profitable trading strategies. Expert knowledge can be used to design a network structure, and available data helps to calibrate parameters conditional to the selected model framework.

DBNs represent graphically a stochastic process using Bayesian Networks that include directed edges pointing in the direction of time [1]. The graph is constructed to represent the a priori expert knowledge expressed in a set of junctions, which facilitate conditional probability calculations. Following Bengtsson work [2], the entire model can be thought of as a compact and convenient way of representing a joint probability distribution over a finite set of variables.

The model assessed in this paper was first presented in [3, 4] as an extension of [5]. It uses a Hierarchical Hidden Markov Model to represent market sentiment dynamics based on features built from transaction prices and volume from orders in the limit order book of the USD/COP, the Colombian foreign exchange rate. Originally, the model performance was compared with results obtained from a Feed-Forward Neural Network model and a random market sentiment classification. It was found that the proposed HHMM model outperformed selected benchmarks yet, trading performance statistical properties were not reported because there were only one testing set in which the trading strategy was executed.

Unfortunately, it is difficult to observe the statistical properties of a trading strategy when there is only access to one testing data set mainly because of certain problems such as the bias introduced to the trading performance statistics by the presence of a high-frequency return correlation structure in the order flow. For example see [6]. One possible solution is the used of data resampling of the testing set where Bootstrapping is the most common technique. However, not all the data correlation structures are undesirable. They may be necessary for the correct assessment of the trading strategy. In this case, it is mandatory to preserve the time structure presented on the LOB.

One of the options to keep the desirable correlation structure on data when finding probability distributions of relevant estimators such as the mean and maximum drawdown of returns is Stationary Bootstrap [7]. Different from the Raw Bootstrap technique presented by Efron [8], stationary bootstrap resamples original series using data blocks of random size to create a new pseudo-time series which are indeed a new stationary series that preserves the original statistical features. However, this method is only applicable for weakly dependent time series and implementation complexity increases as the number of variables involved grows.

Another alternative method is the Random Entry Protocol (REP) as presented by [9]. This technique produces similar results as block-based bootstrapping techniques, yet it does not explicitly resample from the original data. In contrast, the REP directly creates new samples of a trading strategy result randomly choosing several entry points in the original data executing one round-trip trade in each of them. Therefore, return and order flow dependency structure is not transmitted to the final trading result. The REP can manage multivariable entry series as well as variable dependency structures among all the original series. Due to an easy implementation and the ability to deal with different dependency structures, this paper will use the Random Entry Protocol to assess the probability distribution of estimators from the trading strategy selected.

Next section will present the trading strategy model together with a description of the visible feature vector used to feed the HHMM that represent the market structure.

2 The Hierarchical Hidden Markov Model as a Special Case of a Dynamic Bayesian Networks

An HHMM can be represented as a Dynamic Bayesian Network [10]. The state of the whole HHMM can be encoded by the vector $\mathbf{Q_t} = \{Q_t^1, \ldots, Q_t^D\}$ where D means the number of network levels. Changing to a DBN representation also needs to include a set of indicator variables $F_t^d, d = \{1, \ldots, D\}$ which capture whether the HHMM at level d and time t has just finished. If F_t^d is 1, d will mean the hierarchy level the process is currently at. Additionally, an HHMM has termination states for each level. When transforming to DBN, transition matrices should be expanded to include probabilities of level termination.

Our HHMM will have discrete-valued variables. Therefore, conditional probabilities can be encoded as tabular tables. First, we will have conditional probabilities for $Q_t^d, d = \{1, \ldots, D\}$ variables. Formally,

$$P(Q_t^D = j | Q_{t-1}^D = i, F_{t-1}^D = f, Q_t^{1:D-1} = k) = \begin{cases} \widetilde{A}_k^D(i, j) & f = 0 \\ \pi_k^D(j) & f = 1 \end{cases}$$

$$P(Q_t^d = j | Q_{t-1}^d = i, F_{t-1}^d = f, F_{t-1}^{d+1} = b, Q_t^{1:d-1} = k) = \begin{cases} \delta(i, j) & b = 0 \\ \widetilde{A}_k^d(i, j) & b = 1, f = 0 \\ \pi_k^d(j) & b = 1, f = 1, \end{cases}$$

$$(1)$$

where $\widetilde{A}_k^d$ and $\widetilde{A}_k^D$ represent transition probabilities if the process stays in the same level. π_k^d and π_k^D are initial distributions for level d given that parent variables are in state k. It is important to remember that $\widetilde{A}^d(i, j)(1 - \tau_k^d(i)) = A_k^d(i, j)$ where A represents the automaton transition matrix, $\widetilde{A}$ is the DBN transition matrix and $\tau_k^d(i) := A_k^d(i, end)$, the probability of terminating from state i.

Indicator variables F^d are turned on if Q^d enters final state. F^d conditional probabilities are:

$$P(F_t^D = 1|Q_t^{1:D-1} = k, Q_t^D = i) = A_k^D(i, end),$$

$$P(F_t^d = 1|Q_t^{1:d-1} = k, Q_t^d = i, F_t^{d+1} = b) = \begin{cases} 0 & b = 0 \\ A_k^d(i, end) & b = 1, \end{cases} \quad (2)$$

where F_t^d or $F_t^D = 1$ means that the process has signaled an end state in current level and control will be returned to its previous level.

The CPD for the first network slice will be $P(Q_1^1 = j) = \pi^1(j)$ for the top level and $P(Q_1^d = j|Q_1^{1:d-1} = k) = \pi_k^d(j), d = 2, \ldots, D$, elsewhere.

Finally, we will have conditional probabilities for the observed node realizations represented as $P(O_t^m|\mathbf{Q_t}) = \alpha_m, 1 \leq m \leq M$ where M means number of elements in the observed variable's alphabet.

2.1 Proposed Financial Market Representation

Following expert knowledge, we built a financial market model that presents the market as a 2-level structure where the driver variable can enter in 2 different classes that we expect to correspond to traditional bullish and bearish market states. This market states should not be considered long but short trends that realize in intraday data. We should not attend to name a priori any of these states.

We also expect that a simple configuration allows to capture the general market dynamics in each state. Inside each of these trends, markets have two producers of negative and positive observations. These two producers emit unhidden data one at the time from different subfamilies of observations. Thus, this model recognizes that a financial market in very short periods of time is in one of two possible main states and being in that state, they alternately produce positive and negative observations.

Specifically, our DBN assumes $D = 2$, 2 states in the first level and 4 states in the second level.

$$Q^1(1) = \{\text{Market State 1}\},$$
$$Q^1(2) = \{\text{Market State 2}\},$$
$$Q^2(1), Q^2(4) = \{\text{Negative Feature Producer}\},$$
$$Q^2(2), Q^2(3) = \{\text{Positive Feature Producer}\}. \quad (3)$$

Q^1 has two possible market states that have not been previously marked as uptrend or downtrend. Q^2 will represent feature production related to local positive and negative market conditions. It will be guaranteed that F_T^d be clamped to ensure that all models have finished at the end of the data sequence. Additionally, $F_t^1, 1 \leq t < T$ will be always equal to 0 so the model stays in level 1 and 2 for the whole data sequence. We believe that this simple configuration will encapsulate the way expert traders see the USD/COP short term dynamics.

Figure 1 shows HHMM structure as a 2-level automaton and provides a graphical representation of the selected HHMM expressed as a DBN structure. DBN's observed information will be explained in next subsection.

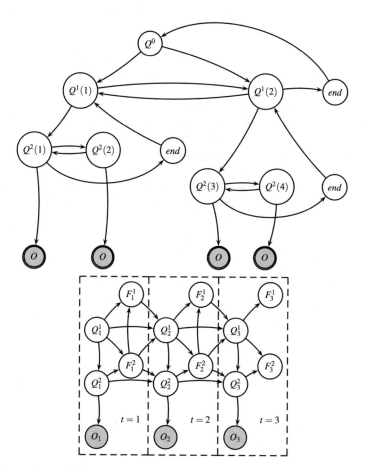

Fig. 1 *Upper* Proposed 2-level automaton. *Lower* Proposed HHMM represented as DBN. Q_n^1 is the market regime and Q_n^2 is the feature producer. Elements of the observed variables are explained in Sect. 3

3 Observed Feature Vector

A feature vector was constructed to capture information from two valuable sources; trades and LOB profile dynamics. This information was used to forecast future trends in financial prices because exact future prices are not needed to create a profitable trading strategy. In the specific case of the LOB profile dynamics, several studies had shown that order books have relevant information to improve financial price direction prediction [11–16]. Therefore, the feature vector will combine elements from the order book and trades.

Assume a fixed time interval $[0, T]$, and let consider $\{P_{t_s}\}_{s \geq 0}$, $t_s \in [0, T]$ the USD/COP trade series. Let also define a new subsequence $\{P_{t_{s_y}}\}_{y \geq 0}$ such that $\{s_y\}$ indexes only reference local extrema from the series P_{t_s}, i.e. $\{P_{t_s} : [P_{t_s} > P_{t_{s-1}} \wedge P_{t_{s-1}} < P_{t_{s-2}}] \vee [P_{t_s} < P_{t_{s-1}} \wedge P_{t_{s-1}} > P_{t_{s-2}}]\}$. This new subsequence allows to sample the trade series every time a market participant is willing to trade on the opposite direction of the current trade trend.

Using the previous series, The zig-zag process, $\{Z_{t_x}\}_{x \geq 0}$, is built recording differences between local extrema as $Z_{t_x} = P_{t_{s_{nx}}} - P_{t_{s_{n(x-1)}}}$, where n is an odd positive integer controlling the number of local extremum price differences accumulated. If $Z_{t_x} \geq 0$, the zig-zag is leading upward or is simply called positive and if $Z_{t_x} < 0$, the zig-zag is leading downward or is simply called negative. If $n = 1$, we do not accumulate zig-zags. n must be a positive odd integer to guarantee that local extremum price differences are calculated in a subseries intercalating maximum and minimum values. This cumulative zig-zag is useful to reduce model's prediction instability. Figure 2 provides a graphical explanation of the above definitions.

The zig-zag and the extremum price series are complemented with information extracted from the order book profile dynamics, LOB_t. A trade indexed LOB series is sampled over the partition $\{t_x\}_{x \geq 0}$, LOB profiles calculated every time the zig-zag

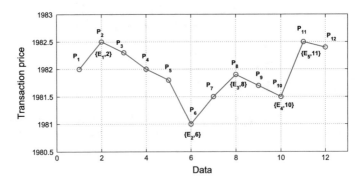

Fig. 2 Transaction price series and the corresponding extremum price and zig-zag processes. Assuming $n = 1$, the first value of the Z process would be $Z_{t_1} = P_{t_6} - P_{t_2}$. The sequence $\{E\}$ is used for reference purpose only and denotes the position at which extremum values were recorded

process is observed. Because of the high sparsity of the order book, the LOB used aggregated price levels every 20 cent.

Based on the raw limit order book process, we calculated two simple-smoothed exponential distance-weighted average volume series, $SEDWAV_{t_x}^b$, $SEDWAV_{t_x}^s$, to capture how volume was concentrated in the buy and offer sides of the order book for every zig-zag x. The closer the volume to the best buy/sell best price, the higher the weight given to that volume. An exponential average was used to expressed this fact. Formally:

$$SEDWAV_{t_x}^{buy} = \frac{1}{s_{nx} - s_{n(x-1)}} \sum_{s_{n(x-1)} < j \le s_{nx}} n_{t_j}^b,$$

$$SEDWAV_{t_x}^{sell} = \frac{1}{s_{nx} - s_{n(x-1)}} \sum_{s_{n(x-1)} < j \le s_{nx}} n_{t_j}^s, \qquad (4)$$

where $n^b(\cdot)$ and $n^s(\cdot)$ are the discretized exponential weighted average volume series that show order book's strength and allow to identify what is commonly known as floors and caps in technical analysis. Figure 3 shows order book evolution for a certain day of the dataset. As described before, there are many price levels with zero volume. Figure 3 also gives visual evidence of bid volume blocks around 1,756 pesos between 9:36:00 am and 10:48:00 am. We set the exponential weighting coefficient at 0.8 so the exponential average used roughly up to 2 Colombian pesos on every side of the LOB, the maximum away of the USD/COP LOB series.

Wavelet Transform of the SEDWAV Series. The raw order book series was denoised before calculating the SEDWAV series using a discrete $2D$-wavelet trans-

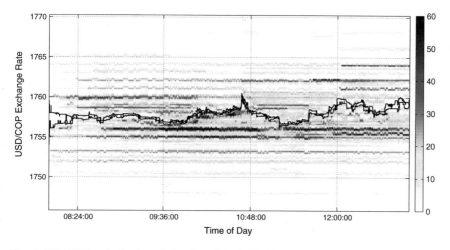

Fig. 3 USD/COP order book evolution from March 21, 2012. *White* spaces correspond to zero volume. Every volume unit is equivalent to 250 thousand US dollars. Maximum volume, 60 in the *color bar*, is equal to 15 million US dollars. *Solid black lines* show best bid/offer price evolution

form of the order book process with a Haar Wavelet over a daily window and a 2-level resolution. The wavelet transform was used to recover a filtered version of the order book series setting all detail coefficients at the second level in all three directions, horizontal, vertical and diagonal, to zero. Thus, we expect to capture relevant changes and leave aside the effect of noisy order book updates. The marginal contribution of the wavelet transform will be assessed on the validation data set. Figure 4 shows the denoised version of the order book information presented in Fig. 3. The wavelet-filtered version of the order book clearly shows the bid volume block previously observed in the raw order book.

Next, we defined volume blocks as wavelet-transformed SEDWAV values that are greater than a threshold α_V. The same threshold is used for buy and sell SEDWAVs. Accordingly, a discrete feature vector was created containing three elements that described zig-zag pattern types, transaction and order book dynamics. In particular,

$$O_x = \{f_x^{(1)}, f_x^{(2)}, f_x^{(3)}\} \quad \text{where,}$$

$$f_x^{(1)} = \begin{cases} 1 & Z_x \geq 0 \text{ local maximum} \\ -1 & Z_x < 0 \text{ local minimum} \end{cases}$$

$$f_x^{(2)} = \begin{cases} 1 & P_{x-4} + \alpha_P < P_{x-2} < P_x - \alpha_P \wedge P_{x-1} < P_{x-1} - \alpha_P \\ -1 & P_{x-4} - \alpha_P > P_{x-3} > P_x + \alpha_P \wedge P_{x-3} > P_{x-1} + \alpha_P \\ 0 & otherwise, \end{cases} \quad (5)$$

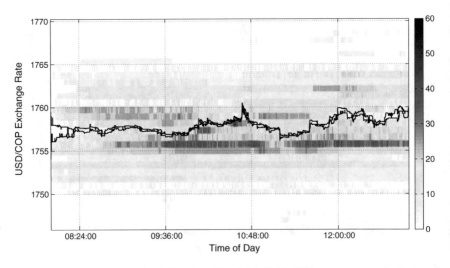

Fig. 4 Denoised order book evolution from March 21, 2012. *White* spaces correspond to zero volume. Every volume unit is equivalent to 250 thousand US dollars. Maximum volume, 60 in the *color bar*, is equal to 15 million US dollars. *Solid black lines* show best bid/offer price evolution

$$f_x^{(3)} = \begin{cases} -1 & SEDWAV_x^{sell} > \alpha_V \\ 1 & SEDWAV_x^{buy} > \alpha_V \\ 0 & otherwise. \end{cases}$$

where α_P is a threshold used to differentiate significant transaction price movements and t_x has been reduced to x to simplify the notation.

Table 1 shows feature vector components and Fig. 5 summarizes feature vector interpretation. The first element in the observed feature vector is the zig-zag type, i.e. maximum or minimum. The second component captures price's momentum comparing current maximum or minimum with its recent historical values. Finally, the third element captures the existence of volume blocks on both sides of the order book. For example, $(1, 1, 1)$ means a local maximum, with a local uptrend and a volume block on the bid side of the order book. $D_{1:9}$ are observations exclusively produced by $Q^2(1)$ and $Q^2(4)$. $U_{1:9}$ are observations only produced by $Q^2(2)$ and $Q^2(3)$, see Fig. 1.

Thus, the HHMM's structure simulates a two-level market in which, first, it enters a market regime and then within each regime, positive and negative features are produced. This structure guaranteed that a positive feature is always followed by a negative feature and vice versa. This structure summarizes expert trader's knowledge of how financial prices evolve when there is no fundamental economic information being released. We expect that $U_{1:4}$ and $D_{1:4}$ features are more probably observed

Table 1 Feature vector observations classified as positive (left) or negative (right)

Symbol	O_x (observation)	Symbol	O_x (observation)
U_1	$(1, 1, 1)$	D_1	$(-1, 1, 1)$
U_2	$(1, -1, 1)$	D_2	$(-1, -1, 1)$
U_3	$(1, 1, 0)$	D_3	$(-1, 1, 0)$
U_4	$(1, 0, 1)$	D_4	$(-1, 0, 1)$
U_5	$(1, 0, 0)$	D_5	$(-1, 0, 0)$
U_6	$(1, 0, -1)$	D_6	$(-1, 0, -1)$
U_7	$(1, -1, 0)$	D_7	$(-1, -1, 0)$
U_8	$(1, 1, -1)$	D_8	$(-1, 1, -1)$
U_9	$(1, -1, -1)$	D_9	$(-1, -1, -1)$

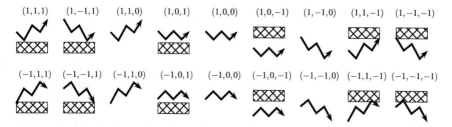

Fig. 5 Graphical interpretation of the feature vector observations classified as positive, $U_{1:9}$ (*upper*) or negative, $D_{1:9}$ (*lower*). *Crosshatched rectangles* represent a order book volume block

during macro uptrends and $U_{6:9}$ and $D_{6:9}$ are more probably found during macro downtrends. U_5 and D_5, defined as no micro trend and no volume block exist associated with a particular market state. These two states will represent what is commonly known as noise in price movement.

Using the previous model, we captured the dynamics of the USD/COP to predict its short-term future behavior. Next section will present the dataset and its characterization.

4 DataSet

The original dataset consisted of three months of tick-by-tick information from the limit order book and transactions of the USD/COP, the Colombian spot exchange rate starting in March 1, 2012. These series were used for calibration, validation, and first testing procedures. This dataset was extended to cover 9 months more to implement the resampling technique starting on June 1, 2012.

The new extended data has also been extracted from the Set-FX market, the local interbank FX exchange market. Dataset covered 120,650 transactions. Transactions with similar time stamp have been aggregated into one observation because it is very likely that the same agent executed them. Additionally, data included 2'242,050 order book updates that helped to reconstruct the original limit order books with an aggregation interval level of 20 cent. Volume was expressed in 250 thousand US dollar units. Due to liquidity issues, the first and last 10 minutes of every day in the available data were not considered. USD/COP spot interbank exchange market opens at 8 am and closes at 1 pm. It is a semi-blind market, participants only know their counterparts after transactions are executed. USD/COP average daily turnover is 1 billion dollars. Figure 6 shows the transaction series during the original revised dataset window. Figures 3 and 4 depicts the raw and filtered order book dynamics of a particular day.

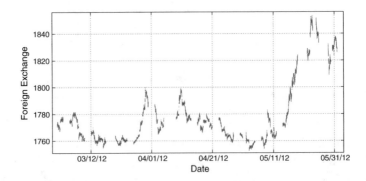

Fig. 6 High-frequency Colombian foreign exchange rate (USD/COP) from March 1, 2012 to May 31, 2012. Series discontinuities represent weekends and holidays

5 Methods and Experiment

Different from other studies [17, 18], the primary goal of this model was to predict market states instead of raw prices or price levels. Original performance results were reported in [3]. The later documented a better performance than results obtained from a standard Feed-Forward Neural Network and random guessing. However, there is no evidence of the probability distribution of the important trading strategy estimators such as the mean of returns. This paper will calculate the trading P/L probability distribution using a Random Entry Protocol technique to overcome the previous problem.

The Random Entry Protocol (REP) consists of creating a new Profit and Loss (P/L) series for a certain trading strategy using round-trip trades executed at random entries of the unique testing dataset. It is not necessary to resample from the original series or to assume synthetic data structures. Of course, this method does not help to calculate confident intervals on the original series but only to construct them for the trading strategy variables (Fig. 7).

The REP was used to generate 1,000 trading results each formed of 290 round-trip trades all extracted from the extended testing set. 290 trades per run were chosen, so the original testing trading P/L series in [3] could be assessed using these new experimental outcomes. This paper will report, mean, standard deviation, max-drawdown, Sharpe Ratio, Hitting Ratio, skewness and kurtosis of trading returns, so it is possible to have a better understanding of the proposed trading strategy (Table 2).

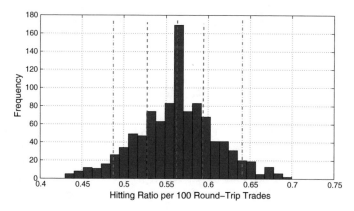

Fig. 7 Probability distribution of the Hitting Ratio calculated over a 290 round-trip trade samples found using the Random Entry Protocol on the extended testing dataset. The outermost lines correspond to 1st, 5th, 95th and 99th % percentiles. The innermost line represents the total return mean

Table 2 First Column: Mean values of relevant trading strategy estimators calculated using the Random-entry Protocol Technique over 1,000 trading P/L paths of 290 trades each

Estimators	Proposed model	Long-and-hold
Round-trip mean (%)	0.016	0.0005
Total return (%)	1.56	0.05
Std round-trip returns (%)	0.09	0.1
Sharpe	0.176	0.0036
MaxDrawDown (%)	0.59	1.97
Hitting ratio (%)	56.30	50.34
Skewness	1.141	−0.509 %
Kurtosis	9.547	11.66 %

Second Column: trading strategy estimators calculated from a Buy-and-Hold during the testing set

Table 3 Relevant percentiles calculated over the probability distribution of P/L key estimators

Estimators	5th p.	25th p.	50th p.	75th p.	95th p.
Mean (%)	0.0023	0.0093	0.0150	0.0218	0.0302
Std (%)	0.07	0.08	0.09	0.09	0.12
Total return (%)	0.23	0.93	1.50	2.18	3.02
Sharpe	0.0303	0.1141	0.1735	0.2391	0.3349
MaxDrawDown (%)	0.37	0.46	0.55	0.66	0.95
Hitting ratio (%)	48.00	53.00	56.00	59.00	64.00
Skewness	−0.804	0.466	0.011	0.016	0.033
Kurtosis	4.31	5.57	7.26	10.21	22.84

The table presents the mean, standard deviation, total return, Sharpe coefficient, Maximum drawdown, hitting ratio, skewness and Kurtosis of 290 round trips trades built using the Random Entry Protocol

6 Model Performance and Results

Simulation results are presented in Table 3. Buy-and-Hold estimators were also calculated for comparison purposes. The first relevant observation is that the mean return of a round-trip trade is positive and significantly different from zero using a two-tailed 95 % confidence interval. The average round-trip trade gives a positive result of 0.016 % which is higher than the average transaction cost for the USD/COP, 0.0018 % (Fig. 8).

The average value of the standard deviation per round-trip trade is 0.09 % in line with standard financial data that shows big differences between standard deviation and mean of return values. Although the proposed strategy yields solid positive returns, they are far away from returns obtained from a risk-free active investment. The trading strategy also has a maximum drawdown (maximum drop in a row) of 0.59 %, six times its round-trip trade standard deviation. A remarkable observation is that the proposed trading strategy clearly outperformed in a Return/Risk framework a simple Buy-and-Hold investment implemented over the same testing set. The studied

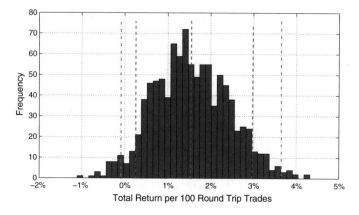

Fig. 8 Probability distribution of the mean of total returns calculated over a 290 round-trip trade samples found using the Random Entry Protocol on the extended testing dataset. The outermost lines correspond to 1st, 5th, 95th and 99th % percentiles. The innermost line represents the total return mean

trading model's mean return per round-trip was 32 times bigger than the mean return of the Buy-and-Hold strategy yet the proposed model shows a similar or even better risk profile. This result is obtained with a hitting ratio that is significantly different from 50 %, the result obtained from pure guessing, see Fig. 7.

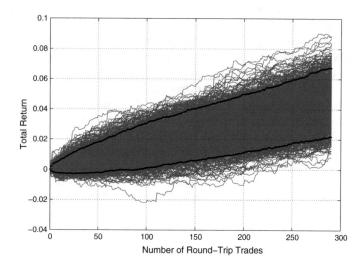

Fig. 9 Probability distribution of the P/L process generated by 290 round-trip trade samples found using the Random Entry Protocol on the extended testing dataset. The *black lines* correspond to 5th, and 95th % percentiles over the complete trading process calculated from 1,000 simulations

Therefore, the use of information coming from the volume evolution of the limit order book and its combination with transaction data, not only helps to create a trading strategy that outperforms other benchmark machine learning techniques [3], but also performs better than a simple passive strategy does. According to Fig. 9, after approximately 290 round-trip trades, there is evidence that supports cumulative positive returns with a 95 % confidence level. Even more, after 290 round-trip trades, the proposed model can deliver a cumulate return between 2 and 6.1 %. This result is enough to cover transaction costs and to produce a higher total return with a lower risk than results obtained from a passive long-strategy.

7 Conclusion

This work proposed a 2-level HHMM that was converted to a DBN for training and assessment purposes. The proposed model assumed that a particular financial market could be viewed as a complex automaton that enters into two main regimes. Then, each regime cycled through two feature producers, throwing negative and positive observations calculated from transaction and wavelet-transformed order book volume data. The former model was tested over the USD/COP foreign exchange rate market using nine months of high-frequency data covering transaction prices and tick-by-tick order book information. The main goal was to construct probability distributions for key estimators of the proposed trading strategy.

This study used the Random Entry Protocol to create resamples of the P/L series of a selected trading model. It was verified that the model produced positive returns significantly different from zero. The mean of trading returns was also much greater than the mean of returns obtained from a pure Buy-and-Hold strategy. The previous result holds even though the passive strategy showed equal or higher risk measured as P/L standard deviation and maximum drawdown.

Finally, through simulations, it was found the optimal number of trades to observe statistically significant positive cumulative returns after accounting for the trading strategy variability. The next step will be to test if this model is exploiting unique patterns just present in the USD/COP market or its results could be extended to other markets.

Acknowledgments We thank Bolsa de Valores de Colombia (BVC) for providing the dataset. This work was partially supported by Colciencias.

References

1. Murphy, K.: Dynamic Bayesian networks: representation, inference, and learning. Ph.D. thesis, University of California, Berkeley (2002)
2. Bengtsson, H.: Bayesian networks—A self-contained introduction with implementation remarks. Ph.D. thesis, Mathematical Statistics, Lund Institute of Technology (1999)

3. Sandoval, J., Hernández, G.: Procedia Comput. Sci. **51**(1593) (2015) (International Conference On Computational Science, ICCS Computational Science at the Gates of Nature). doi:10.1016/j.procs.2015.05.290. http://www.sciencedirect.com/science/article/pii/S1877050915010984
4. Sandoval, J., Hernandez, G.: Science and Information Conference (SAI), pp. 435–442 (2015). doi:10.1109/SAI.2015.7237178
5. Sandoval, J., Hernández, G.: Machine Learning and Data Mining in Pattern Recognition. In: Perner (ed.) Lecture Notes in Computer Science, vol. 8556. Springer International Publishing, pp. 408–421 (2014). doi:10.1007/978-3-319-08979-9_30
6. Covrig, V., Ng, L.: J. Bank. Finan. **28**(9), 2155 (2004)
7. Politis, D.N., Romano, J.P.: J. Am. Stat. Assoc. **89**(428), 1303 (1994). http://www.jstor.org/stable/2290993
8. Efron, B.: The Jackknife, the Bootstrap and Other Resampling Plans. Society for Industrial and Applied Mathematics (1982). doi:10.1137/1.9781611970319, http://epubs.siam.org/doi/abs/10.1137/1.9781611970319
9. Schmidt, A.B.: J. Trading **4**, 62 (2009). doi:10.3905/JOT.2009.4.4.062
10. Murphy, K.P., Paskin, M.A.: Proceedings of Neural Information Processing Systems (2001)
11. Bouchaud, J.P., Farmer, J.D., Lillo, F.: How markets slowly digest changes in supply and demand. Quantitative Finance Papers 0809.0822. arXiv.org (2008). http://ideas.repec.org/p/arx/papers/0809.0822.html
12. Dorogovtsev, S., Mendes, J., Oliveira, J.: Phys. A: Stat. Mech. Appl. **360**(2), 548 (2006). doi:10.1016/j.physa.2005.06.064, http://www.sciencedirect.com/science/article/B6TVG-4GNTCF9-2/2/d9efa60dec0417dacaf9f01a5b4bcc40
13. Eisler, Z., Kertesz, J., Lillo, F.: The limit order book on different time scales. Quantitative Finance Papers 0705.4023, arXiv.org (2007). http://ideas.repec.org/p/arx/papers/0705.4023.html
14. Gu, G.F., Chen, W., Zhou, W.X.: Physica **387**(21), 5182 (2008)
15. Tian, G. Guo, M.: Rev. Quant. Finan. Acc. **28**(3), 287 (2007). http://ideas.repec.org/a/kap/rqfnac/v28y2007i3p287-306.html
16. Weber, P., Rosenow, B.: Quant. Finan. **5**(4), 357 (2005)
17. Hassan, M.R., Nath, B., Kirley, M.: Expert Syst. Appl. **33**(1), 171 (2007). doi:10.1016/j.eswa.2006.04.007, http://www.sciencedirect.com/science/article/pii/S0957417406001291
18. Ortega, L., Khashanah, K.: J. Forecast. **33**(2), 134 (2014). doi:10.1002/for.2270, http://dx.doi.org/10.1002/for.2270

An Adaptive Multi Agent Service Discovery for Peer to Peer Cloud Services

Moses Olaifa, Sunday Ojo and Tranos Zuva

Abstract Cloud computing is evolving into a popular platform that enables on-demand provisioning of computing resources to a growing population of clients. Core to the provisioning of service in the cloud is the discovery of these services in an efficient and timely manner. Centralized and hierarchical approaches to service discovery have exhibited bottlenecks as network load increases and limitation in scalability. Efforts have been made in combining cloud systems and Peer to peer P2P systems to address the problem encountered in the conventional service discovery approaches but not without a new set of challenges ranging from network flooding to poor performance in dynamic networks. This paper presents an efficient and scalable approach for semantic cloud service discovery in a P2P cloud environment. The approach is based on Learning Automata LA and Ant Colony Optimization ACO. The ability of ACO to adapt to changes in real time makes it a better choice in dynamic environments such as cloud. We evaluate this approach against the some existing P2P service discovery approaches, the proposed mechanism showed an improved performance.

1 Introduction

Cloud computing presents a wide range of computing resources on-demand over the internet and dedicated networks to clients across the academic and industrial communities. The computing resources range from applications to hardware infrastructures

M. Olaifa (✉) · T. Zuva
Faculty of ICT, Department of Computer Systems Engineering, Tshwane University
of Technology, Pretoria, South Africa
e-mail: olaifamo@tut.ac.za; newmosesolaifa@yahoo.com

T. Zuva
e-mail: zuvat@tut.ac.za

S. Ojo
Faculty of ICT, Tshwane University of Technology, Pretoria, South Africa
e-mail: ojoso@tut.ac.za

© Springer International Publishing Switzerland 2016   147
L. Chen et al. (eds.), *Emerging Trends and Advanced Technologies
for Computational Intelligence*, Studies in Computational Intelligence 647,
DOI 10.1007/978-3-319-33353-3_8

which are offered via three major categories of services: Software as a Service (SaaS), Platform as a Service (PaaS), and Infrastructure as a Service (IaaS). End-user applications are rendered as services in SaaS; PaaS presents an environment for application development and deployment; IaaS provides computing infrastructures including virtual machines, load balancers, and network connections. The growth in cloud technology usage is not unconnected with benefits derived in terms of flexible on-demand service, broad network access, measured resource usage, pervasive resource access, reduced computing cost, and virtual resource aggregation [1–3]. With an intended pervasive resource provisioning, a seamless interaction between cloud service providers, service requesters and resource brokers is inevitable.

Most essential to the business community is the ability to outsource their computing need to a third party thereby cutting capital expenditure. Four deployment models are presented by the cloud to clients namely: private cloud, public cloud, hybrid cloud and community cloud [3]. The private cloud is available for exclusive use by private clients such as corporations. Resources provided by a private cloud are not available to any other user outside the designated users. Unlike private clouds, public clouds are accessible to the general public. Provisioning is usually by rent for a short or long period of time. Community clouds are dedicated to a particular set of clients based on their shared interests or collaborations. This can be operated and managed by different organizations. Combination of two or all of the above mentioned deployment models results in a hybrid cloud.

Cloud computing promotes two types of providers namely cloud providers and service providers [4]. Cloud infrastructures and platform services are rendered by cloud providers while creation and maintenance of services published in the cloud are offered by the service providers. The utilization of the cloud infrastructures by cloud service providers and requesters requires publication of services by the providers and discovery of the same services by service requesters. Fulfilment of the requirements is subject to the design and implementation of appropriate techniques; building standard tools for supporting publication and discovery activities. Due to the overlapping structure with grid [1, 5], underlying cloud computing technologies can be traced to grid computing, especially those related to resource publication and discovery.

Early cloud, grid computing and other service oriented architecture SOA [6, 7] paradigms were based on centralized and hierarchical approaches to service publication and discovery [8–12]. Services are published in centrally managed or partially distributed storage servers that are queried for discovery of the services. With an increase in the size of services and usage traffic, the centralized approach is faced with the challenge of efficiently managing the constantly scaling distributed and heterogeneous resources. Other issues with centralized approach to publication and discovery of services (resources) include but not exclusive to bottlenecks, single point of failure, inability to scale with growth in services and demand. In order to address these challenges evident in centralized systems, decentralized approaches are considered for efficient service management in vast service networks.

Peer-to-Peer P2P approaches [13, 14] provide an improved substitute to the centralized approach by supporting large scale resource distribution with higher flexibility and scalability levels. Furthermore, due to the involvement of all peers in the discovery activity, reliability of service provisioning is ensured. Core to realizing a successful P2P system is an efficient decentralized discovery of resources in the system [15]. Discovery of services in a peer to peer environment involves the search and location of nodes hosting appropriate candidate services with capabilities to fulfil the service requests. The collaboration between the SOA paradigms and Peer to peer P2P [16, 17] systems was explored to address the problem encountered in the conventional service discovery approaches but not without a new set of challenges. Issues ranging from network flooding to poor performance in dynamic networks.

In this chapter, we present an efficient and scalable approach for semantic cloud service discovery in a P2P cloud environment. The approach is based on Learning Automata LA and Ant Colony Optimization ACO. The ability of ACO to adapt to changes in real time makes it a better choice in dynamic environments such as cloud. The Mobile Ants provided by the ACO locates the shortest paths within the P2P cloud network and the LA selects the optimal path for the mobile ants decision making. The mechanism is evaluated against some existing resource discovery approaches and it showed an improved performance over them. The remaining part of this chapter is arranged as follows: (II) Related work in SOA paradigms; service description and discovery. (III) Theoretical background in Learning Automata LA and Ant Colony Algorithm ACO. (IV) The system design comprising of service description, request attribute linking and service description indexing. (V) Presents the description of the service description model. (VI) Presents the discovery algorithms. (VII) Experimental results.

2 Related Work

2.1 Service Description

Resource provisioning in the cloud is achieved through services that describe the attributes of hosted resources. These service descriptions define the specifications in a manner understandable by human (manual) or machine (automation). Standardization of service description in SOA paradigms such as web services has resulted in languages generally adopted for description of service functional and non-functional attributes. Web service description language WSDL [18] is one notable language for description of web services using six main elements port type, port, message, types, binding, and service. It describes web services as a collection of endpoints or ports that perform different operations on messages containing either document-oriented or procedural-oriented information. WSDL gained prominence for inter-operability across different platforms but exhibits limitations in delivery of semantic support for web services syntactic nature. To improve web service description for automated service discovery and composition, several semantic approaches have been proposed [19–22].

Semantic languages present well-defined expression for exposing service capabilities based on terms and concepts defined in different domain ontologies. With the aid of ontologies with shared vocabularies, improved web service descriptions have been achieved. According to [23], four basic types of semantics for web service description are data semantics, functional semantics, execution semantics and QoS semantics. One of the widely accepted ontology languages for semantic web service description is OWL-S [24]. OWL-S presents an upper ontology for semantic web service description in accordance with W3C OWL standard. It aims at providing a generic framework for semantic description of web services supporting automated service control. Web service description in OWL-S is divided into three parts: service profile, process model and service grounding. Implicit definition of what the service does is realized in the service profile. The process model of the service description models a service as a process or set of processes encompassing the data and control flow. Service grounding defines how to access the web service.

Majority of cloud resource descriptions capture specific aspects of cloud computing with no consideration for inter-operability across cloud platforms, That is, no standardized format that can facilitate inter cloud discovery and composition [25]. A comprehensive review of some of these languages is [25, 26]. Cloud service providers describe resources in different ways and present various interfaces for service discovery. Some descriptions are inclined towards business applications, while others are defined for engineering or technical applications. This poses a challenge to achieving cloud service integration and collaboration. To address the inter-operability gap in cloud service descriptions, different solutions are proposed in [25, 27, 28].

2.2   Service Discovery

To discover appropriate services, service requests are compared to available published services advertised by the different providers. Semantic matching approaches support discovery of services that are semantically related but may be syntactically different. A service producing mobile devices as an output can be regarded as candidate service to a query requesting tablet, if the ontology establishes a close relationship subclass (superClass), between the concepts defining the two terms.

The distributed service management publishes service descriptions for the purpose of reliable service search and discovery. Matching of service descriptions with request requirements is based on the level of similarity between the required and target functionalities. The similarity matching between the required and target capabilities can be redefined as the ability to discover an advertised service outputs that match the request outputs and inputs that match the request inputs. Matching between advertised service and requests input-output parameters can result in full, partial or no match. Full match indicates a complete satisfaction of service request; and a no match implies a failure to identify or non-existence of a matching service. The different levels of similarity define the degree of match between the advertised services and service request. The degrees of matching based on the definitions by [29] are presented as follows:

Let $r = \{r_{in}, r_{pre}, r_{out}\}$ denote a service request with input set r_{in} and output set r_{out} ; $s_i = \{s_{in}, s_{out}\} \in R_i$ denote the set of published services in registry R_i.

Exact match: Service s_i yields an exact match of request r if $\forall r_{out} \in r \exists s_{out_i} \in s_i$.
    That is $s_i \equiv r \Leftrightarrow r_{out} \equiv s_{out_i}$.
Subsume match: Request r subsumes service s_i if $\forall s_i, s_{out} \subseteq r_{out} \wedge s_{in} \subseteq r_{in}$. That
    is, r subsumes $s_i \Leftrightarrow \{s_{in_i}, s_{out_i}\} \subseteq \{r_{in}, r_{out}\}$.
Plugin match: Request r plugs into service s_i if $\forall s_i, s_{out} \supseteq r_{out} \wedge s_{in} \supseteq r_{in}$. That is
    s_i subsumes $r \Leftrightarrow \{s_{in_i}, s_{out_i}\} \subseteq \{r_{in}, r_{out}\}$.
Fail: A match between service s_i and request r fails if $\forall s_i, s_{out} \cap r_{out} = \phi \wedge s_{in_i} \cap$
    $r_{in} = \phi$.

The matchmaking module accepts as input, the set of domain specific concepts, and returns a ranked set of services with the same or related set of concepts. Decision on input and output domain specific concepts are defined by the input and output terms retrieved from the service request. Information about services utilizing the domain specific concepts from multiple repositories is accessible to the matchmaking module. To increase the efficiency and decrease the matching time of the matching module, the service outputs are matched with request output prior to input matching. Services with a fail output match are not considered for the input match. It is assumed that an output match resulting in a failed matching is far from satisfying the need of the service request even if such services produce an exact input match. Ranking of discovered services is based on the number of service request input (output) concepts satisfied by the available services. Services with exact match are ranked highest, followed by the plug in match and subsume match respectively.

As mentioned earlier, services are essential tools to discovery of resources. Approaches to web service discovery are classified into centralized and decentralized. The UDDI [30, 31] is one of the notable centralized approaches to web service discovery that hosts service descriptions in a centralized registry. Unlike the centralized approach, the distributed approaches offer improved scalability. A number of the distributed approaches are based on P2P network structure. Service management is distributed among participating peers thereby ensuring reliable service provisioning. Furthermore, the structured P2P topology guarantees search results in bounded time. Some of the research work in P2P approaches for web service discovery are presented in [32] for a keyword based approach; and [33, 34] for semantic web service discovery to address ambiguity in keyword based service discovery.

In [35], the author categorized resource discovery approaches in grid systems into centralized or hierarchical, P2P and agent based systems. Centralized or hierarchical approaches [12, 36] store information about resources in central or distributed servers that are queried for discovery of these resources. This system provides easy tools for accessing grid services. Like other centralized systems, these approaches suffer from bottleneck, scaling issues, one point of failure among others. Agent based systems [37–39] present improved scalability and reliability properties than the centralized systems. With the agent based approaches, active codes that can be interpreted locally are sent to the servers. Interest in P2P techniques for scalable resource discovery is on the increase. P2P techniques involve every peer in the system in resource management and discovery process. A survey of P2P techniques for grid resource discovery is presented in [40, 41].

3 Theoretical Background

3.1 Learning Automata

LA formalizes a general stochastic system in terms of states, actions, state or action probabilities, and environment responses [15]. The main goal of any automaton is to guide in the selection of an action based on past actions and response from environment in order to improve the total performance function. As depicted in Fig. 1, the automaton randomly selects an action from a set of actions. This is presented to the environment as input and the environment subsequently provides a response as a consequence of the action selected. Thereafter, the learning algorithm updates the action probability vector. For further details on learning automata, readers are referred to [42, 43].

The LA can be formally defined as a 6-tuple $\{\phi, \alpha, \beta, P, A, G\}$, where ϕ denotes a set of internal states, α denotes the set of possible actions, $\beta = [0, 1]$ is the set of responses presented by the environment to the LA, P denotes the probability vector over the set of states, A denotes the reinforcement algorithm for updating the action probability vector for each step n based on the environment response, and G is the

Fig. 1 Learning Automata-environment pair

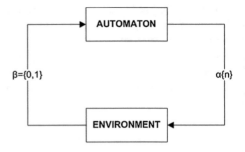

output function $G : \phi \longrightarrow \alpha$. The environment can be defined as a 3-tuple $\{\alpha, s, \sigma\}$ where α represents the finite set of input to the environment s and σ represent the penalty probability vector and the set of output responses obtained by the LA from the environment respectively. The responses are often binary [0,1] where 0 denote penalty and 1 denote reward. Examples of linear updates on the action probability vector for each step are linear reward-penalty, linear reward-in action and linear reward-ε-penalty. In general, the update by the learning algorithm for a reward a at step t for selecting action j is given as follows:

$$P_j(t + 1) = P_j(t) + a(1 - \beta(t))(1 - P_j(t)) - b\beta(t)P_j(t) \tag{1}$$

When an action j is selected at time t with a penalty b, the probability vector is updated as:

$$P_j(t + 1) = P_j(t) + a(1 - \beta(t))P_j(t) + b\beta[(r - 1)^{-1} - P_j(t)] \tag{2}$$

When the value of $a = b$, the algorithm is referred to as linear reward-penalty L_{R-P}; when $b = 0$, reward-inaction L_{R-I} and when $b < a$, it is referred to as linear reward-ε-penalty $L_{R-\varepsilon P}$.

3.2 Ant Colony Optimization

ACO model presents a set of artificial ant systems for solving discrete optimization problems. It is motivated by the manner in which some ant species search for food. In the process of searching, the ants make a deposit of pheromone in the paths followed. The amount of pheromone deposited in a path signifies a favorable path between the source and destination. ACO exploits this approach in addressing optimization problems. The problem is described as a graph where edges are formed by ants represented as mobile agents that can connect the different vertices of the graph. The job assigned to the ants is to find the shortest path connecting two different vertices in the graph. In a way similar to the natural pheromone trail used by ants, a variable τ_{xy} is associated with every edge (x, y), connecting vertex x and vertex y.

The movement of the ants is guided by a probabilistic decision policy that depends on the amount of pheromone trail perceived on the edges of the graph. Positive feedback is thus improved by reinforcing a trail on those edges which are used. To avoid some premature convergence this pheromone trail evaporates over time and ants, transitions to other nodes in the graph happen stochastically. The trail information is communicated locally and indirectly, mediated by physical adjustment of the pheromone trail value on every edge. The ant colony concept has motivated different routing algorithms for network systems. The AntCycle is an ACO routing algorithms used in addressing the Traveling Salesman Problem TSP. Every town has a number of ants positioned on it. Each ant moves to a next town with a probability that is a function of the amount of pheromone trail perceived on edge (x, y) and distance of the town. Ants are only allowed to visit a town not more than once. The selected edge is updated with a value that depends on the tours it participated. To avoid premature convergence, the pheromone trail gradually evaporates. For details on this algorithm and other ant colony based algorithms, readers are referred to [44, 45].

4   System Design

4.1   Service Description

This work adopts the annotated WSDL WSDL-S proposed in [Akkiraju 2005] for describing the cloud services. WSDL-S supports the creation of semantic service description by extending descriptions presented in WSDL with semantic annotations linked to some domain models. Definition of services in WSDL is generally divided into abstract definition and concrete definition. The abstract definition consists of the interface, operation and message constructs. Concrete definition describes the actual service implementation in terms of the binding, service and endpoints. The extension of WSDL by WSDL-S is directed to annotating the abstract definition to achieve automated discovery, composition and invocation. An overview of the extensibility elements proposed includes:

modelReference: this attribute establishes a link between a WSDL entity and a semantic model concept.

schemaMapping: handling of structural dissimilarity between the web service schema elements and the corresponding semantic model concepts is realized by this attribute.

precondition and effect: these are elements defined as child elements of the operation element.

categorization: this attribute defines the category of a service during publication.

4.2   Request Attribute Linking

A two-layered ontology system is proposed, with the first layer attribute ontology, describing the general concepts independent of any domain specific ontology. The second layer involves a set of domain specific ontologies with each defining the concepts and existing relationships between concepts for a specific domain. The ontology system represents the two layers with a tree structure consisting of leaf nodes and roots describing the concepts in the term ontology and domain specific ontologies respectively.

Web service requests are pre-processed into a set of service attributes comprising input, output and precondition elements. Input elements represent the set of available attributes for the operation and output elements represent the expected output after execution. Preconditions elements define the conditions that must be satisfied for successful service invocation. The attribute ontology is searched for concepts related to the set of service request attributes. Identified concepts are linked to the root concepts in domain specific ontologies.

We represent the layered system with a tree composed of different leaves and roots. The leaves and roots denote the attribute concepts in the attribute ontology and concepts in the domain specific ontology respectively. Request attribute related concepts identified in the attribute ontology layer are traced to the tree roots in the domain specific ontologies for domain specific concepts. The system is formally defined as follows:

Definition 1 A service request can be defined as a tuple $r = \{r_{in}, r_{pre}, r_{out}\}$ where r_{in} denotes the set of available inputs, r_{pre} denotes the preconditions that must be satisfied, and r_{out} denotes the set of expected outputs after a successful execution.

Definition 2 A tree is defined as a tuple $t = \{C, \beta, L\}$ where $t \in T$; where $C = \{c_1, c_2, \ldots, c_n\}$ is a set of domain specific concepts in t; $L = \{l_1, l_2, \ldots, l_n\}$ is a set of general concepts in T; and $\beta : C \rightarrow L$ is a mapping function $\forall L \cup C^* \subseteq C$ in where C^* is a sequence of concepts connecting L to C.

Definition 3 $\{c_1, c_2, \ldots, c_n\} \supset \{r_{in}, r_{pre}, r_{out}\} \Leftrightarrow \exists l_i \in L \forall \{r_{in}, r_{pre}, r_{out}\}$. That is $\{c_1, c_2, \ldots, c_n\}$, is a set of root-concepts to $\{r_{in}, r_{pre}, r_{out}\}$ if every input, precondition and output parameter in the service request r has a related concept in the attribute ontology concepts l_i.

4.3   Service Description Indexing

Each peer in the network encodes all the service descriptions hosted by it. A multi-dimensional indexing technique [46] is used since each service description is a combination of input, precondition and output attributes. In the tree, points are partitioned into hierarchical nested Minimum Bounding Rectangles. The tree is made up of nodes that can hold relative number of entries where the leaf nodes hold data points and internal nodes hold the children MDRs. The service descriptions in each peer of the network are represented with three R-Trees. The set of input, precondition and output concepts are indexed with ST_{in}, ST_{pre} and ST_{out} respectively.

5 Multi Agent P2P Cloud Service Discovery Model

The P2P cloud environment is modeled as an undirected graph $G = (V, E)$, where V and E represent the graph vertices and edges respectively. Let $V = \{v_1, v_2, v_3, \ldots v_n\}$ represent the set of nodes within the P2P network and $e_i \; \varepsilon \; E$ represents a path $e_i : v_i \Longrightarrow v_j$ from node v_i to v_j of the P2P cloud network. The $r = \{r_k \mid 1 \le k \le m\}$ represents the set of available services provided by the network peers. For every node $v_i \; \varepsilon \; V$ of the P2P network, an automaton $A_i \; \varepsilon \; A$ is associated for selection of an optimal node of the adjacent nodes to visit in search of the requested service s_k. The trip from the initial node v_i to the destination node v_n produces a sequence of nodes selected by the automata. We assume that the amount of pheromone trail deposited by the ants on each edge is used to calculate the probability of selecting an adjacent node. Each node equally maintains two tables Peer-to-Peer Network Traffic Table (P2PNTT) and Routing Table (RT). P2PNTT maintains frequently updated statistical entries for the network traffic. The RT manages the information on the distances between that node and other nodes in the P2P network. The rows of the table hold the probability vectors of all adjacent peers in the cloud network and each column holds the probability of selecting node v_j from the set of adjacent nodes in search of service s_k. The network controls two types of ant agents namely: forward $F_{i \longrightarrow n}$ and backward $B_{n \longrightarrow i}$ ant agents. The forward agent $F_{i \longrightarrow n}$ maintains a stack of nodes transcended on its way from the requesting node v_i to the destination node v_n. Based on assumption, there exist n varying paths (number of hops) between the source and destination nodes. Each path $e_i : v_i \Rightarrow v_n$ has an associated probability from the probability vectors of all existing paths maintained in the RT. Whenever a service request is received by a node in the network, a Search Ant Agent (SAA) is launched if the required service is not discovered within the node. The ant agent activates the LA in each node that selects an optimal path to be followed by the SAA based on the information from the RT and P2PNTT.

6 Multi Agent LA Algorithm

The algorithm is divided into three separate parts namely: RT Update, service discovery and matching algorithm. The RT Update is designed for the periodic update of the routing table and peer-to-peer network traffic table using Algorithm 1. The service discovery algorithm launches a search agent for the discovery of the requested services using Algorithm 2. At each peer, the matching algorithm matches the services with the service requests using Algorithm 3.

7 Experimental Results

The results of our experiments are presented in this section. The performances of the proposed approach are evaluated by a number of simulations. The performance of our approach Multi Agent LA (MALA) is evaluated against constrained flooding (CF) and the hybrid P2P (H-P2P) approach proposed in [47]. The performances of the algorithms are generally based on two metrics success rate and length of path.

Algorithm 1 Routing Table Update Algorithm

1: Node $v_i \varepsilon V$ releases an ant agent $F_{i \rightarrow n}$ periodically to a randomly selected P2P node hosting service s_k by selecting a node v_j to transcend from the set of adjacent P2P nodes.
2: Node v_j is transcended and pushed on to $F_{i \rightarrow n}$ stack.
3: In node v_j, $F_{i \rightarrow n}$ activates learning automata A_j associated with the node v_j.
4: A_j selects the next node v_j to transcend from the set of adjacent nodes based on its vector probability for service s_k using equation (3):

$$P_{j \rightarrow n} = \frac{P_{j \rightarrow n} + \alpha l_j}{1 + \alpha(|N_{ad}| - 1)} \tag{3}$$

$P_{j \rightarrow n}$ is the probability of selecting node v_j from the set of adjacent nodes N_{ad}. l_j is the heuristic correction value that is equivalent to the length of the queue of the link connecting the current node v_i with its adjacent node v_j. α is a value that weights the importance of the heuristic correction with respect to the information recorded in the routing table. The selection of v_j therefore depends on both a long term learning process and an instantaneous heuristic prediction.
5: Repeat steps 2 to 4 until ant agent $F_{i \rightarrow n}$ reaches the destination node v_n.
6: At the destination node v_n a backward ant agent $B_{n \rightarrow i}$ is created and the stack is transferred to the backward ant agent while $F_{i \rightarrow n}$ is destroyed. $B_{n \rightarrow i}$ follows the path visited by $F_{i \rightarrow n}$ in a reverse direction.
7: At every node v_{n-x} where $1 \leq x \leq (n - 1)$, $B_{n \rightarrow i}$ pops the stack to determine the next node to visit in path $v_n \rightarrow v_i$.
8: At node v_{n-x} in the path $n \rightarrow i$, $B_{n \rightarrow i}$ updates the statistical model for the grid traffic distribution using equations (4) and (5):

$$\mu_n \leftarrow \mu_n + \eta(\tau_{j \rightarrow n} - \mu_n) \tag{4}$$

$$\sigma_n^2 \leftarrow \sigma_n^2 + \eta(\tau_{j \rightarrow n} - \mu_n)^2 - \sigma_n^2 \tag{5}$$

Where μ_n and σ_n^2 are the mean and variance of the grid traffic model respectively. $\tau_{j \rightarrow n}$ represents the newly observed trip from the current node v_{n-x} to node n. η is a factor that compares the recent samples to the mean and variance. The routing table is updated by positive reinforcement by probability using equation (6) or by negative reinforcement by normalization using equation (7).

$$P_{j \rightarrow n} \leftarrow P_{j \leftarrow n} + r(1 - P_{j \leftarrow n}) \tag{6}$$

$$P_{j \leftarrow n} \leftarrow P_{j \leftarrow n} - r P_{j \leftarrow n}) \tag{7}$$

9: Repeat steps 7 and 8 until $B_{n \rightarrow i}$ reaches node v_i.

Algorithm 2 Service discovery Algorithm

1: When a request for service s_k arrives at node v_j, the node is searched for the availability of the service. If service s_k is not discovered in node v_j, the search agent $S_{i \to n}^{r^k}$ is created.

2: A_i is activated and an optimal node $v_j^{\star}$ is selected from the set of adjacent nodes using the probability vector for service s_k updated in route table.

3: v_j is pushed to the stack then $S_{i \to n}^{r^k}$ traverses v_j.

4: Repeat steps 2 and 3 until agent node v_n. The visited stack generates a sequence of nodes.

5: In node v_n, pop stack and follow search path in opposite direction.

Algorithm 3 Matching Algorithm

   C_i denotes the set of concepts corresponding to request attributes.

   S_i denotes the set of concepts corresponding to a service attributes

1: initialize outMatch = exact, inDegree = plugin, preMatch = plugin

2: Forall request attribute concepts Cr do:

3: If Cr_i corresponds to C_{out} then search ST_{out}

4: $C_{out}^{*}[] = C_{out}^{*}[] + C_{out}$

5: forall C_{out} in $C_{out}^{*}[]$ do Compute matchingDegree

6: if matchingDegree < outMatch then return fail.

7: else compute outMatchRank[]

8: forall $C_{out} \subseteq S_i$ in outMatchRank[]

9: forall $C_{in} \cup C_{pre} \subseteq S_i$ repeat Steps 2-7. // for input and precondition attributes.

10: Compute and return inMatchRank[] and preMatchRank[]

11: Rank S_i[outMatchRank, preMatchRank, outMatchRank]

We assume a dynamic cloud environment where services can be added or removed, therefore, the success rate is defined by the true positive rate. The path length defines the number of nodes traversed in the process of searching for a service in the network.

Two factors considered are the network topology and frequency of resources. We generate the topology for the P2P networks using the Inet [48] due to its ability to generate autonomous system level network models. We assume that the H-P2P has cache information on neighboring peers.

In these experiments, we simulated different P2P networks that scale from 1000 nodes to 3000 nodes. Five types of services are introduced with a distribution frequency of 1/1000, 1/500, 1/300, 1/100 and 1/50 respectively.

Figures 2 and 3 present the results of the scalability performance of MALA compared to the other two algorithms for 150 queries. The network sizes for the experiments are 1700, 1900, 2100, 2300 and 2500 nodes respectively. Services are generated using a discrete uniform distribution in order to give every service type a chance of been requested. In the first set of experiments, the RT does not contain entries for all the available services and adjacent nodes. This contributes to the poor performance from the MALA algorithm. Its path length increases as the grid system scales higher. In Fig. 2, both MALA and H-P2P traversed a lower number of nodes than CF. The CF traversed more nodes because of the large number of messages generated during flooding of query messages. Figure 3 shows the best performance from MALA due to the update of its RT. This is followed by H-P2P resulting from the update of cache information of neighboring node.

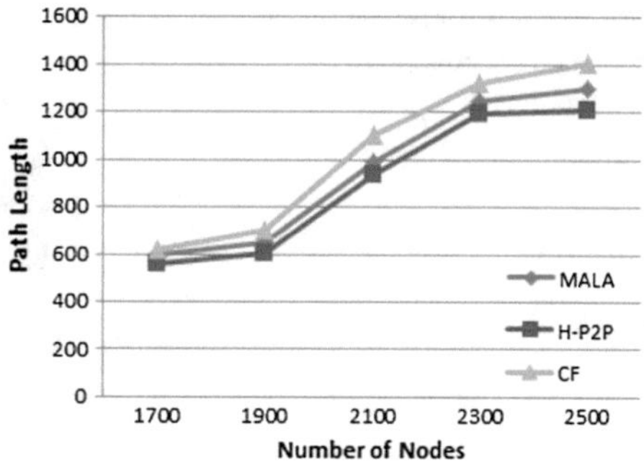

Fig. 2 The path length for partially updated routing table for 150 queries

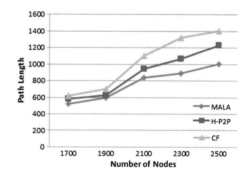

Fig. 3 The path length for a fully updated Routing table for 150 queries

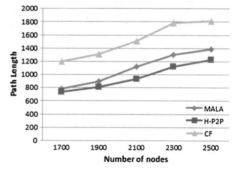

Fig. 4 The path length for partially updated routing table for 300 queries

Figures 4 and 5 describe the results for a RT with entries for all the adjacent nodes and the service types in the cloud network for 300 queries. CF shows a higher number of traversed nodes due to the increase in query size. This shows that for CF, the query size is directly proportional to number of nodes traversed. MALA shows a decrease in the number of nodes traversed due to an update of the RT.

Figures 6 and 7 present the success rate of the algorithm with varying network size but constant service distribution frequency. CF showed a better performance in Fig. 6 because of the size of query messages. However, MALA shows an improved performance in Fig. 7, after the RT is updated.

Fig. 5 The path length for fully updated routing table for 300 queries

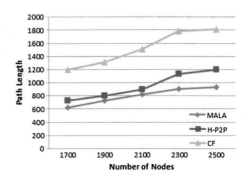

Fig. 6 The success rate for partially updated routing table

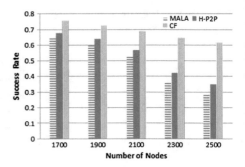

Fig. 7 The success rate for fully updated routing table

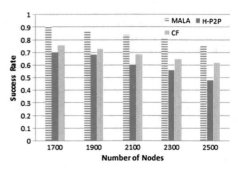

Figures 8 and 9 show the hit rate for constant network size and varying service frequencies before and after RT update respectively. The service frequencies 0.002, 0.003, 0.005, 0.01 and 0.02, are evaluated with a network size of 2500 nodes. Figure 8 shows a better success rate with CF and Fig. 9 shows a better hit rate with MALA due to the fact that the RT manages the complete and updated RT entries. Hence, the ability to select an optimal adjacent node at every node traversed.

Fig. 8 Success rate for varying service frequencies before RT update

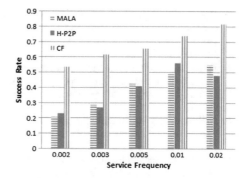

Fig. 9 Success rate for different service frequencies after RT update

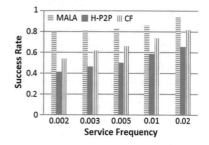

8 Conclusion

In this chapter, we presented a multi agent based reinforcement learning approach to service discovery in a peer-to-peer cloud system. The target is to deploy multiple agents capable of searching for feasible parts through the cloud network traffic. Every node of the system manages a routing table and a network traffic table. Whenever a request for a service is received by a node, an optimal path is followed based on the information provided in the tables. Experimental results show an improvement on the proposed algorithm over two existing approaches.

Future work will consider extending combined theories into the area cloud computing service composition across different clouds and multiple ontology domain.

References

1. Foster, I., Zhao, Y., Raicu, I., Lu, S.: Cloud computing and grid computing 360-degree compared. In: Grid Computing Environments Workshop, 2008. GCE'08, pp. 1–10. IEEE (2008)
2. Sun, L., Dong, H., Ashraf, J.: Survey of service description languages and their issues in cloud computing. In: Eighth International Conference on Semantics, Knowledge and Grids (SKG), 2012, pp. 128–135. IEEE (2012)
3. Mell, P., Grance, T.: The NIST definition of cloud computing. Microsoft Live Mesh (2011). http://www.mesh.com

4. Goscinski, A., Brock, M.: Toward dynamic and attribute based publication, discovery and selection for cloud computing. Fut. Gen. Comput. Syst. **26**(7), 947–970 (2010)
5. Foster, I., Kesselman, C. (eds.): The Grid 2: Blueprint for a New Computing Infrastructure. Elsevier, New York (2003)
6. Perrey, R., Lycett, M.: Service-oriented architecture. In: Symposium on Applications and the Internet Workshops, 2003. Proceedings, pp. 116–119. IEEE (2003)
7. Conner, P., Robinson, S.: Service-oriented architecture. U.S. Patent Application 11/388,624 (2006)
8. Bellwood, T., Clment, L., Ehnebuske, D., Hately, A., Hondo, M., Husband, Y.L., Riegen, C.: The universal description, discovery and integration (uddi) specification. Rapport technique, Comit OASIS (2002)
9. Skoutas, D., Sacharidis, D., Kantere, V., Sellis, T.: Efficient Semantic Web Service Discovery in Centralized and P2P Environments. Springer, Berlin (2008)
10. Czajkowski, K., Fitzgerald, S., Foster, I., Kesselman, C.: Grid information services for distributed resource sharing. In: 10th IEEE International Symposium on High Performance Distributed Computing, 2001. Proceedings, pp. 181–194. IEEE (2001)
11. Kaur, D., Sengupta, J.: Resource discovery in web-services based grids. World Acad. Sci. Eng. Technol. **31**, 284–288 (2007)
12. Molt, G., Hernndez, V., Alonso, J.M.: A service-oriented WSRF-based architecture for metascheduling on computational grids. Fut. Gen. Comput. Syst. **24**(4), 317–328 (2008)
13. Suryanarayana, G., Taylor, R.N.: A survey of trust management and resource discovery technologies in peer-to-peer applications (2004)
14. Meshkova, E., Riihijrvi, J., Petrova, M., Mhnen, P.: A survey on resource discovery mechanisms, peer-to-peer and service discovery frameworks. Comput. Netw. **52**(11), 2097–2128 (2008)
15. Fletcher, G.H., Sheth, H.A., Brner, K.: Unstructured peer-to-peer networks: Topological properties and search performance. In: Agents and Peer-to-Peer Computing, pp. 14–27. Springer, Berlin (2005)
16. Amoretti, M., Zanichelli, F., Conte, G.: SP2A: a service-oriented framework for P2P-based grids. In: Proceedings of the 3rd International Workshop on Middleware for Grid Computing, pp. 1–6. ACM (2005)
17. Wu, C.L., Liao, C.F., Fu, L.C.: Service-oriented smart-home architecture based on OSGi and mobile-agent technology. IEEE Trans. Syst. Man Cybern. Part C: Appl. Rev. **37**(2), 193–205 (2007)
18. Christensen, E., Curbera, F., Meredith, G., Weerawarana, S.: Web services description language (WSDL) 1.1. http://www.w3.org/TR/2006/CR-wsdl20-20060327/wsdl20-z.pdf (2001)
19. Akkiraju, R., Farrell, J., Miller, J.A., Nagarajan, M., Sheth, A.P., Verma, K.: Web service semantics-wsdl-s (2005)
20. Ankolekar, A., Burstein, M., Hobbs, J.R., Lassila, O., Martin, D., McDermott, D., McIlraith, S.A., Narayanan, S., Paolucci, M., Payne, T., Sycara, K.: DAML-S: Web service description for the semantic web. In: The Semantic WebISWC 2002, pp. 348–363. Springer, Berlin (2002)
21. Martin, D., Burstein, M., Hobbs, J., Lassila, O., McDermott, D., Martin, D., McIlraith, S.A., Narayanan, S., Paolucci, M., Payne, T., Sycara, K.: (2004). OWL-S: Semantic markup for web services. W3C member submission, 22 April 2007
22. McGuinness, D.L., Van Harmelen, F.: OWL web ontology language overview. W3C recommendation, vol. 10 (2004)
23. Sheth, A.P.: Semantic Web Process Lifecycle: role of semantics in annotation, discovery, composition and orchestration. In: WWW 2003 Workshop on E.Services and the Semantic Web, Budapest (2003). http://corescholar.libraries.wright.edu/knoesis/33
24. Burstein, M., Hobbs, J., Lassila, O., Mcdermott, D., Mcilraith, S., Narayanan, S., Paolucci, M., Parsia, B., et al.: OWL-S: semantic markup for web services. W3C Member Submission (2004)
25. Nguyen, D.K., Lelli, F., Papazoglou, M.P., Van Den Heuvel, W.J.: Blueprinting approach in support of cloud computing. Fut. Internet **4**(1), 322–346 (2012)

26. Keahey, K., Tsugawa, M., Matsunaga, A., Fortes, J.A.: Sky computing. IEEE Internet Comput. **13**(5), 43–51 (2009)
27. Cardoso, J., Barros, A., May, N., Kylau, U.: Towards a unified service description language for the internet of services: requirements and first developments. In: 2010 IEEE International Conference on Services Computing (SCC), pp. 602–609. IEEE (2010)
28. Sun, Y.L., Harmer, T., Stewart, A., Wright, P.: Mapping application requirements to cloud resources. In: Euro-Par: Parallel Processing Workshops. Springer, Berlin (2011)
29. Paolucci, M., Kawamura, T., Payne, T. R., Sycara, K.: Semantic matching of web services capabilities. In: The Semantic WebISWC 2002, pp. 333–347. Springer, Berlin (2002)
30. Org, U.D.D.I.: Universal Description, Discovery and Integration. UDDI Technical White Paper (2000). http://www.uddi.org/pubs/Iru_UDDI_Technical_White_Paper.pdf
31. Richards, R.: Universal Description, Discovery, and Integration (UDDI). In: Pro PHP XML and Web Services, pp. 751–780. Apress (2006)
32. Schmidt, C., Parashar, M.: A peer-to-peer approach to web service discovery. World Wide Web **7**(2), 211–229 (2004)
33. Basters, U., Klusch, M.: RS2D: fast adaptive search for semantic web services in unstructured P2P networks. In: The Semantic Web-ISWC 2006, pp. 87–100. Springer, Berlin (2006)
34. Paolucci, M., Sycara, K.P., Nishimura, T., Srinivasan, N.: Using DAML-S for P2P discovery. In: ICWS, pp. 203–207 (2003)
35. Hameurlain, A., Morvan, F., Samad, M.E.: Large scale data management in grid systems: a survey. In: 3rd International Conference on Information and Communication Technologies: From Theory to Applications, 2008. ICTTA 2008, pp. 1–6. IEEE (2008)
36. Antonioletti, M., Atkinson, M., Baxter, R., Borley, A., Hong, N.P.C., Collins, B., Westhead, M.: The design and implementation of Grid database services in OGSA-DAI. Concurr. Comput.: Pract. Exp. **17**(2), 357–376 (2005)
37. Cao, J., Jarvis, S.A., Saini, S., Kerbyson, D.J., Nudd, G.R.: ARMS: an agent-based resource management system for grid computing. Sci. Programm. **10**(2), 135–148 (2002)
38. Tan, Y., Han, J., Wu, Y.: A multi-agent based efficient resource discovery mechanism for grid systems. J. Comp. Inf. Syst. **6**(11), 3623–3631 (2010)
39. Han, L., Berry, D.: Semantic-supported and agent-based decentralized grid resource discovery. Fut. Gen. Comput. Syst. **24**(8), 806–812 (2008)
40. Trunfio, P., Talia, D., Papadakis, H., Fragopoulou, P., Mordacchini, M., Pennanen, M., Haridi, S.: Peer-to-Peer resource discovery in grids: models and systems. Fut. Gen. Comput. Syst. **23**(7), 864–878 (2007)
41. Ranjan, R., Harwood, A., Buyya, R.: Peer-to-peer-based resource discovery in global grids: a tutorial. IEEE Commun. Surv. Tutor. **10**(2), 6–33 (2008)
42. Narendra, K.S., Thathachar, M.A.: Learning automata: an introduction2012: Courier Dover Publications (2012)
43. Narendra, K.S., Thathachar, M.: Learning automata—a survey. IEEE Trans. Syst. Man Cybern. **4**, 323–334 (1974)
44. Dorigo, M., Maniezzo, V., Colorni, A.: Ant system: optimization by a colony of cooperating agents. IEEE Trans. Syst. Man Cybern. Part B: Cybern. **26**(1), 29–41 (1996)
45. Mullen, R.J., et al.: A review of ant algorithms. Expert Syst. Appl. **36**(6), 9608–9617 (2009)
46. Kocak, T., Lacks, D.: Design and analysis of a distributed grid resource discovery protocol. Cluster Comput. **15**(1), 37–52 (2012)
47. Zhou, J., Abdulla, N. A., Shi, Z.: A hybrid P2P approach to service discovery in the cloud. Int. J. Inf. Technol. Comput. Sci. **3**(1), 1 (2011)
48. Winick, J., Jamin, S.:. Inet-3.0: internet topology generator. Technical Report CSE-TR-456-02, University of Michigan (2002)

Modelling and Detection of User Activity Patterns for Energy Saving in Buildings

Jose Luis Gomez Ortega, Liangxiu Han and Nicholas Bowring

Abstract Recently, it has been noted that user behaviour can have a large impact on the final energy consumption in buildings. Through the combination of mathematical modelling and data from wireless ambient sensors, we can model human behaviour patterns and use the information to regulate building management systems (BMS) in order to achieve the best trade-off between user comfort and energy efficiency. Furthermore, streaming sensor data can be used to perform real-time classification. In this work, we have modelled user activity patterns using both offline and online learning approaches based on non-linear multi-class Support Vector Machines. We have conducted a comparison study with other machine learning approaches (i.e. Linear SVM, Hidden-Markov and K-nearest models). Experimental results show that our proposed approach outperforms the other methods for the scenarios evaluated in terms of accuracy and processing speed.

1 Introduction

Buildings are one of the main contributors to CO_2 of the atmosphere with a 40 % of the total emissions in the UK [1]. It has been noted that occupant presence and behaviour have a significant impact on building energy performance [2], therefore efficient energy management needs to be able to reduce energy consumption costs without neglecting occupant comfort which ultimately can affect productivity [3].

J.L.G. Ortega (✉) · L. Han
School of Computing, Mathematics and Digital Technology,
Manchester Metropolitan University, John Dalton Building, Manchester M1 5GD, UK
e-mail: jose.l.gomez@mmu.ac.uk; jose.l.gomez@stu.mmu.ac.uk

L. Han
e-mail: l.han@mmu.ac.uk

N. Bowring
School of Engineering, Manchester Metropolitan University,
John Dalton Building, Manchester M1 5GD, UK
e-mail: n.bowring@mmu.ac.uk

© Springer International Publishing Switzerland 2016
L. Chen et al. (eds.), *Emerging Trends and Advanced Technologies for Computational Intelligence*, Studies in Computational Intelligence 647,
DOI 10.1007/978-3-319-33353-3_9

Today's building management systems (BMS) are usually operated based on a fixed seasonal schedule, maximum design occupancy assumption, and pre-defined comfort levels (constant) to ensure satisfactory temperatures, ventilations and luminance at all times, but fail to capture dynamic information. This is both costly and inefficient. For example, typical public areas such as office buildings or shopping centres tend to have illuminated all the spaces yet many times all of them are empty. Moreover, depending on which activities occupants might be performing, different systems or devices must be controlled (e.g. a user performing an exhausting physical activity would want to turn off the heating, to decrease the air conditioning temperature or to open a window). This is particularly true in domestic buildings where the activities users might perform are really diverse. Therefore, it is crucial for a BMS proper operation to dynamically incorporate user occupancy and activity information. This information will provide the system with the means to establish the right balance between energy consumption and user satisfaction.

To dynamically capture information about the occupants and their surroundings, wireless sensing techniques have shown great potential, taking advance of the fact that these sensor devices are easy to install, non-intrusive and inexpensive [4]. This explains their increasing popularity of their use for activities related to human behaviour detection tasks in buildings such as energy saving, smart homes or health monitoring. Through sensed information from users and indoor conditions, it is feasible to create a mathematical model, which will process that data and will be capable of recognising human behaviour and ambient parameters [5] in order to manage BMS more effectively (Fig. 1). Machine learning (ML) approaches and probabilistic models are specially suitable for the development of human behaviour pattern modelling as they can successfully extract relevant information from datasets as well as handle human behaviour randomness.

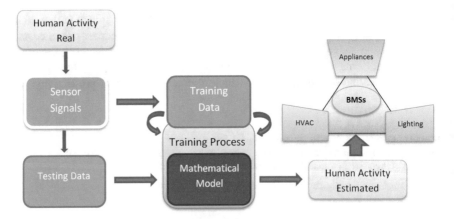

Fig. 1 Human activities trigger sensor events which are recorded and used to train a model. The same sensors also provide the data which will be for model testing and its normal performance. The information we retrieve from the model will help regulate BMSs more efficiently according to real occupant needs

In addition to explore human occupancy patterns from data, sensor information can be collected and introduced into the models as time sequences in a streaming data fashion and used to perform activity detection in real-time. To achieve this, these models need to address some specific issues [6]:

- To be able to handle the continuous flow of data and give a prediction within a certain amount of time.
- To find the best balance between performance and computational costs (time and memory).
- To update the model as new (unbounded) datapoints are being included.

In this work, we have proposed a novel non-linear multiclass Support Vector Machine (SVM) (MCSVM) to perform activities of daily living (ADL) classification in three different scenarios. We have evaluated the performance of our model based on an offline and an online approach and compared our results with other ML techniques. The rest of the paper is organised as follows: Sect. 2 outlines the related work in literature, Sect. 3 gives details about the methodologies adopted in this work, Sect. 4 explains the experiments carried out, and Sect. 5 presents the discussion based on the experimental results and we conclude in Sect. 6 where we also point future lines of research.

2 Previous Work

This section presents related previous works, which covers approaches to human behaviour patterns modelling in buildings and an overview of related works performing real-time user activities classification.

2.1 Occupancy and Activity Models Based on ML and Sensor Data

The exploitation of different ambient sensors to monitor occupancy and ambient conditions for dynamically adaptive energy control and comfort management has initially shown promising results [7, 8]. Various sensors (e.g. temperature or air quality such as CO_2, or a combination of them) were deployed in building scenarios and data was collected and processed for understanding the relationship between occupant behaviour patterns and energy consumption. With the introduction or real-time data collection techniques, new studies demonstrated how 'live' sensing approaches could be used to build data-driven models for occupancy detection [9, 10]. These models however did not consider occupant comfort levels, and could not correctly predict occupancy (e.g. maximum of 70 % accuracy prediction in [11]). Recently, many authors have opted for SVM as the building block for their model design [9, 11, 12], showing why it is considered one of the best choices for pattern recognition applications.

Singla et al. [13] considered a hidden Markov model (HMM) based algorithm for ADL detection from data acquired by means of a combination of motion, temperature and analogue sensors for water and cooking hobs usage monitoring. The combination of sensors was intended to overcome some of the limitations when using just one sensor as also noted in the work conducted by Dong et al. [9], where another mixture of sensors including motion, temperature, humidity and CO_2 levels were used. Other contributions which addressed ADL estimation from sensors were [14, 15]. All these works pointed the significance of the activity estimation and proposed methods to achieve this by using multi-sensor data and different ML techniques based on supervised learning.

Other authors used k-Nearest Neighbour models, as in Li et al. [16], where they developed a model for occupancy detection in order to build a demand driven heating, ventilation and air conditioning (HVAC) control showing how kNN performed better than other strategies in their application. Hajj et al. [17] and Scott et al. [18] also used a kNN to model user/building interactions.

2.2 Online Modelling Approaches

Human activity data can be analysed to extract conclusions such as typical household occupancy patterns or energy consumption profiles. However, this data can also be used to make decisions in real-time. Models devoted to the improvement of energy efficiency can take advance of streaming sensor data to regulate building systems accordingly at the exact time. The aforementioned works in [9, 11] or in [8, 19] tried to present methodologies which could address this problem and use the information as streaming data to make 'live' human behaviour patterns classification.

Recently, many efforts have been devoted to the development of a kernel SVM which can be updated when partial data is included. This is the case of successful contributions as the passive-aggressive algorithm [20], the incremental SVM [21] or other similar approaches including multi incremental SVM [22]. Most of these approaches are based on the online gradient descent approach to preserve the KKT conditions while updating the model. However, in spite of all the efforts, the actual proposed methods suffer from a computational complexity making the updating process even more costly where the models were trained with the whole datasets [23], which recently made researchers to look into online solutions for less complex models as an alternative [24, 25].

3   The Proposed Methodology

Inspired by the work in [26], the aim of this research is to develop a model to monitor ADLs in buildings more accurately using the same dataset. We have developed two different approaches to perform activity identification as follows:

Offline Approach Based on Non-Linear Multiclass SVM (MCSVM):
We have developed a pattern recognition system to identify human activity based on non-linear multiclass SVM-based approach trained and tested using ambient sensor data from a real world scenario. For the purpose of the comparison study, we have also applied other methods including HMM, KNN and Linear SVM to our case and conducted experimental evaluation.

Online Approach Based on Non-Linear Multiclass SVM (MCSVM):
We have proposed a methodology to perform 'live' detection using SVM approach. This method increases the speed of the classifier training and can handle unbound streaming data more efficiently.

3.1   Data Description

Wireless ambient sensors represent a powerful means to get information from our surroundings in a simple and efficient way with no intrusive issues. Different sensors can be used to monitor various indoor parameters such as motion or temperature which are relevant for the tasks we aim to achieve. Some researchers have made their datasets publicly available (i.e. WSU Casas [15] or DomusLab [27]), which is extremely helpful in two major ways: Firstly, it allows other researchers to use the datasets to conduct their own research. Secondly, it gives researchers a ground truth to evaluate their models and compare the results. In our work we have used the dataset described in [26] for ADL recognition purposes.

3.1.1   Sensors and Labelling

The dataset consists of data divided into days and collected by sensors deployed in three different house scenarios, namely **House A**, **House B** and **House C**. The various sensors used are motion (PIR), pressure (piezo) and contact sensors. As an example, **House A** has a total of 14 sensors and the labels were annotated by the occupants using a bluetooth voice detector device rather than other scenarios in which the activities were just written down on paper. As an example, a list of the sensors used and the different labels for **House A** can be found in Table 1. For more detailed information about the datasets, please refer to: https://sites.google.com/site/tim0306/.

Table 1 Detail on the sensors used and the labels for each activity corresponding to *House A*

Activity number	Activity description	Sensor number	Sensor description
1	Idle*	1	Microwave
2	Leave house	2	Hall-Toilet door
3	Use toilet	3	Hall-Bathroom door
4	Take shower	4	Cups cupboard
5	Brush teeth	5	Fridge
6	Go to bed	6	Plates cupboard
7	Prepare breakfast	7	Frontdoor
8	Prepare dinner	8	Dishwasher
9	Get snack	9	Toilet flush
10	Get drink	10	Freezer
–	–	11	Pans cupboard
–	–	12	Washing machine
–	–	13	Groceries cupboard
–	–	14	Hall-Bedroom door

* Represents the data that is unlabelled

3.1.2 Data Pre-processing

One of the main tasks before data processing is discretisation. The dataset we are using consists of annotations on the sensors fired and the activities performed at an exact time with seconds precision. However, discretising the data in such small amounts of time is inefficient since the sensor sampling rate is of at least a few seconds (since the action that triggered the sensor happened until the moment the information is available in the dataset). Moreover, if we want to use this data for online learning we need to carefully consider the amount on time between samples as we need to perform any updating operations before the new datapoint arrives.

Based on all these reasons, the time granularity will be determined by the machine capacity and the activities we are modelling. For example, if we want to model presence of people in a room for lighting control, we don't want to wait more than an instant of time for the system to operate [5]. But for longer activities such as *sleeping* or *leave home* we can afford a more coarse-grained discretisation. A choice of 60 s per time slot has shown to be fine enough with an acceptable error rate as shown by [28, 29]. For this particular dataset, timeslices of 60 s incur in a performance error below 2 % [26], which proves to be a good trade-off between loss of information and efficiency.

The three different scenarios **House A, House B** and **House C**, comprise 25, 14 and 19 days of data respectively. We conducted a pre-processing step on the data in which we created two types of matrices of features. In one of them we created a binary mxn matrix in which the columns n represented each of the sensors and each row m represented each sample. For each day we had 1440 samples (60 min × 24 h) instances per day spaced 60 s between them. A 1 in any position would represent the activation of the sensor corresponding to the row (more than one sensor might be active at any point), and the number of columns in which this value is repeated, would represent the time that sensors are activated. The label matrix consisted of an array with a numerical value k ($k = num_{activities}$) for each activity and also the 1440 timeslices per day.

3.2 SVM Based User Activity Pattern Modelling and Detection

In our work, we have proposed to apply non-linear multiclass SVM (MCSVM) for human activity identification through ambient sensor data. The main reasons include (1) the data is non-linear (2) our task is a typical multiclassification task. We have evaluated its performance against three other state-of-the-art ML algorithms namely HMM, kNN and linear SVM techniques.

3.3 Support Vector Machines

The SVM model learns by representing feature points in space and establishing a hyperplane separating the points of different classes. The most representative points in the boundaries are called Support Vectors and their position determines the final plane which will maximise the distance between the mentioned support vectors of different classes. The features will then be separated into regions by the hyperplane and the new points tested will be classified based on their region location. In order to go beyond linear classification, SVM uses the kernel trick method, which consist of modelling feature relations (from input space to feature space) instead of the features themselves, allowing the use of multidimensional hyperplanes. With these hyperplanes, SVM learners can handle multivariate data (such as from occupancy typical datasets with many sensors) which can be thus processed and classified.

We call our data $(\mathbf{x_i}, y_i)$, $i = 1, \ldots m$, where m is the total number of samples, $\mathbf{x_i} \in \chi \subseteq \mathbb{R}^d$ is the input of sensor readings and $y_i \in \{+1, -1\}$ represent each of the labels (multiclass SVM also keeps as will be explained below). This model seeks to learn from data to find a hyperplane function such as:

$$f(x) = \mathbf{w^T x} + b. \tag{1}$$

where $f(x)$ represents the plane as a function of the training samples $\mathbf{x}$, b is the bias term and $\mathbf{w^T}$ is the *weight vector*. In order to learn the model parameters b and $\mathbf{w}$, we want to optimise the function:

$$\frac{1}{2}||w||^2, \quad s.t. \quad y_i(w^T \mathbf{x_i} + b) \geq 1, \tag{2}$$

which is solved by means of a Lagrangian multiplier, yielding the optimisation expression:

$$\min_{\alpha} \frac{1}{2} \sum_{i=1}^{m} \sum_{j=1}^{m} y_i y_j \alpha_i \alpha_j \mathbf{x_i^T} \mathbf{x_j} - \sum_{i=1}^{m} \alpha_i. \tag{3}$$

When (3) is solved using quadratic programming, returns a matrix of α coefficients for each $\mathbf{x_i}$. Substituting the values in α into one of the Lagrangian constraints we can solve the value for $\mathbf{w}$ and for the bias term b.

Also, the coefficients in α can be separated into zero and non-zero values. All the non-zero values in the α matrix will correspond to what we call the support vectors, which are the values in the dataset which determine the final shape of the hyperplane we are learning. For more information about SVM models, we refer the reader to the work in [30].

3.3.1 Kernel Trick and Soft Margin

What we have seen so far about SVM is the linear hard margin version of the SVM. For the soft margin version, which allows to have 'noisy' points within the marginal region, the main optimisation problem (2) changes to:

$$\frac{1}{2}||w||^2 + C \sum_{i=1}^{m} \xi_i, \quad s.t. \quad y_i(w^T \mathbf{x_i} + b) \geq 1 - \xi_i, \xi_i > 0, \tag{4}$$

where the term C regulates the penalty for each point in the margins.

Also, to handle non linear data, the SVM model can be further adapted to map the inputs x into another plane. Changing the input space into an feature space $\Phi(\mathbf{x})$ through a kernel function $K(x_i, x_j)$. So, the function that we will need to learn change to:

$$f(x) = \mathbf{w^T} \Phi(\mathbf{x}) + b. \tag{5}$$

In terms of the kernels used for the *kernel trick*, the most common functions are:

- Linear: $K(x_i, x_j) = x_i^T x_j$,
- Polynomial: $K(x_i, x_j) = (\gamma x_i^T x_j + c)^d$ for $\gamma > 0$,
- RBF: $K(x_i, x_j) = e^{(\gamma||x_i - x_j||^2)}$ for $\gamma > 0$,
- Sigmoid: $K(x_i, x_j) = tanh \gamma x_i^T x_j + c$.

This kernel method enables the SVM to separate non-linear data as a linear model would do. Although this technique significantly increases the flexibility and the potential to successfully generalise with more complex data, it also increases the

Fig. 2 One versus one multiclass SVM approach. We construct a classifier (hyperplane) for each pair of classes. For prediction, each pair is tested and the most voted option is chosen

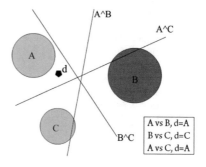

| A vs B, d=A |
| B vs C, d=C |
| A vs C, d=A |

complexity of the quadratic programming optimisation, which in turn results in a much more computational intensive task.

3.3.2 Multi-class SVM

The SVM is inherently a model to perform binary classification by nature. However, many of the real-world applications need to perform multi-class classification such as the present one. The SVM function used for our work performs the *one versus one* [31] multiclass inference approach, which has been noted as the most convenient technique for multi classification tasks. The *one versus one* involves creating a different classifier for each pair of classes, evaluating and finally selecting the most recurrent. When two or more classes are tied as the most recurrent, a random one is selected.

As shown in Fig. 2, we see an example of the *one versus one* classification method for three classes (A, B and C), which also needs to learn three SVM classifiers. For a new point d, the pair *A versus B* yields A, the pair *B versus C* gives C and the pair *A versus C* will predict class A. From the results sequence ABA, we predict the new inferred label would be A.

3.4 MCSVM Versus Alternative ML Techniques for Performance Evaluation

We have developed our MCSVM based model on MATLAB using the popular libsvm libraries [32]. All the evaluations have been conducted using a leave-one-day-out k-fold cross validation, where the k parameter corresponded to the length in days of each scenario. All the accuracies reported were calculated using the total of correct predictions over the total of predictions, from the confusion matrix generated. We tested the same data with other three state-of-the-art classification algorithms (i.e. HMM, kNN and LSVM) to compare their performance.

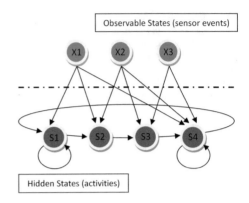

Fig. 3 HMM basic structure. The observable states are the sensor readings and the hidden states that trigger those sensors are the activities. The probability of each state depend on the observations X and the transition probability

3.4.1 Hidden Markov Model

Hidden Markov Models are generative probabilistic models based on the Markov sequences which assume interdependence between the current and the previous occurrence of classes. This usually results in a good way for modelling real events as sequences of states. However, to prevent unmanageable computing loads, this model computes only the probability of being in a determined state for consecutive times-lices (first order Markov chain). The results for this model were produced using the HMM MATLAB function included in the statistical toolbox of the product [33]. A more detailed description on this model can be found in [34]. Figure 3 shows a basic HMM structure.

3.4.2 K-Nearest Neighbours

The kNN model is a simple classification method extensively used for pattern recognition. It merely computes the distance between a new point and its closer neighbours to infer the label of the most repeated label amongst the neighbouring points. The most recurring label will be the chosen for the new point's classification. For this model we used our own kNN function based on the euclidean distance, selecting the number of neighbours which reported better accuracies. A more detailed description of this model can be found in [35].

3.4.3 Linear SVM

The technicalities of this model were covered in previous sections. This linear SVM (LSVM) is a simpler approach than the non-linear SVM as this does not apply the input into feature space mapping. This model has generally shown great potential for classification (unless the data is highly non-linear), and yields results that can challenge those obtained by other more complex models such as MCSVM, but faster and more efficient. This algorithm is specially indicated when dealing with large quantities of sparse data with high number of features and samples. To apply this approach, we used the liblinear MATLAB libraries [36].

4 Experimental Evaluation

In this section we give details on the experiments we have conducted to evaluate the performance of our two approaches: (1) Offline MCSVM and (2) Online MCSVM.

4.1 Offline MCSVM Performance Evaluation

When using a SVM classifier, we need to select some parameters beforehand, namely a kernel (the different options were discussed above), a degree parameter in the case of polynomial and RBF, and an error coefficient C which will determine the soft margin constraint. We made preliminary experiments to see the MCSVM performance using the most commonly used kernels: linear, polynomial, radial basis and sigmoid. For parameter estimation, we used a grid search of exponential growing values. From the experimental results we can conclude that the best overall kernel to be used with this dataset was the RBF (see Fig. 4), which is also the best choice for the majority of the real world classification tasks.

In any case, MCSVM performed significantly better than the other methods (kNN, LSVM and HMM) evaluated in our experiment. For example, in the *House A*, MCSVM had 84 % of accuracy with much more consistency with some days achieving more than 90 %. Also in *House B*, this model clearly outperforms all the methods before proposed as in Table 2. In the latter scenario *House C* the values reached only

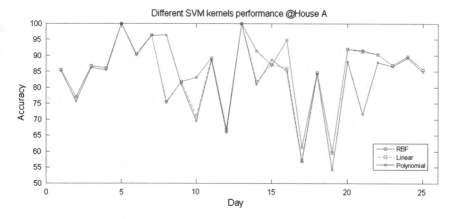

Fig. 4 Example of performance of the MCSVM model using different kernels for each day of the scenario House A

Table 2 Results from the performance evaluation experiment

	MCSVM	LSVM (%)	HMM (%)	KNN	CRF*(%)
House A	85.60	80.63	68.49	73.06	89.80
House B	84.74	73.53	63.51	65.12	78.00
House C	44.57	43.21	36.54	40.23	46.30
Average	71.64	65.79	56.21	59.47	71.37

CRF*: The last column shows the best accuracies reported for the team who made this dataset available from their own experiments

a 44.57 % accuracy, however this is still the best overall option amongst our three methods.

As an alternative for the MCSVM, LSVM obtained good accuracy results which, although still not as accurate as MCSVMs, demonstrated that this approach can have its uses when constraints such as time or memory allocation have an impact on the design of the model. With an average of 65.79 % it is the best option among the rest of the models we evaluated in this experiment.

Regarding the kNN, results were a 73 % average accuracy for *House A* and slightly lower in the other two, which is not a bad performance for such a simple classifier. We conducted the kNN inference by modifying the neighbour parameter until finding the peak performance when 87–91 neighbours were considered. Apart from the typical euclidean, other preliminary measurements for distance where evaluated (i.e. mahalanobis and correlation), yet the variations were too small to be significant and never above the values mentioned. We kept the euclidean approach. in case of the HMM, we solved the 0 emissions/transitions probability by applying a smoothing technique to give a small probability to every transmission and emission even if it never happened throughout the dataset. In our experiment we used the *Pseudoemissions* and *Pseudotransitions* option in the MATLAB's HMM function, although the results obtained using HMM is our experiments were significantly lower than other techniques.

4.2 MCSVM Online Approach

MCSVM model is complex and computationally intensive, and one of the main characteristics an online algorithm has to have is speed and efficiency. We compared the time (Machine: IOS WIN7, @2.40GH, 4GB RAM, MATLAB 2013a) each of our models took to perform the training and testing phases in our previous offline evaluation experiment (see Table 3).

Based on these times, it can be argued that the MCSVM might be too slow to perform real-time predictions.[1]

---

[1]Note that in this work we have discretised our data in timeslices of 60 s. Therefore, we need to develop an online approach fast enough to be able to incorporate a new point and give an estimated class prediction within 1 min.

Table 3 Time from the performance evaluation experiments

	MCSVM	LSVM	HMM	KNN
House A	420.85	20.58	0.91	36.76
House B	47.31	10.84	0.55	18.55
House C	829.21	31.34	0.744	27.79

The times are expressed in seconds. MCSVM has much higher computational times than the rest

Here, we propose a method that improves significantly the MCSVM training times by removing all non-support vectors from data, which in addition allows to integrate large quantities of new datapoints with a much more constrained growth of parameters.

4.2.1 Online MCSVM Based on Feature Reduction

As discussed in previous sections, one characteristic of the SVM algorithm is that it optimises the maximum margin distance between the hyperplane learned and the datapoints, giving the final function after training based upon just a portion of the total of samples; these are called the support vectors. The major drawback about this SVM approach is to solve the quadratic optimization, which requires a significant amount of memory and time compared to other methods, specially in the case of the non-linear SVM.

To successfully process real-time activity data, we need our algorithm to perform any model update operations to include new datapoints before the next one arrives. Moreover, if we keep on continuously introducing and memorising new points the memory necessary for the data storage will become unmanageable. The real challenge for our MCSVM approach is to be able to process and re-train the model fast and accurately enough to be ready for the next sample and to be able to handle unbounded data.

Here, we have proposed a simple feature reduction of the conventional MCSVM algorithm and we have used this approach to perform online ADL classification. To do this, we based our idea in the fact that the SVM models, once trained, use only a part of the data for their parametrisation called support vectors. As a natural characteristic of the SVM, when the training stage is finished, we can use all the non-zero coefficients from the matrix α obtained after the quadratic optimisation to identify which are the points (SVs) that are actually contributing to the final function. Once these SVs are identified, we can remove the rest of the points from the set and re-train the model with just the previous SVs as training set. The accuracy of the model after being re-trained in this fashion is exactly the same as it was when the whole dataset was used, however the processing times are largely lower (Table 4).

Table 4 Time of the evaluation model experiment for a MCSVM trained with the whole dataset against the same approach but trained with the SVs obtained from the model used with the whole dataset

	MCSVM-Full data	MCSVM-SVs	Num. Datapoints	Num. of SVs
House A	420.85	118.27	35486	9390
House B	47.31	10.14	19968	5756
House C	829.21	334.63	26236	12980

The first two columns compare the training times using the whole dataset or just the selected SVs. The other two columns indicate the number of data-points used for training in each case

4.2.2 Online MCSVM Evaluation

To validate the online approach, we have conducted an experiment in which we trained a MCSVM model using a third of the data (approximately 33 %) for the initial model training. Once the model has been initialised $t = 0$, our algorithm removes all the non-SVs from the data, makes a prediction of the $t + 1$ next label based on the sensor readings. Regardless the model is right or wrong in its prediction,[2] we include the new point in the training set and re-train the model with the previous SVs and the new point. As we have reduced dramatically the number of datapoints, the updating times are faster. Once the model has been re-trained, we test another datapoint and we train again including the new datapoint at $t + 2$ and we keep only the SVs after each re-training. This process continues throughout the test set which includes the rest 66 % on the points. In Fig. 5 we can see an graphical explanation of the process and we have included the pseudocode of our algorithm in Appendix 1.

This method was applied to the three scenarios and the predictions made step by step were compared with a batch testing of the remaining 66 % as a whole. The results of the online approach are much better than the offline evaluation as shown in Figs. 6, 7 and 8.

5   Discussion

Based on our experimental results, the offline MCSVM approach gives the more accurate performance among all the models evaluated, reaching values above 80 % accuracy in two out of three scenarios. In Figs. 9, 10 and 11 we can see how MCSVM outperforms LSVM, HMM and kNN significantly for *House A*, *House B* and *House C*. Comparing with the results reported in [26], in all scenarios SVM accuracies are better than all the previous generative models proposed (NB, HMM and HSMM); and also above the discriminative one (CRF) in the *House B* scenario. When averaging

---

[2]Our approach also differs from traditional online methods in which we do not apply any penalisation from a loss function, nonetheless this feature could be easily incorporated in the future versions of this algorithm.

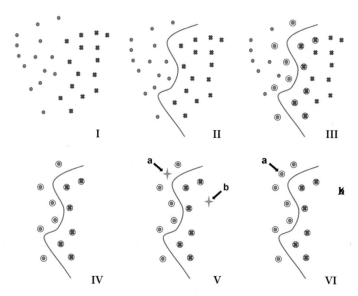

Fig. 5 Removing non-SVs online algorithm → Step I: we process the data. Step II: we train a model to separate the data. Step III: the model created identifies the SVs. Step IV: we keep only the SVs. Step V: we evaluate a new point which can be a potential new SV (**a**) or not (**b**). Step VI: if the new point is identified as a SV, we use it in the next iteration (**a**). If it is not, we discard it (**b**)

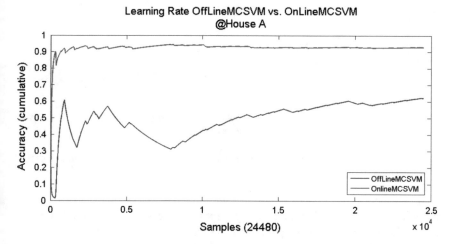

Fig. 6 The online learning versus offline learning in House A after 25000 samples/retraining (66 % of the data)

the results for the three scenarios, SVM presents the best overall result of all the previous models proposed, with a 71.64 %.

Furthermore, we have shown that this MSCSVM approach can be adapted to perform online classification by increasing the model efficiency. Results are promising

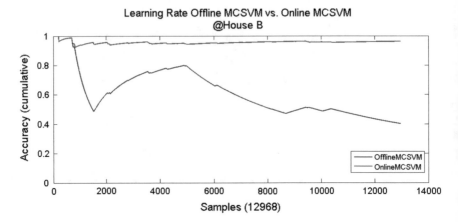

Fig. 7 The online learning versus offline learning in House B after 12000 samples/retraining (66 % of the data)

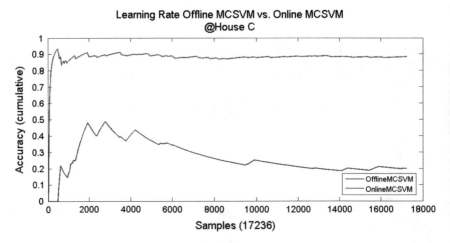

Fig. 8 The online learning versus offline learning in House B after 12000 samples/retraining (66 % of the data)

and have shown great potential, clearly improving the traditional batch approach. An specific constraint in this application was the time, but our approach made each prediction and update iteration under 2 s in House A and B, and in less than 7 s in House C; which means that, for this particular case in which our data is sampled in every 60 s, the model has plenty of time to make its updates and predictions within the time frame.

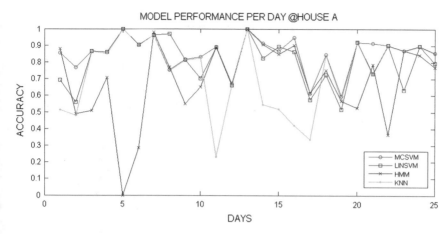

Fig. 9 Model performance @House A

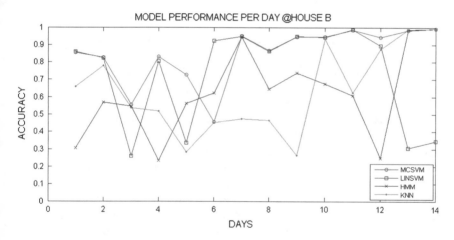

Fig. 10 Model performance @House B

6 Conclusions and Future Work

Accurate occupancy and activity pattern recognition play an important role in energy saving in buildings. This mainly include CO_2 reduction, cost savings and occupant comfort management. In this work, we have modelled user occupancy and activities based on an offline and an online machine learning approaches, using a publicly available dataset to perform activity detection more accurately. We have presented our proposed human activity pattern detection system based on a nonlinear multi-class SVM classifier (MCSVM), which has shown great potential for handling non-linear and complex features. We have conducted a multi-activity estimation based on labelled data collected by means of wireless ambient sensors using this approach. Further-

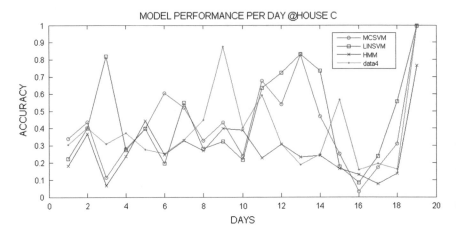

Fig. 11 Model performance @House C

more, we have compared our results with the levels of performance achieved with other state-of-the-art machine learning approaches (i.e. LSVM, HMM and kNN). The MCSVM model has achieved results over 80% accuracy in two of the three scenarios, significantly improving classification accuracy. In our online approach, we have trained the initial model with a third of the available data, and we have made predictions for the remaining two thirds of the data both in a batch and in an online fashion. Results show that our method achieves high standards of performance compared to the batch approach. Also, each iteration of the online approach is made in a few seconds so this method can be used for fast sampling rates of data acquisition.

Further work can be done to assess and quantify how pre-processing techniques may affect the final results for a model given, and also how different sampling criteria might modify the performance and help finding the best trade-off between overall and low repetitive class prediction performance. In the case of the online approach, the training samples still tend to grow while new data is incorporated (around a 5, 10 and 16% of the new datapoints have been added to the initial training data for each scenario) but this could be addressed by limiting the number training points if memory becomes a constraint after a certain number of iterations. Also, larger datasets could be used to study if the increasing number of suport vectors reach a saturation point when all the points available are able to represent the target function accurately.

Acknowledgments This work has formed a part of the funded project "Occupancy Pattern Detection for Energy Efficiency and Comfort Management in Buildings Based on Environmental Sensing", funded by Manchester Metropolitan University.

Appendix 1

Data is divided into 2 parts. First part is used for $trainLabels$ and $trainFeatures$.
The rest is used for $testLabels$ and $testFeatures$.
Initialisation: $SVMmodel := SVM\ training\ (trainLabels, trainFeatures)$.
$trainFeatures \leftarrow SVpoints, trainLabels \leftarrow SVlabels$; SVs from the previously
trained model become the training set.
for i:=1 to m; where m is the number of testing samples, **do**
    $activityprediction_i := SVM\ prediction\ (testLabels_i;\ testFeatures_i;\ SVMmodel)$.
    $trainLabels \leftarrow trainLabels + testLabels_i$.
    $trainFeatures \leftarrow trainFeatures + testestLabels_i$.
    $SVMmodel := SVM\ training\ (trainLabels; trainData)$;
    $trainFeatures \leftarrow SVpoints\ \&\ trainLabels \leftarrow SVlabels$, we update the datapoints
    with the SVs from the last SVM model.
end for

References

1. Gov.uk.: Energy performance of buildings-improving the energy efficiency of buildings and using planning to protect the environment-policies-gov.uk (2012). https://www.gov.uk/government/policies/improving-the-energy-efficiency-of-buildings-and-using-planning-to-protect-the-environment/supporting-pages/energy-performance-of-buildings
2. Nguyen, T.A., Aiello, M.: Energy intelligent buildings based on user activity: a survey. Energy Build. **56**, 244–257 (2013)
3. Jazizadeh, F., Marin, F.M., Becerik-Gerber, B.: A thermal preference scale for personalized comfort profile identification via participatory sensing. Build. Environ. **68**, 140–149 (2013)
4. Teixeira, T. Dublon, G., Savvides, A.: A survey of human-sensing: methods for detecting presence, count, location, track, and identity. ACM Comput. Surv. **V** (2010)
5. de Dear, R.J., Akimoto, T., Arens, E.A., Brager, G., Candido, C., Cheong, K.W.D., Li, B., Nishihara, N., Sekhar, S.C., Tanabe, S., Toftum, J., Zhang, H., Zhu, Y.: Progress in thermal comfort research over the last twenty years. Indoor Air **23**(6), 442–461 (2013)
6. Gaber, M.M., Zaslavsky, A., Krishnaswamy, S.: Mining data streams: a review. SIGMOD Rec. **34**(2), 18–26 (2005). http://doi.acm.org/10.1145/1083784.1083789
7. Dong, B., Andrews, B.: Sensor-based occupancy behavioral pattern recognition for energy and comfort management in intelligent buildings. In: Proceedings of Building Simulation, pp. 1444–1451 (2009)
8. Han, H., Jang, K., Han, C., Lee, J.: Occupancy estimation based on co2 concentration using dynamic neural network model. aivc.org (2013)
9. Dong, B., Andrews, B., Lam, K.P., Höynck, M., Zhang, R., Chiou, Y.-S., Benitez, D.: An information technology enabled sustainability test-bed (ITEST) for occupancy detection through an environmental sensing network. Energy Build. **42**(7), 1038–1046 (2010)
10. Ekwevugbe, T., Brown, N., Pakka, V., Fan, D.: Real-time building occupancy sensing using neural-network based sensor network. In: 2013 7th IEEE International Conference on Digital Ecosystems and Technologies (DEST), pp. 114–119 (2013)
11. Mamidi, S., Chang, Y.H., Maheswaran, R.: Improving building energy efficiency with a network of sensing, learning and prediction agents. In: International Conference on Autonomous Agents and Multiagent Systems (AAMAS) (2012)

12. Howard, J., Hoff, W.: Forecasting building occupancy using sensor network data. In: Proceedings of the 2nd International Workshop on Big Data, Streams and Heterogeneous Source Mining Algorithms, Systems, Programming Models and Applications—BigMine '13, pp. 87–94 (2013)
13. Singla, G., Cook, D.J., Schmitter-Edgecombe, M.: Recognizing independent and joint activities among multiple residents in smart environments. J. Ambient Intell. Hum. Comput. 1(1), 57–63 (2010)
14. van Kasteren, T., Englebienne, G., Kröse, B.J.A., van Kasteren, T.L.M.: Transferring knowledge of activity recognition across sensor networks. Pervasive Comput. (2010)
15. Cook, D., Crandall, A., Thomas, B., Krishnan, N.: CASAS: A smart home in a box. Computer (2013). http://www.ncbi.nlm.nih.gov/pmc/articles/PMC3886862/
16. Li, N., Calis, G., Becerik-Gerber, B.: Measuring and monitoring occupancy with an RFID based system for demand-driven HVAC operations. Autom. Constr. 24, 89–99 (2012)
17. Hajj, H., El-Hajj, W., Dabbagh, M., Arabi, T.: An algorithm-centric energy aware design methodology. In: IEEE Transactions on Very Large Scale Integration (VLSI) Systems (2014)
18. Scott, J., Brush, A.B.: PreHeat: controlling home heating using occupancy prediction. In: Proceedings of the 13th International Conference on Ubiquitous Computing, pp. 281–290 (2011)
19. Krishnan, N., Cook, D.: Activity recognition on streaming sensor data. In: Pervasive and Mobile Computing (2014). http://www.sciencedirect.com/science/article/pii/S1574119212000776
20. Crammer, K., Dekel, O., Keshet, J., Shalev-Shwartz, S., Singer, Y.: Online passive-aggressive algorithms. J. Mach. Learn. Res. 7, 551–585 (2006). http://dl.acm.org/citation.cfm?id=1248547.1248566
21. Cauwenberghs, G., Poggio, T.: Incremental and decremental support vector machine learning. In: Advances in Neural Information Processing Systems 13: Proceedings of the 2000 Conference, vol. 13, p. 409 (2001)
22. Karasuyama, M., Takeuchi, I.: Multiple incremental decremental learning of support vector machines. Trans. Neur. Netw. 21(7), 1048–1059 (2010). http://dx.doi.org/10.1109/TNN.2010.2048039
23. Kivinen, J., Smola, A., Williamson, R.: Online learning with kernels. Trans. Signal. Process. 52(8), 2165–2176 (2004). http://dx.doi.org/10.1109/TSP.2004.830991
24. Tsai, C.-H., Lin, C.-Y., Lin, C.-J.: Incremental and decremental training for linear classification. In: Proceedings of the 20th ACM SIGKDD International Conference on Knowledge Discovery and Data Mining, ser. KDD '14, pp. 343–352. ACM, New York, NY, USA (2014). http://doi.acm.org/10.1145/2623330.2623661
25. Karampatziakis, N., Langford, J.: Importance weight aware gradient updates. CoRR abs/1011.1576 (2010). http://arxiv.org/abs/1011.1576
26. van Kasteren, T., van Kasteren, T.L.M.: Human activity recognition from wireless sensor network data: benchmark and software. Activity Recogn. Pervasive Intell. Environ. 4, 165–186 (2011)
27. Gallissot, M., Caelen, J., Bonnefond, N., Meillon, B., Pons, S.: Using the Multicom Domus Dataset. LIG, Grenoble, France, Research Report RR-LIG-020 (2011)
28. Hoque, E., Stankovic, J.: AALO: Activity recognition in smart homes using Active Learning in the presence of Overlapped activities. In: Proceedings of the 6th International Conference on Pervasive Computing Technologies for Healthcare, pp. 139–146. IEEE (2012)
29. Alemdar, H., van Kasteren, T., Ersoy, C., van Kasteren, T.L.M.: Using active learning to allow activity recognition on a large scale. Ambient Intell., 105–114 (2011)
30. Burges, C.: A tutorial on support vector machines for pattern recognition. Data Mining Knowl. Discov. 43, 1–43 (1998). http://link.springer.com/article/10.1023/A:1009715923555
31. Pal, M.: Multiclass approaches for support vector machine based land cover classification (2008). arXiv:0802.2411
32. Chang, C.-C., Lin, C.-J.: LIBSVM: a library for support vector machines. ACM Trans. Intell. Syst. Technol. 2, 27:1–27:27 (2011). http://www.csie.ntu.edu.tw/~cjlin/libsvm

33. MATLAB and Statistics Toolbox Release: The MathWorks Inc. Natick, Massachusetts, United States (2012)
34. Rabiner, L.R.: A tutorial on hidden Markov models and selected applications in speech recognition. Proc. IEEE **77**(2), 257–286 (1989)
35. Indyk, P., Motwani, R.: Approximate nearest neighbors: Towards removing the curse of dimensionality. In: Proceedings of the Thirtieth Annual ACM Symposium on Theory of Computing, ser. STOC '98, , pp. 604–613. ACM, New York, NY, USA (1998). http://doi.acm.org/10.1145/276698.276876
36. Fan, R.-E., Chang, K.-W., Hsieh, C.-J., Wang, X.-R., Lin, C.-J.: LIBLINEAR: a library for large linear classification. J. Mach. Learn. Res. **9**, 1871–1874 (2008)

A New Architecture to Guarantee QoS Using PSO in Fixed WiMAX Networks

Eden Ricardo Dosciatti and Augusto Foronda

Abstract The sharing of communication networks, especially with multimedia services such as IPTV, video conferencing and VoIP has increased in recent years. These services require more resources and generate a great demand on the network infrastructure, requiring the guarantee quality of services. For this, scheduling mechanisms, call admission control and traffic policing should be present to guarantee quality of service. The networks of communication for wireless broadband, based on the IEEE 802.16 standard, called WiMAX are used in this work, because this standard only specify the mechanisms of how these policies should be implemented. Based on these factors, a new architecture was developed in order guarantee the quality of service, using the meta-heuristic Particle Swarm Optimization for fixed WiMAX networks, presenting a method for calculating the duration of the time frame, which allows a control of queues in the scheduler in order to uplink traffic from the base station.

1 Introduction

The number of users who use communication networks has increased significantly, particularly with the use of multimedia services such as IPTV, video conferencing and VoIP, that require more resources and generate a large demand on the infrastructure of the network, making access to broadband service had been continued growth and global demand, the major challenge that must be faced is the democratization of access, considered as the great frontier of communication nowadays.

E. Ricardo Dosciatti (✉)
Federal University of Technology Paraná, UTFPR,
Via do Conhecimento, km 01, Pato Branco, Paraná, Brazil
e-mail: edenrd@utfpr.edu.br

A. Foronda
Federal University of Technology Paraná, UTFPR,
Avenida Sete de Setembro, 3165, Curitiba, Paraná, Brazil
e-mail: foronda@utfpr.edu.br

© Springer International Publishing Switzerland 2016
L. Chen et al. (eds.), *Emerging Trends and Advanced Technologies for Computational Intelligence*, Studies in Computational Intelligence 647,
DOI 10.1007/978-3-319-33353-3_10

With the creation of technological devices of last generation that use wireless networks, allowed a change in access profile to computer networks, generating a strong demand for applications involving voice, video and data traffic [1]. Thus, the connection quality has become an essential parameter to modern life and the great goal is to provide broadband access in an efficient manner, with low cost of deployment and maintenance and enable the user to have the information available anytime and anywhere.

A possible and viable solution to reduce the costs of deployment of broadband Internet access in areas where infrastructure is not present is the use of wireless technologies, such as radio, WiMAX (Worldwide Interoperability for Microwave Access), LTE (Long Term Evolution), satellite, 3G, 4G, that does not use cables, reducing deployment costs, maintenance and upgrade.

This work uses WiMAX technology, a communication technology that uses broadband high speed wireless metropolitan networks [2]. The IEEE 802.16 standard [3], which describes the rules and recommendations to support the development and deployment of WiMAX networks, specifying scheduling policies, traffic control and connection control, however, does not establish how to implement them.

For this, a new architecture is proposed for the scheduler of the BS (Base Station) in the direction of uplink traffic. The main goal of this new architecture is the provision of QoS to the classes of service specified by the IEEE 802.16 standard, with the guarantee of resources for classes with real-time traffic and guaranteed minimum resources required for classes with no real-time traffic. The new architecture consists of three processing steps performed in a hierarchical manner, and called *Layer*1, *Layer*2 and *Layer*3.

In the following sections, metaheuristic particle swarm optimization (PSO) and the IEEE 802.16 standard are presented. The new architecture is specified showing results and a comparison with other schedulers in this area. The end of this chapter, the conclusions are presented.

2   Particle Swarm Optimization (PSO)

PSO is a heuristic optimization method based on autonomous agents or population of individuals, which is inspired by the experiences of the model of social behavior observed in many species of birds and fish, and even in human social behavior [4].

In PSO, the population, called swarm, is composed of particles. This particle swarm performs research within a multi-dimensional search space. These searches are candidate solutions of each particle to a given problem. This search space in a given time interval, each particle occupies a specific position and moves with a speed, which is modified based on the experience obtained by the particle itself and the experience of other particles of the swarm.

Each individual can be the solution to the problem being investigated, being assigned to each individual a value that is related to the adequacy of the particle with the solution of the problem, which is called the fitness function, and also a variable that represents the speed moving the individual (particle). With the development of the solution, individuals will adjust their speed on the best solution (best fitness) found by himself and also the best solution group (swarm), performing this process until particles find the best overall solution or next her. The value of the fitness function is calculated based on the objective function you want to optimize (maximize or minimize). Thus, the value of the fitness function is defined by the nature of the optimization problem and is computed by an objective function that evaluates the solution.

2.1 Mathematical Model of the PSO

In PSO algorithm, each particle, treated as a point in the search space represents a potential solution to the problem and follow the best positions found so far by particle group.

A key characteristic of the PSO approach is its simplicity, since the model consists of only two equations. In these equations, the particles move in the search space looking for the best possible solution. At each iteration, the velocity is updated by

$$v_i^{(k+1)} = (w * v_i^{(k)}) + (c_1 * r_1 * [pbest_i^{(k)} - x_i^{(k)}])$$
$$+ (c_2 * r_2 * [gbest^{(k)} - x_i^{(k)}]). \tag{1}$$

The new position of the particle is determined by the sum of your current speed and the new position, expressed by

$$x_i^{(k+1)} = x_i^{(k)} + v_i^{(k+1)}. \tag{2}$$

In Eqs. (1) and (2), i is a particle that is a candidate solution; k is the counter of iterations; v_i^k represents the current velocity of the particle i, and the update speed of each particle depends on parameters that must be adjusted to each problem to be optimized; x_i^k represents the current position of the particle i; c_1 and c_2 represent the parameters of confidence, also called parameters **cognitive** (c_1), which is how a particle trusts itself, also called **memory**, and **social** (c_2), which is how a particle swarm trusts, also called **cooperation**; w is the factor of inertia or inertial weight, which is a parameter used to control the exploration and exploitation of the search space; r_1 and r_2 represent two vectors with random numbers uniformly distributed in the interval [0,1]. r_1 is the random variable of the cognitive part and r_2 is the random variable of the social part; $pbest_i^k$ (*personal best*) is the best position ever achieved by the particle i based on their own experience; $gbest^k$ (*global best*) is the best position found by the particle swarm, based on best *pbest* positions obtained so far; s is the swarm size.

3 Standard IEEE 802.16

The purpose of the networks specified by the IEEE 802.16 standard [3] is to provide wireless broadband high-speed metropolitan area networks (WANs), supporting the interconnection between heterogeneous networks.

3.1 Topology of a Network Standard IEEE 802.16

WiMAX network consists of BSs, which provide access to subscriber stations (SS) of a particular area. Depending on the configuration, a WiMAX network is composed of one BS and one or more SSs, as shown in Fig. 1. BS is the central node that coordinates all activities involving the SSs. The BSs are kept in towers to optimize the coverage area of the network. Each of the BSs that are part of the network are interconnected by a network, and connected to the Internet, allow the SSs also have access to the Internet.

IEEE 802.16 standard defines two types of topologies for communication between network components, PMP (point-to-multipoint) and Mesh. In this new architecture, PMP topology is used because it controls network parameters needed to evaluate the performance of the proposed architecture.

3.2 Structure of a Frame in the IEEE 802.16 Standard

In IEEE 802.16 standard, the physical layer operates in the frames format. The standard specifies that frames have a duration between 2.5 and 20 ms [3]. In this

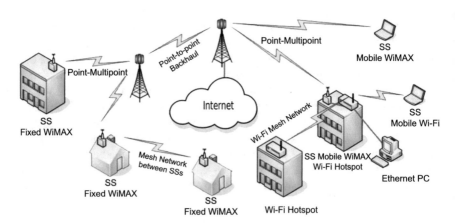

Fig. 1 Topology of a WiMAX network

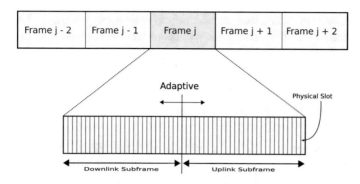

Fig. 2 A frame structure in TDD transmission scheme

new architecture, the duration of the frame is calculated according to the network parameters. Thus, in each iteration of the system, if there is a change in the parameters of the network, the duration of frame is calculated using an optimization algorithm based on PSO.

Frames are constituted by time intervals called physical slot (PS). The number of PSs in a frame is a function of the duration of an OFDM symbol, and a time frame.

Each frame consists of two subframes, called downlink subframe and uplink subframe. The BS uses the downlink subframe to send data and control information for the SSs. The uplink subframe is used for the SSs to send data and messages request bandwidth to the BS.

IEEE 802.16 standard specifies two modes of media access, time division duplexing (TDD) and frequency division duplexing (FDD). In this work, the TDD mode access is used.

Figure 2 shows the structure of a frame. The duration of the downlink and uplink subframes is adaptative according with the parameters used by the system.

4 A New Architecture Proposal

The scheduling in uplink direction, located at the BS, is complex to be implemented and should guarantee QoS for all requests from SSs. Thus, it is expected that implementation of a scheduler uses the available bandwidth efficiently, allowing a larger number of users to be allocated in the network.

This work aims to optimize the use of network by calculating the time frame (TQ), for service classes ertPS (extended real-time Polling Service) and rtPS (real-time Polling Service). In each iteration of the scheduling algorithm, if necessary, with the network parameters used at that time, a recalculation of TQ runs, with the possibility of allocating more system users, allowing the network to be more dynamic and more efficient.

This section presents a new architecture to guarantee QoS using PSO in fixed WiMAX networks. This architecture will act on the BS in scheduling traffic in the uplink direction.

4.1 Model of the New Architecture

The model of the new architecture is shown in Fig. 3 and in Fig. 4. SSs send messages request bandwidth, which are received at the BS uplink subframe and stored in virtual queues, according to service classes. For this to be possible, the BS must be synchronized with multiple SSs to manage the operations of requisition and allocation of bandwidth.

The strategy for the development of the new architecture is a hierarchical approach, with the use of three layers that are executed in sequential order, depending on the traffic. This division is performed in accordance with the purpose and functionality of each layer, with the possibility of modify parameters independently on each layer.

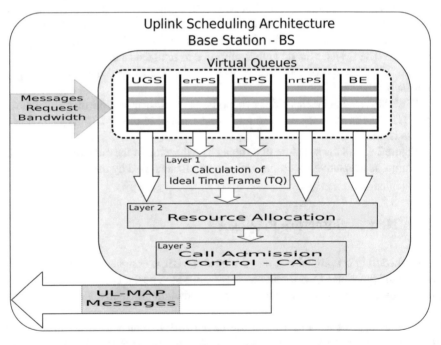

Fig. 3 Model of the new architecture: Base Station—BS

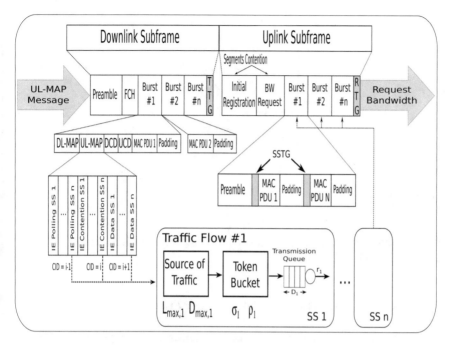

Fig. 4 Model of the new architecture: uplink and downlink subframes

4.2 Layer 1—Calculating the Duration of the Time Frame

The new architecture follows the trend of the latest solutions, such as the work of [5–8], for schedulers that operate in the BS for uplink traffic in IEEE 802.16 networks. The difference in relation to the related works is that the new architecture implements the calculation for the duration of the time frame (TQ) for the classes of service ertPS and rtPS.

The calculation of TQ is considered ideal for the duration of the time frame and is performed in each iteration of the scheduler, if necessary, for data transmission. Thus, the first layer of the new architecture is responsible for calculating the duration TQ. In this section, the analytical model for calculating the TQ is discussed and equations of constraint, to make the most efficient scheduling are presented.

To implement the new architecture, the model based on *LR* server [9] was used to calculate the ideal *TQ*, which is determined by two parameters: the latency and rate allocated by the server. Incoming traffic to the classes of service ertPS and rtPS are modeled by the token bucket algorithm. The idea is that the token bucket sets a limit for the input traffic and the scheduler *LR* establish a rate of allocation for each user. Thus, if the rate allocated by the scheduler *LR* is greater than the token bucket rate, the maximum delay is calculated, represented by D_i. In Fig. 5 is shown as D_i

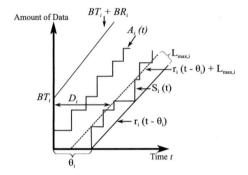

Fig. 5 *LR* server

is calculated, namely, the delay is based on the upper slope which results in a tighter bound delay in the network.

4.2.1 Analytical Model to Calculate the Time Frame

The performance of the new architecture is determined by two parameters for each session i ($i = 1, \ldots, N$): latency (θ_i) and allocated rate by the server (r_i). Thus, for each session i, the latency is a time frame period ($T_{DL} + T_{UL}$), the time to transmit a packet of maximum size ($\frac{L_{max,i}}{R}$), and the parameters of the intervals separating the subframes of *downlink* and *uplink*, T_{TTG} and T_{RTG}, expressed by

$$\theta_i = T_{DL} + T_{UL} + \frac{L_{max,i}}{R} + T_{TTG} + T_{RTG}. \tag{3}$$

The value r_i must be greater than the service rate $S_i(t)$ offered by the scheduler over a time interval $(0, t)$ (see Fig. 5).

The application using the session i declare the maximum packet size $L_{max,i}$ and requests a maximum delay that is allowed, represented by $D_{max,i}$, which is used by the scheduler in the new architecture to calculate the service rate of each individual session to ensure the required delay and optimize the number of stations on the network. In (3), R is the channel physical capacity.

The incoming traffic A_i, in a given period of time t, in session i, goes through a token bucket, each session during the interval $(0, t)$, the incoming traffic passes by the token bucket, limited by token bucket size, BT_i, and the token bucket rate, BR_i, in period t, expressed by

$$A_i(t) \leq BT_i + BR_i t. \tag{4}$$

Thus, the packet is queued in the station until it accesses the wireless medium and transmit. The queuing delay is measured from when the last bit of the packet is received and queued at the station until it accesses the wireless medium and transmit.

In the new architecture, the delay in the queue depends on the parameters of the token bucket, network latency and allocated rate by the server.

As incoming traffic $A_i(t)$ is modeled by a token bucket and the scheduler allocates a server rate r_i, a maximum delay D_i can be provided by

$$D_i \leq \frac{BT_i}{r_i} + \theta - \frac{L_{max,i}}{r_i}, \tag{5}$$

where $\frac{L_{max,i}}{r_i}$ is the difference between the lower and upper slope in Fig. 5, that is, the slightest inclination of the $(r_i (t - \theta_i))$ is the smallest valid limit for the service at any point the busy period and the steepening of the server $(BT_i + BR_i t)$ is valid at the point where a packet leaves the system. Thus, the delay can be calculated based on this and the result is a maximum delay limit of the network, called the upper limit of delay bound D_{bound}, which is represented by $\frac{BT_i}{r_i} + \theta_i - \frac{L_{max,i}}{r_i} \leq D_{max,i}$. This upper limit of delay must be less than or equal to the maximum delay that is required by the user and can be accepted by the system and is defined by

$$\frac{BT_i}{r_i} + \theta_i - \frac{L_{max,i}}{r_i} \leq D_{max,i}. \tag{6}$$

For each user that enters the system and will make the transmission of your data, should be considered an overhead. In the new architecture, this overhead is composed of contention segments in the uplink subframe and expressed by

$$\Delta = RI_{inic} + BW_{request}, \tag{7}$$

where RI_{inic} is reserved for initial maintenance opportunity in which admission requests can be met on the network, as well as changing the burst profile. After that, with $BW_{request}$ begins the period in which the SSs can perform requests bandwidth for transmission in the next uplink subframe.

The allocated rate by the server to schedule each user is defined as r_i. However, taking into consideration the overhead Δ in (7), is to add this value to the transmission of each user, determined by

$$r_i' = r_i + \frac{\Delta * R}{TQ_i}. \tag{8}$$

Then, from (3), (5) and (8), the maximum delay D_i in the new architecture is boudend by

$$D_i \leq \frac{(BT_i' - L'_{max,i}) * TQ_i}{r_i' * TQ_i - \Delta * R + L'_{max,i}} + T_{DL} + T_{UL} + \frac{L'_{max,i}}{R} + T_{TTG} + T_{RTG}, \tag{9}$$

where $L'_{max,i}$ is the $L_{max,i}$ with the overhead in (7) and BT'_i is the token bucket size with the same overhead. How to obtain (9) is shown in the appendix.

Thus, the first restriction condition to calculate ideal TQ, which is related to the delay, is given by

$$\frac{(BT'_i - L'_{max,i}) * TQ_i}{r'_i * TQ_i - \Delta * R + L'_{max,i}} + T_{DL} + T_{UL} + \frac{L'_{max,i}}{R} + T_{TTG} + T_{RTG} \leq D_{max,i}. \quad (10)$$

In the new architecture, the values of R, $L_{max,i}$ and BT_i, are declared by the application. The value of r'_i is calculated by (10), and the value of TQ_i is initialized as the IEEE 802.16 standard [3].

In (11), the second constraint condition for the new architecture can perform the calculation of the ideal TQ, is defined by

$$BR_i + \frac{\Delta R + L'_{max,i}}{TQ_i} \leq r'_i, \quad (11)$$

where the token bucket rate plus the rate to transmit the overhead and a maximum packet size must be smaller than the server rate to bounded delay.

The new architecture calculates a value for the TQ that is ideal, allocating a maximum number of SSs under the restriction conditions shown in (10) and (11). Thus, the maximum number of SSs is achieved when the value of r'_i, for each SS, is the minimum to guarantee the delay bound D_{bound}. Then, is necessary to find the minimum value for the following function:

$$f(TQ) = \frac{\sum_{i=1}^{n} TU_i}{TQ} \rightarrow min, \quad (12)$$

where TU_i is the time each SS has to transmit the data packet and is calculated by (13). In this calculation, the value of overhead Δ is not considered, since only the data load is what concerns for the system.

$$(TU_i) = \frac{r'_i * TQ_i}{R} - \Delta. \quad (13)$$

The values of TQ_i and TU_i are recalculated when: (i) require access new SSs to the BS; (ii) SSs leave the network; or (iii) if a SS exchange their QoS parameters.

4.2.2 TQ Optimization Using PSO

In this paper, the PSO metaheuristic [4] was chosen because it is a heuristic optimization method that aims to search for values within a set of parameters to maximize or minimize a given objective function subject to constraints imposed by the problem. The restrictions mean that has to reduce the number of possible solutions, and if a solution satisfies all the constraints, it is called a viable solution. To optimize the TQ, in each iteration of the system, if necessary, a value for the duration of ideal TQ is calculated. The use of the PSO reduce the number of executions for the calculation of TQ, and hence the execution time will be shorter, reducing the computational complexity. This is due to the fact that only viable solutions are analyzed.

The mathematical formulation of the optimization problem is described below:

Fitness Function:

$$Minimize \quad f(TQ) = \frac{\sum_{i=1}^{n} TU_i}{TQ} \tag{14}$$

Subject to:

$$TQ_{min,i} \leq TQ_i \leq TQ_{max,i} \tag{15}$$

$$\frac{(BT_i' - L_{max,i}') * TQ_i}{r_i' * TQ_i - \Delta * R + L_{max,i}'} + T_{DL} + T_{UL}$$
$$+ \frac{L_{max,i}'}{R} + T_{TTG} + T_{RTG} \leq D_{max,i} \tag{16}$$

$$BR_i + \frac{\Delta R + L_{max,i}'}{TQ_i} \leq r_i' \tag{17}$$

$$(Equation\ (16) \quad AND \quad Equation\ (17)) \leq R \tag{18}$$

In (15), the values of $TQ_{min,i}$ and $TQ_{max,i}$ are 2.5 ms and 20 ms, respectively, as determined by IEEE 802.16 standard [3].

In the new architecture, the values of TQ_i and TU_i are recalculated when: (i) a new SS request access to the BS; (ii) SSs leave the network; or (iii) if a SS change its QoS parameters. This is a very important feature and it is an advantage of the new architecture as compared to the schedulers of the work presented in [5–8]. In these works, the duration of TQ_i is declared as a fixed value at the beginning of scheduling, thus the duration of the TU_i, being related to the value of TQ_i also become fixed throughout the process scheduling.

4.3 Layer 2: Allocation of Resources

The task of *Layer* 2 is to allocate the available resources for data transmission in the uplink direction in a WiMAX network. These resources should be divided considering the transmission capacity of each segment in the uplink subframe. For this reason, this layer performs the calculation of an OFDM symbol, which is a small piece of information, it is possible to determine the value of the PS, which is the transmission capacity allocation unit used in the uplink subframe. Details for calculating the duration of an OFDM symbol can be found in [10].

4.4   Layer 3: Call Admission Control (CAC)

The *Layer* 3 performs the CAC to manage the QoS provisioning in a WiMAX network, namely which requests should be attended. Therefore, the new architecture makes the processing of service classes that are in the queue scheduling, using four rows of scheduling, with the aggregation of priorities for the five classes of services specified by the IEEE 802.16 standard [3].

After the execution of the three layers, the transmission needs of each class are organized to meet the next subframes by providing a message, called UL-MAP, which specifies the data regions allocated to each station in the next uplink subframe (see Fig. 4).

5   Results

The results achieved with the new architecture are demonstrated in two steps. In the first step, the calculation of the duration of TQ will be performed using PSO and compared in relation to the work of [11]. In the second step, with the ideal *TQ* that was calculated in the first step, the new architecture is compared with the work of [7] to check if the calculation of the ideal *TQ* has an effective gain in terms of delay and throughput.

5.1   First Step: Calculation of Ideal TQ

Figure 6 shows the case that is developed in this work, using the PSO metaheuristic, in which two constraint conditions must be satisfied for the search of the solution.

For the execution of performance tests of the new architecture, an IEEE 802.16 network was simulated, consisting of a BS that communicates with eighteen SSs, with only one traffic type for each SS and the fate of all flows is BS. The simulation environment used was developed in C programming language [12], for the process to be as reliable as possible to the environment used by [11].

Table 1, shows the type of traffic sent by SSs eighteen. Six sent to the BS, audio CBR traffic; six SSs send to BS, VBR video traffic; and six SSs send MPEG4 video traffic to BS.

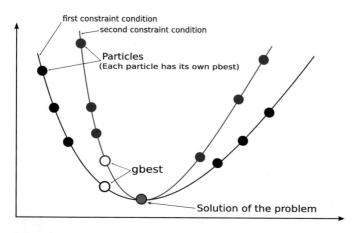

Fig. 6 Model constraint conditions used in PSO to calculate the ideal TQ

Table 1 Description of traffic types

SS	Application	Arrival period (ms)	Package size (max) (bytes)	Sending rate (kb/s) (average)
$1 \rightarrow 6$	Audio	4.7	160	64
$7 \rightarrow 12$	VBR video	26	1024	≈ 200
$13 \rightarrow 18$	MPEG4 Video	2	800	3200

Table 2 Parameters of PHY and MAC layers

Description	Value
Bandwidth	20 MHz
Duration of the OFDM symbol	13.89 μs
Delay required by user	5/10/15 e 20 ms
Δ (Initial ranging and BW request) $\rightarrow$ 9 OFDM symbols	125.10 μs
TTG + RTG $\rightarrow$ 1 OFDM symbol	13.89 μs
UL subframe (header) $\rightarrow$ 10 % OFDM Symbol	1.39 μs
Maximum data rate	70 Mbps

Table 2 summarizes all the parameters of the PHY and MAC layers used in this simulation.

Table 3 presents the token bucket parameters that are estimated of according to the characteristics of the incoming traffic.

The performance of the new architecture is evaluated by the required delay by the user and by allocated stations. Compared to the work of [11], the value of the ideal TQ is practically equal to optimal TF in the work of [11]. This proves that the calculation of the value of the ideal TQ, with the use PSO, was validated as shown at the graph of Fig. 7.

Table 3 Token bucket parameters

	Audio	VBR video	MPEG4 video
Bucket size (bits)	3000	18000	10000
Bucket rate (kb/s)	64	500	4100

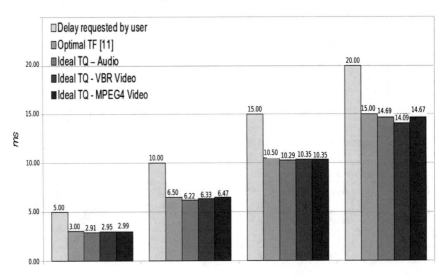

Fig. 7 Duration of the ideal *TQ* for audio traffic, video VBR traffic and video MPEG4 traffic

The difference between this work and that of [11] is the number of runs to calculate the ideal *TQ*, which allows a gain of runtime to perform the calculation. Figure 8 shows the number of executions for the traffic types used with the delay required by the user of 5, 10, 15 and 20 ms. The graph in Fig. 8 shows that there was a reduction of around 95 % in the number executions to find the ideal *TQ* compared to [11].

There is a relationship between the size swarm (also called the number of particles) and the number of executions to calculate the ideal *TQ* using the PSO metaheuristic. Thus, Fig. 9 presents the simulation results to find the ideal *TQ*, with the particles varying according to (19), which determines the initial size of the swarm.

$$s = \delta + [2 * sqrt(D)], \tag{19}$$

where, s is the swarm size, δ is the property that makes the change to find the ideal swarm size for solution of the problem and D is the number of dimensions determined by problem.

In Fig. 9 the audio traffic was simulated, where the delay required by the user is 15 ms, where the ideal *TQ* found is 10.29 ms, where the swarm size is 27.83 particles, and the relationship between the swarm size and the number of executions is 417.43.

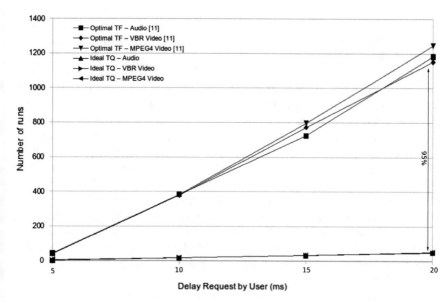

Fig. 8 Number of executions for the three types of traffic

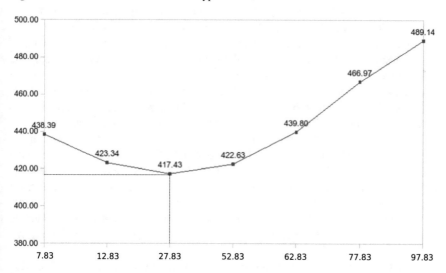

Fig. 9 Runtime versus number of particles

5.2   Second Step: Use of Ideal TQ with PSO

The second step shows the simulation results using the ideal *TQ*. The environment is modeled and evaluated by the network simulator NS-2 (Network Simulator-2) [13], together with the module that implements the MAC layer of the IEEE 802.16

standard, developed in [14]. The simulations are designed to verify the properties of the new architecture and analyze its behavior in a network with channel under ideal conditions without loss.

In this second step, the simulation scenario consists of a BS and 62 SSs, being 15 SSs with UGS traffic, 20 SSs with nrtPS traffic, 20 SSs with BE traffic and with rtPS traffic, SSs ranging from 1–7 during the simulation. In this scenario, each SS has only one application. Applications are represented by sources of voice traffic, video traffic, FTP traffic and Web traffic, which are served by UGS, rtPS, nrtPS and BE services, respectively. The voice traffic is modeled by an on/off source whose duration follows an exponential distribution with mean equal to 1.2 and 1.8 s respectively, with packets of 66 bytes are generated every 20 ms during the period on. Video traffic was obtained from traces of actual videos. The traffic file transfer via FTP is generated using message lengths exponentially distributed with a mean of 512 kbytes. The Web traffic is generated from a distribution Lognormal/Pareto hybrid whose body distribution corresponding to an area of 0.88 and is modeled as a lognormal distribution with mean equal to 7247 bytes, whereas the tail is modeled by a Pareto distribution with average of 10558 bytes.

The interval between periodic data grants for UGS service is 20 ms. The interval between unicast polling for sending request messages band is 20 ms for the rtPS service and for nrtPS flows of 1 s. The limited maximum delay requirement for the flow of rtPS class is 100 ms and the requirement for minimum bandwidth varies for each connection. The nrtPS flows have requirement of minimum bandwidth of 200 kbps and BE flows has no QoS requirement. For each value tested regarding the arrival rate of connections, 10 replications with different seeds were performed to generate the confidence interval of 95 %. Each simulation lasted 1000 s.

The results are compared with [7] and the goal is to assess how new architecture behaves using an ideal TQ, when the number of connections rtPS ranges from 1 to 7. In [7], the duration of time frame is fixed, namely, has no variation during the execution of the simulations. One of the properties of the new architecture is used to calculate the duration of TQ through optimization with PSO to improve network performance. Thus, every time the scheduler is used, a new calculation is performed for the duration of ideal TQ, if necessary. Thus, the calculation of the ideal TQ is made, using the same parameters of the experiment [7]. The duration of the time frame is 5 ms for all scenarios. The calculated value of the ideal TQ, in this simulation, was 4.6 ms.

As shown in Fig. 10 in both schedulers, the delay in the class of UGS service was not affected by increasing the offered load. The delay in rtPS class of service increased when new connections are allowed, but keeping below 100 ms required. It is noticed that there is a maximum delay for rtPS connections, below 30 ms, and the requested was 100 ms, occurring a waste of bandwidth, since the scheduler is always ensuring 100 ms.

Figure 11 shows the results for nrtPS and BE connections using the new architecture and the work of [7]. The new architecture has better performance because it uses the ideal TQ. Thus, as shown in Fig. 11, until the system admits five rtPS traffic connections, the throughput of traffic connections with BE is higher compared

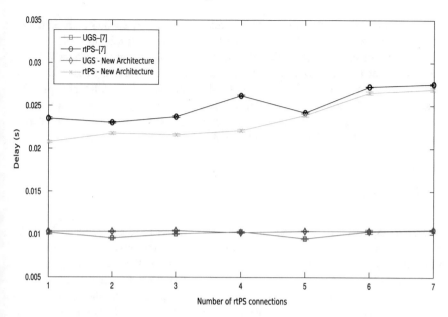

Fig. 10 Delay of UGS and rtPS connections

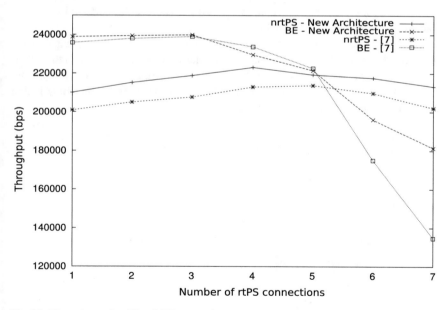

Fig. 11 Throughput of nrtPS and BE connections

to connections with nrtPS traffic. This is due to the fact that when there is plenty of resources, the scheduler is designed for the low-priority traffic, to avoid wasting bandwidth. The nrtPS traffic which generates a load 200 kbps are all satisfied. When the number of rtPS connections is more than five connections, the resources allocated to the BE connections are used to meet the needs of classes with higher priority. Thus, the results demonstrate that with an ideal TQ, the performance is improved in the new architecture because it avoids transporting empty frames sectors, allocating more users and reduce the waste of bandwidth.

6 Conclusion

In *Step* 1, the result obtained to calculate the duration of ideal TQ using the PSO metaheuristic, was very efficient, and had his calculation, regarding the length of time the frame, compared to the work of [11] and has been validated, reaching practically the same values of the optimal TF in relation to the ideal TQ. However, the amount of executions to find the value of ideal TQ, regarding the calculation of the optimum TF in [11] was decreased on average 95 % of the number of executions. This demonstrates that a decrease in the computational complexity, since the new architecture makes it faster calculation of ideal TQ and enables better utilization of frame for transmission of user data.

In *Step* 2, a comparison was performed with [7], relative to the throughput and delay, showing that the new architecture becomes more efficient, since in the calculation of the ideal TQ, there is an improvement in network parameters, given the fact that the frames are being transported with less voids, because the tendency is that the entire space of the frame is occupied.

Acknowledgments The authors would like to thank all researchers and collaborators at the Advanced Nucleous of Communication Technology of UTFPR.

Appendix

Steps to obtain (9):

   Substituting (3) into (5) yields

$$D_i \leq \frac{BT'i - L'_{max,i}}{r_i} + T_{DL} + T_{UL}$$

$$+ \frac{L'_{max,i}}{R} + T_{TTG} + T_{RTG}, \tag{20}$$

where $L'_{max,i}$ is the $L_{max,i}$ with the overhead in (7) and BT'_i is the token bucket size with the same overhead.

Substituting (8) into (20) yields

$$D_i \leq \frac{BT'i - L'_{max,i}}{r'_i - \frac{\Delta * R}{TQ_i}} + T_{DL} + T_{UL}$$

$$+ \frac{L'_{max,i}}{R} + T_{TTG} + T_{RTG}, \tag{21}$$

where r'_i is total allocated server rate.

Then, considering that at least one packet $\frac{L'_{max,i}}{TQ_i}$ must be transmitted in a TQ:

$$D_i \leq \frac{BT'i - L'_{max,i}}{r'_i - \frac{\Delta * R}{TQ_i} + \frac{L'_{max,i}}{TQ_i}} + T_{DL} + T_{UL}$$

$$+ \frac{L'_{max,i}}{R} + T_{TTG} + T_{RTG}. \tag{22}$$

And then, (9) is

$$D_i \leq \frac{(BT'_i - L'_{max,i}) * TQ_i}{r'_i * TQ_i - \Delta * R + L'_{max,i}}$$

$$+ T_{DL} + T_{UL} + \frac{L'_{max,i}}{R} + T_{TTG} + T_{RTG}. \tag{23}$$

References

1. Rosa, E.C., Guardiero, P.R.: Cac and Uplink Scheduling Algorithms in WiMAX Networks. Telecommun. Mag. **13**(2), 32–39 (2011). ISSN 1516-2338
2. Taghipoor, M., Mjafari, S., Hosseini, V.: Quality of service and resource allocation in WiMAX. In-Tech Croatia (2012). doi:10.5772/2454
3. IEEE Standard for Air Interface for Broadband Wireless Access Systems. IEEE Std 802.16-2012 (Revision of IEEE Std 802.16-2009), pp. 1–2542 (2012). doi:10.1109/IEEESTD.2012.6272299
4. Kennedy, J., Eberhart, R.C.: Particle Swarm optimization. In: IEEE Proceedings of the IEEE International Joint Conference on Neural Networks, vol. 4, pp. 1942–1948 (1995). doi:10.1109/ICNN.1995.488968
5. Sayenko, A., Alanen, O., Hämäläinen, T.: Scheduling Solution for the IEEE 802.16 Base Station. Comput. Netw. **52**, 96–115 (2008). doi:10.1016/j.comnet.2007.09.021
6. Masri, M., Abdellatif, S., Juanole, G.: An Uplink Bandwidth Management Framework for IEEE 802.16 with QoS Guarantees. NETWORKING. Lecture Notes in Computer Science, vol. 5550, pp. 651–663. Springer, Berlin (2009). doi:10.1007/978-3-642-01399-7_51
7. Borin, J.F., Fonseca, N.L.S.: Uplink Scheduler and Admission Control for the IEEE 802.16 Standard. In: Global Telecommunications Conference, GLOBECOM 2009, IEEE, vol 1, issue no. 1, pp. 1–6 (2009). doi:10.1109/GLOCOM.2009.5425779
8. Ferreira, F.A.: Uma Proposta de Escalonamento Baseado na Disciplina Priority Queuing (PQ) para Redes IEEE 802.16. UFU, 2011, master thesis (in portuguese). http://repositorio.ufu.br/handle/123456789/405. Accessed 17 Oct 2014

9. Stiliadis, D., Varma, A.: Latency-rate servers: a general model for analysis of traffic scheduling algorithms. IEEE/ACM Trans. Netw. **6**, 611–624 (1998). doi:10.1109/90.731196
10. Msadaa, I.C., Camara, D., Filali, F.: Scheduling and CAC in IEEE 802.16 Fixed BWNs: a comprehensive survey and taxonomy. IEEE Commun. Surv. Tutor. **12**(4), 459–487 (2010). doi:10.1109/SURV.2010.033010.00038
11. Dosciatti, E.R., Godoy Jr, W., Foronda, A.: An efficient approach of scheduling with call admission control to fixed WiMAX networks. IEEE Latin Am. Trans. **10**(1), 1256–1264 (2012). doi:10.1109/TLA.2012.6142470
12. Kernighan, B.W., Ritchie, D.M.C.: Teh C—Programming Language, 2nd edn. Prentice Hall, Englewwod Cliffs (1988)
13. NS-2. Network Simulator 2. http://www.isi.edu/nsnam/ns. Accessed 17 Oct 2014
14. Borin, J.F., Fonseca, N.L.S.: Simulator for WiMAX networks. Simul. Modell. Pract. Theory **16**(7), 817–833 (2008). doi:10.1016/j.simpat.2008.05.002

Colour-Preserving Contrast Enhancement Algorithm for Images

J.A. Ojo, I.D. Solomon and S.A. Adeniran

Abstract Conventional contrast enhancement techniques often fail to produce satisfactory results for low-contrast images, and cannot be automatically applied to different images because their processing parameters must be specified manually to produce a satisfactory result for a given image. This work presents a colour-preserving contrast enhancement (CPCE) algorithm for images. Modification to images was performed in the HSV colour-space. The Hue component is preserved (unchanged), luminance modified using Contrast Limited Adaptive Histogram Equalization (CLAHE), while Saturation components were up-scaled using a derived mapping function on the approximate components of its discrete wavelet transform. Implementation was done in MATLAB and compared with CLAHE and Histogram Equalization (HE) algorithms in the RGB colour space. Subjective (visual quality inspection) and objective parameters (Peak-signal-to-noise ratio (PSNR), Absolute Mean Brightness Error (AMBE) and Mean squared error (MSE)) were used for performance evaluation. The method produced images with the lowest MSE, AMBE, and highest PSNR when tested, yet preserved the visual quality of the image.

1 Introduction

A greyscale image may be defined as a two-dimensional function, $f(x, y)$, where x and y are spatial (plane) coordinates, and the amplitude f at any pair of coordinates (x, y) is called the intensity of the image at that point [1]. The image is said to be

J.A. Ojo (✉) · I.D. Solomon
Department of Electronic and Electrical Engineering,
Ladoke Akintola University of Technology, Ogbomoso P.M.B. 4000, Nigeria
e-mail: jaojo@lautech.edu.ng

I.D. Solomon
e-mail: soleade2@yahoo.com

S.A. Adeniran
Department of Electronic and Electrical Engineering, Obafemi Awolowo University,
Ile Ife, Nigeria
e-mail: sadenira@oauife.edu.ng

© Springer International Publishing Switzerland 2016
L. Chen et al. (eds.), *Emerging Trends and Advanced Technologies
for Computational Intelligence*, Studies in Computational Intelligence 647,
DOI 10.1007/978-3-319-33353-3_11

digital if the values of x, y and f are all finite. A digital image can be defined as one which is composed of a finite number of elements, pixels, each of which has a particular location and value.

A digital image is represented as an $M \times N$ matrix:

$$f = \begin{pmatrix} f_{0,0} & f_{0,1} & f_{0,N-1} \\ f_{1,0} & f_{1,1} & f_{1,N-1} \\ f_{M-1,0} & f_{M-1,1} & f_{M-1,N-1} \end{pmatrix} \tag{1}$$

An image $f(x, y)$, have values that range from 0 to $L - 1$, where $L = 2^x$ and x is the class unit of the image.

RGB colour image may be viewed as a stack of three different 2-dimensional images, each of which corresponds to red, green and blue channel images. CMYB image may be viewed as one that is comprised of four different 2-dimensional images, each of which corresponds to cyan, magenta, yellow and black channel images.

Assuming that the image is generated from a physical process, the intensity values 'f' of the image are proportional to energy radiated by the physical source (capturing device). Hence, image formation can be modelled as the multiplication of source illumination 'i' by the scene object reflectance 'r'.

$$f(x, y) = i(x, y) \times r(x, y) \tag{2}$$

where 'i' is the source illumination that depends on the capturing device and the capturing area, 'r' is the reflectance that depends on the characteristics of the imaged objects.

$$0 < i(x, y) < \infty \quad and \quad 0 < r(x, y) < 1 \tag{3}$$

$r = 0$ indicates total absorption and $r = 1$ indicates reflectance. The same mathematical expressions can be applied to images formed through transmission of illumination through a medium such as an x-ray image. In this case, transmissivity replaces reflectivity [2].

Sometimes an image may be too dark or too bright, which makes it difficult to recognize the different objects or scenery contained in the image. Earlier studies proved that digital images face the problem of low contrast due to various factors such as poor illumination, storage format, low lens power and operator proficiency. To improve recognition of details in such an image, it is necessary to compensate for the loss of image quality, and one of the possible ways suggested [3] is to employ contrast enhancement algorithms.

Many procedures for contrast enhancement such as Histogram Equalisation have been proposed, especially for aviation, medical digital radiography and infrared imaging systems where the initial images are of poor quality [4]. This work presents a contrast enhancement algorithm that preserves the colour of an image.

The other sections of this chapter are organised as follows: Sect. 2 gives a brief review of existing contrast enhancement techniques, while Sect. 3 describes the newly proposed colour-preserving contrast enhancement (CPCE) technique.

2 Contrast Enhancement Using Histogram Processing Techniques

Image enhancement processes consist of a collection of techniques that seek to improve the visual appearance of an image or to convert the image to a form better suited for analysis by humans or machine. The existing contrast enhancement techniques for mobile communication and other real time applications falls under two broad categories that is contrast shaping based methods and histogram equalization based methods [5]. These methods may lead to over enhancement and other artefacts such as flickering and contouring.

The contrast shaping based methods are based on calculating an input-output luminance curve defined at every luminance level. The shape of the curve must depend on the statistics of the image frame being processed. For example, dark images would have a dark stretch curve applied to them. Although contrast shaping based methods are the most popular methods used in the consumer electronics industry, they cannot provide the desirable localized contrast enhancement. For example, when a dark stretch is performed, bright pixels still become brighter. However, a better way to enhance darker images is to stretch and enhance the dark regions, while leaving brighter pixels untouched [6].

Histogram equalization is a technique that generates gray map which change the histogram of image and redistributes all pixel values to be as close as possible to user specified desired histogram. A very popular technique for contrast enhancement of image is histogram equalization technique [7]. The histogram equalization (HE) is a method to obtain a unique input to output contrast transfer function based on the histogram of the input image, which results in a contrast transfer curve that stretches the peaks of the histogram (where more information is present) and compresses the troughs of the histogram (where less information is present) [5]. Therefore it is a special case of contrast shaping technique.

As a standalone technique, histogram equalization is used extensively in medical imaging, satellite imagery and other applications where the emphasis is on pattern recognition and bringing out hidden details. Thus histogram equalization results in too much enhancement and artefacts like contouring, which is unacceptable in consumer electronics [8].

During the last decade a number of techniques have been proposed to deal with these problems. One of such is Brightness Preserving Bi-histogram Equalization. The histogram is divided into two parts based on the input mean, and each part is equalized separately. This preserves the mean value of image to a certain extent [9].

An adaptation of HE, termed as Contrast Limited Adaptive Histogram Equalization (CLAHE), divides the input image into a number of equal sized blocks and then performs contrast limited histogram equalization on each block [8]. The contrast limiting is done by clipping the histogram before histogram equalization. This tends to tone down the over enhancement effect of histogram equalization and gives a more localized enhancement. However it is much more computationally intensive than histogram equalization. If the blocks are non-overlapping, an interpolation scheme is needed to prevent blocky artefacts in the output picture.

Another method is histogram specification (HS), which takes a desired histogram by which the expected output image histogram can be controlled [10]. However, specifying the output histogram is not a smooth task as it varies from image to image. During past years various researchers have also focused on improvement of histogram equalization based contrast enhancement techniques such as mean preserving Bi-Histogram Equalization (BBHE) [9], dualistic sub-image histogram equalization (DSIHE) [11] and minimum mean brightness error bi-histogram equalization (MMBEBHE) [12].

The DSIHE method uses entropy value for histogram separation. The MMBEBHE is the extension of BBHE method that provides maximal brightness preservation. Though these methods can perform good contrast enhancement, they also cause more annoying side effects depending on the variation of gray level distribution in the histogram. Recursive Mean-Separate Histogram Equalization (RMSHE) was proposed, which provides better contrast results over BBHE [12]. Furthermore another technique that is Brightness Preserving Dynamic Fuzzy Histogram Equalization (BPDFHE) has been proposed. This technique modifies brightness preserving dynamic histogram equalization technique to improve its brightness preserving and contrast enhancement abilities while reducing its computational complexity. The technique uses fuzzy statistics of digital images for their representation and processing. Therefore, representation and processing of images in the fuzzy domain enables the technique to handle the inexactness of gray level values in a better way and provide improved performance [13].

Contextual and variational contrast enhancement for image algorithm enhances the contrast of an input image using inter-pixel contextual information. This algorithm uses a 2D histogram of the input image constructed using a mutual relationship between each pixel and its neighbouring pixels. A smooth 2D target histogram is obtained by minimizing the sum of frobenius norms of the differences from the input histogram and the uniformly distributed histogram. The enhancement is achieved by mapping the diagonal elements of the input histogram to the diagonal elements of the target histogram. This algorithm produces better enhanced images compared to other existing state-of-the-art algorithms [14].

2.1 Other Contrast Enhancement Methods

Other proposed algorithms for contrast enhancement include, a Discrete Cosine Transform (DCT) based compressed domain known as Alpha Rooting (AR) [15], Multi Contrast Enhancement with Dynamic Range Compression (MCEDRC) [16] and wavelet based domain contrast enhancement [17]. An enhancement of colour images by high-frequency spatial information feedback from the saturation component into the luminance component was proposed by [18]. A colour contrast enhancement method which adopts xy-chromaticity diagram and consist of two steps: (i) maximally saturating all the chromatic colours within a certain gamut, and (ii) desaturation operation based on centre of gravity law for colour mixture was proposed by [19]. Likewise a genetic algorithm approach to colour images in which the enhancement problem is formulated as an optimization problem was suggested by [20].

In recent years, multi-scale technologies have been widely used in image processing. The multi-scale methods mainly enhance the edge information of the image. In [21] a contrast enhancement method based on multi-scale gradient transformation was proposed, while an adaptive strategy for wavelet based image enhancement was proposed by [22]. These methods have been used for colour image enhancement. However, manual parameters specification and colour shifting problem are often associated with some of these methods. The hue component indicates the colour, hence, it is critical to consider hue in colour enhancement method. The colour shift problem has been solved in earlier works [23],while a hue preserve colour image enhancement method without gamut problem was proposed by [24]. This method keeps hue preserved in order to avoid colour shifting problem.

A perceptually motivated automatic colour contrast enhancement algorithm was developed in [25]. The enhancement was inspired by a luminance-reflectance modelling. Segmentation was done to remove halo effect, while illumination and reflectance extraction was performed. Colour cast was removed and the illumination-image was converted to LAB colour space, L-component image was remapped and recombined with A and B to get new LAB image. Inverse conversion to RGB was performed to get the enhanced image.

Compressed domain contrast and brightness improvement algorithm for colour image through contrast measuring and mapping of DWT coefficients was proposed by [26]. The Input image in RGB was converted to HSV colour space. S was enhanced by HE, DWT was used to decompose V and local and global enhancement techniques were used to enhance the decomposed V images. IDWT was used to synthesis new HSV image. The new HSV image was converted back to RGB to get the enhanced image.

A hybrid colour image enhancement technique based on contrast stretching and peak based histogram equalization was proposed in [27]. The research was carried out on acute leukaemia images. A 3×3 Gaussian filter was convolved with R, G, and B components of the input image. PDF and CDF were calculated for the three layers and partial contrast enhancement was applied to the three layers modified by multiple peaks.

Undecimated Wavelet Transform (UWT) based contrast enhancement was proposed by [28]. The contrast enhancement algorithm was applied to value (V) component of the HSV colour space, which is regarded as the intensity image. UWT is applied to the intensity image using the "à trous" algorithm to get the decomposed images. Contrast enhancement was realized by tuning the magnitude of approximation coefficients at each level with respect to the approximation coefficients of one higher level. Synthesis was performed by inverse UWT to get the enhanced intensity image and the new HSV was converted to RGB to get the enhanced image.

An adaptive contrast enhancement of coloured foggy images was proposed by [29]. The input foggy image was converted to HIS colour space. Gamma correction was applied on the S-image, morphological operation and transmission ratio were used to enhance the I-image. The new HIS image was converted back to RGB to get the enhanced image.

Menotti et al. [30] proposed a fast HUE-preserving histogram equalization methods for colour image contrast enhancement. Two fast hue-preserving HE methods based on 1D and 2D histograms of the RGB colour space for image contrast enhancement were proposed in this work. The first method estimates a RGB 3D histogram equalized using R-red, G-green, and B-blue 1D histograms, while the second method employs RG, RB, and GB 2D histograms. The histogram equalization was performed using shift hue-preserving transformations to avoid unrealistic colours.

3   Colour-Preserving Contrast Enhancement Algorithm

This section describes the newly proposed colour-preserving contrast enhancement (CPCE) algorithm. The image enhancement is achieved through modifications in both spatial and frequency domains on the HSV colour model, since HSV has attributes very close to the ways humans recognize colours.

The input colour image in RGB was converted to HSV. The flowchart for the CPCE algorithm is shown in Fig. 1. The H-component was unaltered throughout to preserve the colour, while discrete wavelet transform was applied to the S-component to decompose it to approximate and detailed components. The approximate components were enhanced using a mapping function derived from a 'scale-up' triangle function. The enhanced approximate components were combined with the unaltered detail components and Inverse Wavelet transform was performed to obtain the enhanced S-components. V-component in the HSV colour space was enhanced using CLAHE. Unchanged H-component, enhanced S-component and CLAHE-enhanced V-component were combined to give the new HSV image. New HSV image was converted to RGB to give the enhanced output image. The algorithm was proposed using MATLAB 7.8 version.

Five low contrast colour kodak png images (dark and bright) were acquired and used as test images. These images were obtained from the internet. These images were used as the input images. The input RGB colour image was converted to HSV colour image. The discrete wavelet transform was used to decompose the saturation into approximate components and detail components. A specified mapping function is applied on the approximate coefficients to modify the values and reconstruction of the saturation information is performed by applying Inverse Discrete Wavelet transform (IDWT). Thus, the above mentioned process consists of three steps: Discrete wavelet transform, approximate coefficients enhancement and inverse discrete wavelet Transform.

Approximate component enhancement was done by using the mapping functions derived from the scale-up triangle as shown in Fig. 2. Equation 4 was derived from Fig. 2, which was further simplified to Eq. 7 and used to map the approximate coefficients to new values. The new approximate component was combined with the unaltered detail components using Inverse discrete wavelet transform to give the new values for the saturation components.

The purpose of saturation adjustment is to make the colour image soft and vivid. The saturation component often contains more high frequency spectral -energy/image details than its luminance counterpart [31]. CLAHE was adopted for the enhancement of the V component in the HSV colour space. The V-image was divided into 8×8 tiles and the clip-limit used was 0.01. Uniform distribution was used as the histogram shape for the image tiles.

Algorithm 1

1. **Input**: *Image* of size m, n
2. Convert *Image* from RGB to HSV
3. Compute Discrete Wavelet Transform of the S-component to give *DS*
4. Extract approximate component *A* from *DS*
5. Compute upper threshold *newmax* and lower threshold *newmin*
6. **for** $i = 1 : m$ **do**
7.   **for** $j = 1: n$ **do**
8. $newA(i, j) = new\max - \frac{(new\max - oldA(i,j)) \times (new\max - new\min)}{old\max - old\min}$
9.   **end for**
10. **end for**
11. Compute Inverse Discrete Wavelet Transform to get *nS*
12. Apply CLAHE to *V* to get *nV*
13. Convert *H*, *nS*, *nV* to *nRGB* to get *newImage*
14. **Output**: *newImage*

$$\frac{old\max - oldA(i, j)}{old\max - old\min} = \frac{new\max - newA(i, j)}{new\max - new\min} \tag{4}$$

$$new\min = (old\min \times 2.5289) \tag{5}$$

$$new\max = (old\max \times 0.9) \tag{6}$$

$$newA(i, j) = new\max - \frac{(new\max - oldA(i, j)) \times (new\max - new\min)}{old\max - old\min}, \tag{7}$$

Fig. 1 The flowchart of the
proposed method

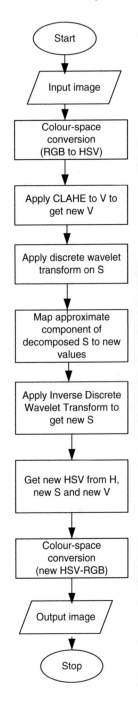

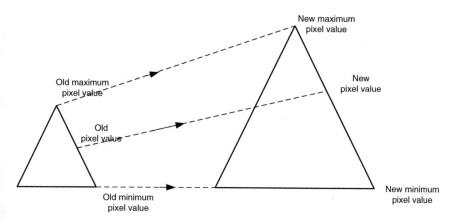

Fig. 2 Scale-up triangle model

where i, j represent the row and column respectively which shows the location of a particular saturation approximate component pixel. Scale-up triangle model is shown in Fig. 2.

3.1 Evaluation Parameters

To check the performance of the proposed method there is need for quality evaluation metrics. Subjective and objective parameter metrics were used. Objective Evaluation Parameters used are Absolute Mean Brightness Error (AMBE), Mean Squared Error (MSE), Peak Signal to Noise Ratio (PSNR), expressed in Eqs. 8–10 [32]. Low AMBE indicates a better brightness preservation of the method; a lower MSE indicates a better performance, while the higher the PSNR value, the better the reconstructed image.

$$AMBE = |E(y) - E(x)| \qquad (8)$$

where: $E(x)$ = average intensity of input image
$E(y)$ = average intensity of enhanced image.

$$MSE = \frac{1}{MN} \sum_{i=0}^{M-1} \sum_{j=0}^{N-1} \left[\left(r(i, j) - r''(i, j)\right)^2 + \left(g''(i, j) - g''(i, j)\right)^2 + \left(b(i, j) - b''(i, j)\right)^2 \right],$$

$$(9)$$

where: $r(i, j)$, $g(i, j)$, $b(i, j)$ are image pixels of original image and $r''(i, j)$, $g''(i, j)$, $b''(i, j)$ are image pixels of enhanced image of size $M \times N$

$$PSNR = 10\log_{10} \frac{R^2}{MSE_R + MSE_G + MSE_B},\tag{10}$$

where $R = 255$ for an 8-bit/class 8 image, $R = 1$ for a double-precision image and MSE_R, MSE_G, MSE_B are the MSE values between the R component, G component and B component of the original and the enhanced image respectively

4 Results and Discussions

MatLab was used for the implementation of the proposed algorithm simulated on a Dell inspiron 3420 with Windows 7 operating system, 4 GB RAM and core i3 processor at 2.4 GHz. The proposed method was tested with five of the kodak test images in 'png' format. Figures 3, 4, 5, 6 and 7 shows the experimental results of the images used for testing. The original image, results for CPCE, HE, and CLAHE methods are shown in (a), (b), (c) and (d) respectively.

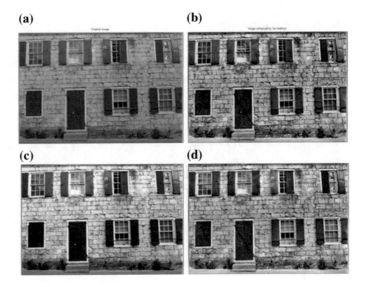

Fig. 3 Wall

(a)   (b)

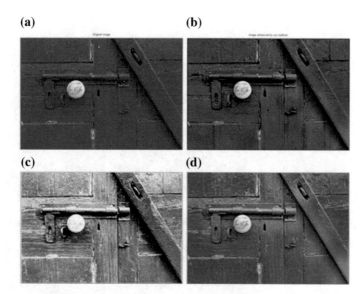

(c)   (d)

Fig. 4 Door surface

(a)   (b)

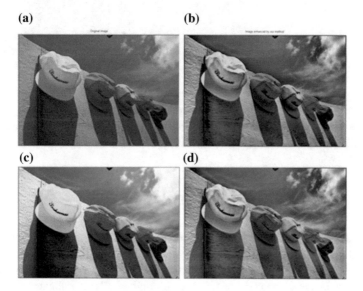

(c)   (d)

Fig. 5 Caps on a wall

Fig. 6 Boat

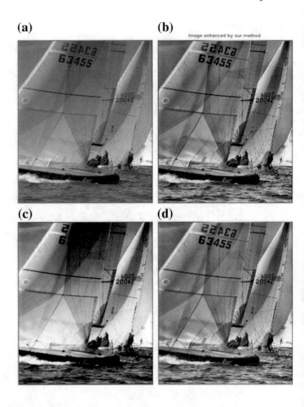

To test the performance of the proposed method for colour image contrast enhancement, the results were evaluated subjectively by visual inspection and objectively by three evaluation parameters (AMBE, MSE, and PSNR).

4.1 Results of Objective Evaluation Parameters

The values in Table 1 show that CPCE method produced images with the lowest MSE and MBE and the highest PSNR values for all the test images. Lowest AMBE indicates that CPCE gave the best brightness preservation, lowest MSE also indicates lowest error in the enhanced images, and the highest PNSR indicates that CPCE produced the image with highest quality. These results show that the proposed method has the best objective performance for the three methods compared.

Fig. 7 Lady

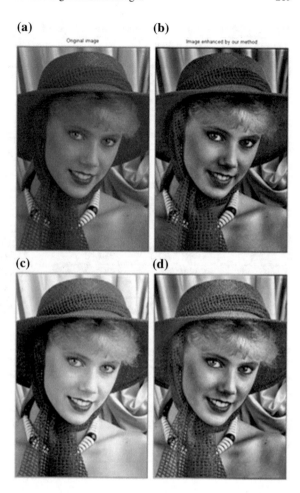

4.2   Results of Subjective Evaluation Parameter

Subjective parameter (visual inspection) was carried out by twenty people selected at random. The original and enhanced images were presented to the observers. The enhanced images were placed side by side and each inspector was told to pick the preferred image with respect to the original image (images with the better contrast and colour preservation of the enhanced images). The summary of the results is as shown in Table 2. The results showed that CPCE was perceived to give the best performance of the three methods. CPCE has average opinion score of 60%, CLAHE, 30% and HE, 10%.

Table 1 Objective evaluation parameter results for test images

Image	Evaluation parameter	HE	CLAHE	CPCE
Wall	AMBE	21.890	18.881	5.651
	MSE	0.112	0.103	0.085
	PSNR (dB)	21.8892	22.774	24.679
Door	AMBE	50.528	19.066	0.251
	MSE	0.112	0.061	0.048
	PSNR (dB)	21.918	27.889	30.440
Caps on the wall	AMBE	30.993	11.141	1.565
	MSE	0.112	0.073	0.0590
	PSNR (dB)	21.885	26.207	28.298
Boat	AMBE	7.960	7.168	1.381
	MSE	0.1120	0.095	0.083
	PSNR (dB)	21.890	23.547	24.933
Lady	AMBE	23.369	18.191	3.927
	MSE	0.112	0.086	0.062
	PSNR (dB)	21.889	24.494	27.846

Table 2 Results of subjective evaluation parameter

Image	HE	CLAHE	CPCE
Wall	1	5	14
Door	1	4	15
Caps on a wall	–	8	12
Boat	8	6	6
Lady	–	7	13
Total opinion score	10	30	60

5 Conclusion

A colour-preserving contrast enhancement (CPCE) technique for colour images has been presented in this chapter. The proposed algorithm converts RGB input image to HSV colour space. The V and S components of HSV were enhanced while H-channel remains unchanged. S-image was converted into wavelet domain and scale-up triangle mapping was used for its enhancement, while the V component was enhanced using CLAHE. The new HSV image was converted back to RGB to give the enhanced image. Unlike some methods, CPCE does not require the user to specify or adjust parameters for different images. The experimental results also showed that CPCE successfully enhanced the contrast while it still preserves the colour in the original image. In other words CPCE gave brigher details in low energy image regions and does not degrade the details in the high energy image regions. CPCE also gave a more quality result when compared with conventional methods, such as HE and CLAHE.

References

1. Jiang, Q.W.: Theory Based on Fuzzy Image Enhancement Technology. Research and Implementation. East China Normal University, Software Engineering Institute, Shanghai (2009)
2. Laminita, V., Todd, W.: An Introduction to Mathematical Image Processing. Park City Institute, Utah Undergraduate Summer School (2010)
3. Stark, J.A.: Adaptive contrast enhancement using generalization of histogram equalization. IEEE Trans. Image Process. 9(5), 889–906 (2000)
4. Cheng, H.D., Xue, M., Shi, X.J.: Contrast enhancement based on novel Homogeneity measurement. Pattern Recogn. 36, 2687–2697 (2003)
5. Gonzalez, R.C., Woods, R.E., Eddins, S.L.: Digital Image Processing Usin MATLAB. Prentice Hall, New Jersey (2003)
6. Agaian, S., Silver, B., Panetta, K.: Transform coefficient histogram based image enhancement algorithms using contrast entropy. IEEE Trans. Image Process. 16(3), 741–757 (2007)
7. Caselles, V., Lisani, J.L., Morel, J.M., Sapiro, G.: Shape preserving local histogram modification. IEEE Trans. Image Process. 8(2), 220–230 (1998)
8. Zuiderveld, K.: Contrast Limited Adaptive Histogram Equalization. Graphics Gems IV, pp. 474–485. Academic Press Professional Inc (1994)
9. Kim, Y.T.: Contrast enhancement using brightness preserving bi histogramequalization. IEEE Trans. Consum. Electron. 4(1), 21 (1997)
10. Coltuc, D., Bolon, P., Chassery, J.M.: Exact histogram specification. IEEE Trans. Image Process. 15(5), 1143–1151 (2006)
11. Wang, Y., Chen, Q., Zhang, B.: Image enhancement based on equal area dualistic sub-image histogram equalization method. IEEE Trans. Consum. Electron. 45(1), 68–75 (1999)
12. Pizer, S.M., Amburn, E.P., Austin, J.D., Cromartie, R., Geselowitz, A., Greer, T., Romeny, B.T.H., Zimmerman, J.B., Zuiderveld, K.: Adaptive histogram equalization and its variations. Comput. Vis. Graph. Image Process. 39(3), 355–368 (1987)
13. Sheet, D., Garud, H., Suveer, A., Mahadevappa, A.M., Chatterjee, J.: Brightness preserving dynamic fuzzy histogram equalization. IEEE Trans. Consum. Electron. 56(4), 24752480 (2010)
14. Celik, T., Tjahjadi, T.: Contextual and variational contrast enhancement. IEEE Trans. Image Process. 20(12), 3431–3441 (2011)
15. Aghagolzadeh, S., Ersoy, O.K.: Transform image enhancement. Opt. Eng. 31(3), 614–626 (1992)
16. Lee, S.: An efficient content based image enhancement in the compressed domain using retinex theory. IEEE Trans. Circuits Syst. Video Technol. 17(2), 199–213 (2007)
17. Lal, S., Chandra, M., Upadhyay, G.K.: Contrast enhancement of compressed image in wavelet based domain. In: Proceedings of International Conference on Signal Recent Advancements in Electrical Sciences,pp. 479–489, Tiruchengonde, India, Jan 08–09 2010
18. Thomas, B.A., Strickland, R.N., Rodriguez, J.J.: Colour image enhancement using spatially adaptive saturation feedback. In: Proceedings of 4th IEEE Conference on Image Processing, pp. 30–33, SantaBarbara, CA, USA (1997)
19. Lucchese, L., Mitra, S.K., Mukherjee, J.: A new algorithm based on saturation and desaturation in the xy chromaticity diagram for enhancement and re-rendition of colour images. In: Proceedings of 8th IEEE Conference on Image Processing, pp. 1077–1080, Thessaloniki, Greece (2001)
20. Shyujin-Jang, M.S., Leou, J.J.: A genetic algorithm approach to colour image enhancement. Int. J. Pattern Recogn. 31(7), 871–880 (1998)

21. Lu, J., Hearly, D.M.: Contrast enhancement via multi-scale gradient transformation. In: Proceedings of SPIE Conference on Wavelet Application, , pp. 345–365, Orlando, FL, USA (1994)
22. Brown, T.J.: An adaptive strategy for wavelet based image enhancement. In: Proceedings of IMVIP Conference on Irish Machine Vision and Image Processing, pp. 67–81, Belfast, Northern Ireland (2000)
23. Gupta, A., Chanda, B.: A hue preserving enhancement scheme for a class of colour images. Pattern Recogn. Lett. **17**(2), 109–114 (1996)
24. Naik, S.K., Murthy, C.A.: Hue-preserving colour image enhancement without Gamut problem. IEEE Trans. Image Process. **12**(12), 1591–1598 (2003)
25. Choudhury, A., Medioni, G.: Perceptually motivated automatic colour contrast enhancement. In: Proceedings of 12th IEEE Conference on Computer Vision, pp. 1893–1900 (2009)
26. Prabukumar, M., Cristopher, Clement J.: Compressed domain contrast and brightness improvement algorithm for colour images through contrast measuring and mapping of DWT coefficients. Int. J. Multimedia Ubiquitous Eng. **8**(1), 55–69 (2013)
27. Balachandra Reddy, A., Manjunath, K.: A hybrid colour image enhancement technique based on contrast stretching and peak based histogram equalization (2012)
28. Numan, U., Samil, T., Suleyman, D.: Undecimated wavelet transform based contrast enhancement. Int. J. Comput. Inf. Syst. Control Eng. **7**(9), 571–574 (2013)
29. Mohanram, S., Joyce, S.H.T., Aarthi, B., Sakthivel, P.: An adaptive contrast enhancement of coloured foggy images. Proc. Int. Conf. Glob. Innovations Comput. Technol. **2**(1), 208–213 (2014)
30. Menotti, D., Najman, L., Jacques, F., de Arnaldo, A.: Fast hue preserving histogram equalization method for colour image contrast enhancement. Int. J. Comput. Sci. Inf. Technol. **4**(5), 243–259 (2012)
31. Pan, Q., Zhang, L., Dai, G., Zhang, H.: Two denoising methods by wavelet transform. IEEE Trans. Signal Process. **47**(12), 3401–3406 (1999)
32. Anish, K.V., Agya, M.: Colour image enhancement techniques: a critical review. Indian J. Comput. Sci. Eng. **3**(1), 39–45 (2012)

Sequential Pattern Discovery for Weather Prediction Problem

Almahdi Alshareef, Azuraliza Abu Bakar, Abdul Razak Hamdan,
Sharifah Mastura Syed Abdullah and Othman Jaafar

Abstract This study proposes the Sequential Pattern Discovery algorithms to solve weather prediction problem. A novel weather pattern discovery framework is presented to highlight the important processes in this work. Two algorithms are employed; namely episodes and sequential pattern mining algorithms. The episodes mining algorithm is introduced to find frequent episodes in rainfall sequences and sequential pattern mining algorithm to find relationship of patterns between weather stations. Real data are collected from ten rainfall stations of Selangor State, Malaysia. The sequential pattern algorithm is applied to extract the relationship between ten rainfall stations in 33 years periods of time. The patterns are evaluated experimentally by support and confidence values while some specific rules are mapped to the location of stations and analysed for more verification. The proposed study produces valuable patterns of weather and preserves important knowledge for weather prediction.

Keywords Episodes mining · Sequential patterns · Pattern discovery

A. Alshareef · A.A. Bakar (✉) · A.R. Hamdan
Center for Artificial Intelligence Technology,
Faculty of Information Science and Technology,
Universiti Kebangsaan Malaysia, 43600 Bangi, Selangor Darul Ehsan, Malaysia
e-mail: azuraliza@ukm.edu.my

A. Alshareef
e-mail: sheriftsm@gmail.com

A.R. Hamdan
e-mail: arh@ukm.edu.my

S.M.S. Abdullah · O. Jaafar
Institute of Climate Change, Universiti Kebangsaan Malaysia,
43600 Bangi, Selangor Darul Ehsan, Malaysia
e-mail: pghikp@ukm.my

© Springer International Publishing Switzerland 2016   223
L. Chen et al. (eds.), *Emerging Trends and Advanced Technologies
for Computational Intelligence*, Studies in Computational Intelligence 647,
DOI 10.1007/978-3-319-33353-3_12

1   Introduction

The dynamic nature of the weather has led to many researches relating to weather predictions such as weather forecasting warning system [1], rainfall and flood forecasting [2, 3]. The condition of the weather gives impact to many sectors, for example in the power usage of industry, residential facilities, agriculture [4, 5]. The continuous change in climate highlights the need of weather prediction systems that is up to par with the current technology existed [6], including the Malaysian weather prediction system. Rainfall prediction is one of the areas being actively researched all over the world [7–9]. The rainfall data in Malaysia have been used previously in researches such as gauging the size of the rainfall cells [10, 11]. Neural network modeling has been a popular method in weather and climate prediction modeling [1, 8]. Sequential pattern mining is one of data mining techniques that potentially give new insights to the weather prediction problems. To date, the sequential and association pattern mining has yet to be employed in weather prediction problem. Previously it has been used in other application such as alarm log analysis, financial events, and stock trend relationship analysis [11]. Several works by [12] propose algorithms that find frequent episodes from the input sequence. In this study we propose a sequential pattern discovery algorithm for prediction of Malaysia rainfall.

2   Methodology

The pattern discovery framework for weather data involves five main phases as depicted in Fig. 1. Firstly, the rainfall data are collected from the Institute of Climate Change University Kebangsaan Malaysia (UKM) consisting data of amount of rainfall collected hourly from 10 stations around Selangor state. Then, the raw data is preprocessed and represented in symbolic form using Symbolic Approximation representation (SAX) [13], Thirdly, algorithm is employed to identify the rainfall sequences. The frequent episodes and sequential rule generation algorithms are employed in phase four and five respectively.

2.1   Data Collection

The time series rainfall data is gathered from Institute of Climate Change, University Kebangsaan Malaysia, Malaysia. The data set comes from 10 rainfall stations in Selangor state of Malaysia. The period covered is about 35 years from 1950 to 2006, with a stamped daily time and one variable of values. Table 1 shows the description of the rainfall data sets and Table 2 shows examples of the dataset.

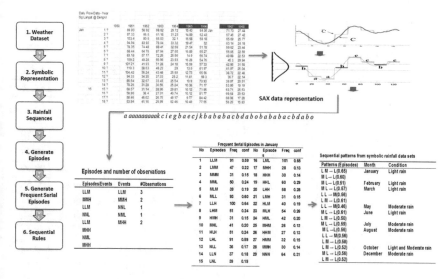

Fig. 1 Weather sequential pattern discovery framework

Table 1 Rainfall time series data characteristics

Code	Stations	Period of record	Time series length
1	Serdang: 5R-Mardi	1975–2003	10248
2	Ampang: 17R-JPS	1953–2003	18300
3	Janaletrik: 21R-Ponsoon	1953–2003	18300
4	Lenggeng_NS: 33R-Ldg	1947–2000	19398
5	Kajang_Sel: 28R	1975–2004	9882
6	Mantin_NS 32R-Setul	1960–2003	15738
7	UKM_Sel 51R-	1991–2004	4758
8	PemasokanAmpang_Sel 18R-	1947–1995	17568
9	Dominion 53R-Ldg	1970–2006	13176
10	LdgCairo NS 40R	1965–2000	12810

2.2 Symbolic Data Representation

The rainfall time series data is represented in symbolic using the Symbolic Approximation algorithm called SAX [13–15]. SAX allows distance measurements to be less than delineated in the symbolic space, so one can move from a common representation of time series (a sequence of data points interpolated by a line) in a symbolic way. More specifically, SAX performs a discretization in two stages (see Fig. 2). First, a time series is divided into segments of equal size, w, the values of each segment are approximate, and then segments are replaced by a single coefficient, which is in its way. Aggregating these w coefficients form the general approach of the PAA [14]

Table 2 Original rainfall dataset

Daily flow data—year

Sg Langat@Dengkil

		1960	1961	1962	1963	1964	1965	1966	1967	1968
Jan	1	?	69.93	58.92	68.62	29.72	15.43	64.05	71.73	27.44
	2	?	87.33	66.6	61.18	31.23	14.89	62.43	67.45	27.42
	3	?	103.4	80.6	65.03	32.1	16.58	58.18	65.09	25.77
	4	?	94.94	83.92	78.04	33.32	19.47	52	63.14	24.18
	5	?	78.35	74.48	88.41	32.68	21.54	51.78	59.62	23.44
	6	?	68.44	64.76	87.94	27.65	16.89	55.27	55.05	22.59
	7	?	83.18	57.17	72.26	25.56	14.9	56.74	49.88	22.53
	8	?	109.2	49.28	56.96	23.93	16.28	54.76	45.3	29.94
	9	?	131.21	41.93	51.28	24.18	15.59	57.33	42.96	31.58
	10	?	119.3	38.63	48.23	29	13.6	61.97	41.97	25.04
	11	?	104.42	36.24	43.48	25.99	12.73	65.56	38.72	22.48
	12	?	94.33	34.35	37.03	25.2	11.81	68.3	36.7	22.14
	13	?	86.54	32.67	33.43	25.54	10.9	70.93	39.97	20.51
	14	?	78.26	31.28	30.56	25.24	10.38	71.17	53.95	19.19
15	15	?	68.57	31.14	28.96	29.81	10.12	71.96	63.71	25.53
	16	?	58.96	38.4	27.01	40.74	10.12	81.77	69.64	20.63
	17	?	58.46	46.02	26.75	49.17	9.77	84.42	68.08	17.28
	18	?	53.94	41.16	25.99	52.46	10.49	77.55	58.26	15.83

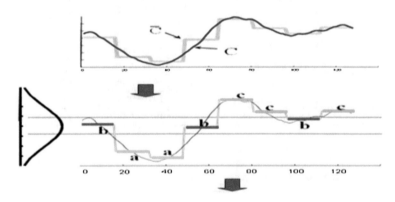

Fig. 2 Example of a time series converted with SAX method

representation of time series. Second, to convert the Piecewise Aggregate Approximation (PAA) coefficients to symbols, breakpoints that divide the distribution space in an area equal probabilities are calculated, where a is the alphabet size specified by the user. In other words, the breakpoints are determined so that the probability that a segment in any of the regions is approximately the same. If the symbols are not

Table 3 Alphabet and word size obtained by SAX algorithm

Data sets	Raw data		SAX		
	Word size	Alphabet size	Run	Word size	Alphabet size
Data 1	90	13	4	47	13
Data 2	90	20	5	39	14
Data 3	90	17	10	41	14
Data 4	90	28	8	34	16

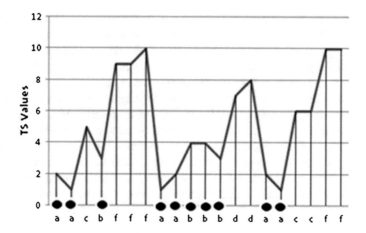

Fig. 3 Alphabet size by SAX algorithm

equally likely, some channels are more likely than others, hence probabilistic bias in the process is added.

The alphabet obtained using the SAX algorithm on data from 10 rainfall stations. Table 3 shows the optimal alphabet sizes that are generated by the SAX. Those alphabet sizes were chosen as optimal alphabets using the algorithm in [15]. It should be noted that the SAX has the ability to keep information loss as low as possible. For example, in station1 the compression rate for the alphabet size is 0.16 and 0.5 for word size, which indicates that 0.84 of the alphabet was kept and 0.5 of the time series length was kept and this is a clear advantage in that the algorithms managed to retain more of the important information.

Figure 3 shows a sequence of rainfall which is represented by improved SAX [16]. There are some important points (shown by the black circles) which are missing in original SAX [13–15] (a and b). Those points might be important in helping analysts to explain the trend in the rainfall and the loss of those points might cause distraction upon the rainfall sequence.

The original data were transformed into symbolic form, as the following steps:

1. Input the data (rainfall data)
2. Find the optimal alphabet and word size [15]

3. Apply piecewise aggregate approximation (PAA) and SAX to reduce time series length and transfer numerical values of rainfall data into symbolic values based on the optimal alphabet and word size
4. Output is symbolic representation.

2.3 Rainfall Data Sequences

A basic sliding window process is employed that moves through the time series sequences until they pass a change point. While the window goes through the time series, the segmentation process is performed. The rainfall database contains data collected over a period of time using the sliding window approach. Each window represents some sequence of time stamped data where the sequence is a single time stamp. The amount of data contains in the window may therefore vary as the window is progressed along the time series. Figure 4 shows an example of the segmentation of a rainfall sequence using the proposed dynamic window size. There are n windows with different sizes. The figure shows that each window contains two parts; the first part on the left side represents points of days in equal distribution, whereas the second part on the right side represents the change point (1 day) that is completely different from the first part in term of value. It should be noted that each window represents one rainfall case.

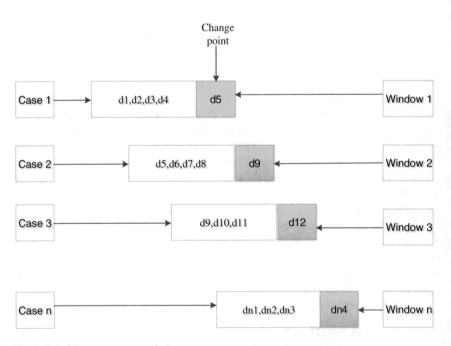

Fig. 4 Rainfall sequence on n windows

Table 4 Example of episodes and number of observations

Episodes/events	Events	# Observations
LLM	LLM	3
MMH	MMH	2
LLM	NNL	1
NNL	NML	1
LLM	MHH	2
MHH		
NML		
MHH		

The rainfall data sequences is classified by experts into four main events, namely No rain (N) for 0–2 mm/day, Light (L) for 3–10 mm/day, Moderate (M) for 11–30 mm/day and Heavy (H) for 31–60 mm/day. Let $C = [e_1, \ldots, e_m](m \geq 2)$ be a finite alphabet with total order that denotes the rainfall events $\leq$ over N. Each element $e \in C$ is called an event. $S = s_1, \ldots, s_w (w \geq 2)$ is a serial of events where each s is numbers of events (e), for example $(L \rightarrow M \rightarrow H)$ where $e_1 = L$, $e_2 = M$, $e_3 = H$ and the width of this episode is 3. Table 4 shows an example of serial episodes in the rainfall data and the number of observations.

2.4 Episodes Generation

The frequent pattern algorithm is employed to find the most repeated episodes among those that were generated in the Allen operation process [17]. The output of the algorithm is the most frequent patterns, i.e., those that satisfy the minimum support predefined by the user. The frequent algorithm is shown as Algorithm 1. The frequent episode is counted based on number of occurrence and it is evaluated based on its confidence level (minimum confidence is set to 0.1). The numbers of frequent episodes are produced and adapted as new patterns for rainfall data sets.

Input: rainfall sequences
Output: Frequent Episodes
Step 1. Read the rainfall data sequences
Step 2. Generate episodes for every 3 events (use Allen Interval concept)
Step 3. Read each episodes and count the frequency
Step 4. Calculate the confidence of the episodes (conf_e)
Step 5. Compare with the min_conf = 0.1.
Step 6. If the conf_e >= min_conf then input to frequent episode list
Step 7. Generate frequent episodes

Algorithm 1. Episodes Generation

2.5   Sequential Rules Generation

This algorithm is used to test the relationship between the data from the 10 rainfall stations. The algorithm by [17] is employed to find the association rules between attributes. The proposed algorithm accepts as input the attributes that represent the rainfall stations. The output of the algorithm is the most sequential patterns that satisfy the minimum support and minimum confidence predefined by user. The apriori based algorithm for sequential patterns generation is shown in Algorithm 2.

Input: rainfall dataset
Output: Sequential rules
Step 1. Scan the rainfall data base to get the support for each rainfall event.
Step 2. Generate set of frequent rainfall event.
Step 3. Generate a set of candidate k-rainfall events.
Step 4. Use Apriori property to prune the infrequent k-rainfall events from this set.
Step 5. Scan the rainfall database to get the support (*Sup*) with each candidate k-rainfall event in the final set.
Step 6. Compare *Sup* with min-sup, and get a set of frequent k-rainfall events.
Step 7. Generate all nonempty sub events.
Step 8. Generate rules where each rule satisfies the min-conf.

Algorithm 2. Sequential Rules Generation

2.6   Experiments and Result Discussion

The proposed algorithms are employed in two experiments. Experiment 1 for episodes and sequential patterns are discussed in next section, and Experiment 2 for the association rules algorithm employed to find association patterns between stations explained Sect. 4.

2.7   Expert Validation

A common way to achieve the data mining results is to let the domain expert perform this task, and several data mining systems support this capability. In our approach, patterns discovered during the data mining stage are validated by the expert, and, depending on how well they represent the actual behaviours of the rainfall data, some patterns are valid and valuable for further use while some patterns are very common and not interesting to the climate expert. Then the accepted patterns are obtained from the experts. One of the main issues about validating patterns of rainfall by a human expert is the real relation between the knowledge obtained by the system and the means of the knowledge experts. In this section, two experts are involved to evaluate the performance of frequent pattern algorithm and sequential pattern algorithm on rainfall data sets. Experts have divided the discovered patterns into two types (normal and abnormal). Experts noted that there are interesting patterns in abnormal patterns. They measured the performance of the patterns based on the locations of the stations and the distance between each station and taking into account the time as main factor to evaluate both patterns frequent and sequential patterns.

3 Experiment 1: Episodes and Sequential Patterns

This section presents and discusses the results of the application of the episodes and sequential patterns algorithms. We give the experimental results for the combinations of the algorithm when applied to real events. The rainfall produced 23228 events from different 10 rainfall stations. Table 5 shows an example of frequent serial episodes in the month of January for the 10 rainfall stations. Those episodes are extracted based on min-confidence (0.1). In the table, high confidence was found in extracted episodes 'LML' which occurs 101 times in January, which means that when L comes before M and L comes after L with confidence 0.65 this indicates new rules for that specific month (January) and the given data sets. Another frequent set of episodes that can be seen is 'LLH', which indicates another rule which says if the pattern is L and L then H and the month is January with confidence 0.64, we can conclude that based on the association rules facts that (the rules are believable if the min-confidence satisfies 0.1).

Table 5 Frequent serial episodes in January

No	Episodes	Freq	Conf	No	Episodes	Freq	Conf
1	LLM	91	0.59	16	LML	101	0.65
2	LMM	47	0.22	17	MHH	28	0.13
3	MMM	31	0.15	18	HHH	30	0.14
4	MML	50	0.24	19	HHL	60	0.29
5	MLM	39	0.19	20	LHH	58	0.28
6	MLL	93	0.60	21	LMH	31	0.15
7	LLH	100	0.64	22	HLM	40	0.19
8	LHM	51	0.24	23	MLH	54	0.26
9	HMH	31	0.15	24	HML	42	0.20
10	MHL	41	0.20	25	MHM	26	0.12
11	HLH	51	0.24	26	HHM	27	0.13
12	LHL	91	0.59	27	HMM	32	0.15
13	NLL	36	0.17	28	MMH	30	0.14
14	LLN	37	0.18	29	NNN	64	0.31
15	LNL	39	0.19				

Tables 5 and 6 present and summarize the extracted patterns from the 10 rainfall stations in different time periods. It is important to remark that the patterns are frequent episodes, each episode denotes three events, and each event is a period of rainfall points. Those points are represented in four different classes (either N, L, M or H). The length of each class is dynamic, for example class N represents rainfall points from 5 to 15 days, class L from 4 to 10 days, class M represents rainfall points from 3 to 7 days and class H represents rainfall points from 2 to 4 days. The patterns are divided into normal patterns which denote the rainfall in class L and M, abnormal denote the patterns with more N and H classes.

3.1 Normal Patterns

Normal patterns are patterns which more frequents occur and usually known by the experts of the Institute of Climate Change. The patterns are represented in Table 6 according to the time period (by month). The first column of the table shows the episodes for each month and each month has at least three patterns as normal rules. The patterns are represented with more advanced information, where each pattern is stamped with a time period and confidence as a measurement for the pattern. The second column contains the months of the rainfall patterns and the third column shows the condition regarding each month patterns.

Normal patterns can be seen in 9 months of the year. As can be seen, the patterns for January tell us that in this month there are more light rain, because the three patterns have L as the conclusion with the highest confidence (between 0.60 and

Table 6 Extraction normal patterns from symbolic rainfall data sets

Patterns (Episodes)	Month	Condition
L M → L(0.65) M L→ L(0.60)	January	Light rain
M L→ L(0.51)	February	Light rain
M L → L(0.57) L L → M(0.56) L M → L(0.61)	March	Light rain
L L → M(0.46)	May	Moderate rain
M L → L(0.61) L M → L(0.50)	June	Light rain
M L → L(0.59)	July	Moderate rain
M L →L(0.56) L L → M(0.56) L M → L(0.58)	August	Moderate rain
L M → L(0.52)	October	Light and moderate rain
M L → L(0.56) L M → L(0.52)	December	Moderate rain

0.65), whereas the remaining patterns have confidence levels below 0.6. Hence a conclusion regarding the rainfall pattern in January can be drawn that patterns are normal. Other month has normal patterns is February with confidence (0.51). Those patterns are evaluated by the experts and specified as normal patterns because they used their experience to analysis them and conclude that whenever patterns contains L and M then L is treated as normal and will not affect the peck of the river.

3.2 Abnormal Patterns

Abnormal patterns are unusual patterns that rarely occur and are very interesting for the experts. The abnormal patterns can be seen in Table 7 in eight different months of the year. It shows that, No rain (N) is only indicated in June and considered as abnormal patterns because through those patterns experts can see when and which period of the year is dry. This month has two patterns that satisfied the confidence, namely, two patterns indicate No rain (N) with confidence levels (0.50 and 0.53). Another abnormal pattern is Heavy rain. Those patterns are indicated at 7 months of the year. The most interesting patterns are seen in April, which points to this con- clusion when rain is L and followed by L then ending with H with confidence level (0.50). It should be noted that heavy rain (H) is indicated in six different months (Jan- uary, February, May, July, August and September). The content of the patterns for these months is almost the same with only a slight difference among some months. Moreover, abnormal patterns cannot be seen in 2 months of the year, namely March and June. April is the most similar to the reality and April patterns are very inter- esting. The patterns show useful knowledge that can be used for weather prediction.

Table 7 Extraction of abnormal patterns from symbolic rainfall data sets

Patterns (episodes)	Month	Condition	Patterns (episodes)	Month	Condition
L L→ H(0.64)	January	Heavy rain	L L→ H(0.43) H L→ L(0.44)	August	Heavy rain
L L→ H(0.69) H L→ L(0.65)	February	Heavy rain	L H→ L(0.61) H H→ L(0.55) L L→ H(0.57)	September	Heavy rain
H L→ L(0.53) H L→ H(0.51) H H→ L (0.50)	April	Heavy rain	H L→ L(0.55) L L→ H(0.51)	October	Heavy rain
H L→ L(0.50) L L→ H(0.49)	May	Heavy rain	H L→ L(0.60) L L→ H(0.61) L H→ L(0.53)	November	Heavy rain
L L→ N(0.53) L N→ L(0.50)	June	No rain	HH→H (0.53)	December	Heavy rain
L L→ H(0.55) L H→L(0.56)	July	Heavy rain			

Furthermore, abnormal and useful patterns are extracted in December, which shows very heavy rain as H and H then H with confidence level (0.51). The experts classify these patterns as very rare and can be an anomaly pattern.

4   Experiment 2: Sequential Rules of Rainfall Stations

This section describes the experimental results from the Apriori based algorithm for sequential patterns generation. The same size of rainfall data is used in experiment 1. As mentioned in the previous Sect. 10 different rainfall stations are involved in the experiment. The locations of those stations are shown in the Fig. 5.

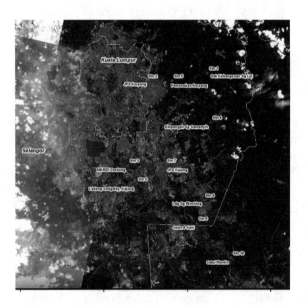

Fig. 5 Map of the Selangor rainfall stations

Table 8 Example of experiment with 100 rules

Conf	Min.sup	# of rules	# of rainfull stations
0.1	0.1	100	10
0.1	0.2	100	10
0.1	0.3	100	10
0.1	0.4	100	9
0.1	0.5	100	7
0.1	0.6	100	6
0.1	0.7	100	10
0.1	0.8	100	5
0.1	0.9	100	5

The experiment is conducted based on the rainfall events with different time periods. The number of rules is set from 100, 300 and 500 rules. In addition the value of Conf is set to 0.7 and the minimum support used is from 0.1 until 0.9. The result from the experiment (with 100 rules) is shown in Table 8. It shows an example of generated rules (fixed at 100) for 10 rainfall stations. The confidence is also fixed (at 0.6) and the support is changed nine times from 0.1 as min-sup to 0.9 as max.sup. It indicates that for all the attributes (rainfall stations) almost all appeared in the rules; the best rules are chosen where confidence equals 0.6, support equals 0.7 and the number of rules equal 100. The experiment is performed on 10 rainfall stations for different months. The Apriori algorithm is run with a fixed number of rules (100) and with confidence (0.1) and support (0.6). The rules are divided by using the time period as the class, where each month is specified as the class. In this section, 2 months (January and February) are explained in detail to show how the representation phase performed efficiently with respect to the time series data. The sequential patterns are divided into two types of patterns: one is normal patterns and second is abnormal patterns.

4.1 Normal Patterns

Normal patterns are patterns frequently occur and usually known by the experts of the Institute of Climate Change. It should be noted that the Tables 9 and 10 explain the output of the sequential patterns algorithm. Stations are represented as St# where # is the station number. In Table 9, it can be seen that ten normal sequential patterns in January. Ten patterns satisfied the min-conf and min-sup, and all ten rainfall stations can be seen appeared in the patterns. Pattern 1–4 in Table 9 indicate the normal relation of the three stations with high confidence.

Table 9 Example of the normal rainfall patterns in January

No	Patterns				Conf	Sup
1	St1 = M	St4 = L	==>	St2 = M	1.00	0.6
2	St2 = M	St6 = M	==>	St1 = M	1.00	0.6
3	St6 = H	St7 = H	==>	St1 = M	1.00	0.6
4	St1 = L	St8 = M	==>	St2 = L	1.00	0.6
5	St5 = L	St10 = L	==>	St4 = L	0.69	0.6
6	St4 = M	St7 = M	==>	St1 = L	0.68	0.6
7	St2 = L	St4 = L	==>	St1 = L	0.68	0.6
8	St3 = L	St4 = L	==>	St1 = L	0.67	0.6
9	St4 = L	St5 = L	==>	St1 = L	0.67	0.6
10	St1 = L	St8 = M	==>	St2 = L	0.67	0.6

Table 10 Example of the normal rainfall patterns in February

Patterns				Conf	Sup
St5 = L	St8 = M	==>	St4 = L	0.67	0.6
St6 = L	St10 = L	==>	St4 = L	0.67	0.6
St6 = L	St8 = L	==>	St4 = L	0.67	0.6
St3 = L	St6 = L	==>	St4 = L	0.66	0.6
St2 = L	St8 = L	==>	St4 = L	0.66	0.6
St5 = L	St7 = L	==>	St4 = L	0.65	0.6
St5 = L	St8 = L	==>	St4 = L	0.65	0.6
St5 = L	St7 = L	==>	St1 = L	0.65	0.6

It also shows that when all are L in January then St1 (Semenyih station) is L. These patterns appear to be related to reality because the experts specified that station1 is main station among all stations. This is because station1 is situated beside the Semenyih River, whereas the other stations are situated on other side of the river, so the rain data collected by those stations flows to the Semenyih River where station1 is situated. From those patterns the conclusion is drawn that whenever nine rainfall stations show a Light pattern then station1 is always light (L) with confidence 0.67 to 0.69 whenever the time period is January. Another set of normal sequential patterns is shown in Table 10 where patterns are for the month of February. The patterns fall into different classes; L and M and it can be seen that ten patterns satisfied the min-conf and min-sup, and almost all of the ten rainfall stations appeared in the patterns.

Table 10 shows that whenever the month is February and stations are light (L) then three rainfall stations (St2, St5 and St8) are always Light (L) with confidence between 0.61 and 0.69. Other indications from the patterns are that whenever stations St4 and St5 are light (L) then St8 is moderate (M) with high confidence (0.67).

4.2 Abnormal Patterns

Abnormal patterns are unusual patterns that rarely occur and are very interesting for the experts. The abnormal patterns can be seen in Table 11. Abnormal patterns are shown in February with five different patterns. It should be noted that only seven rainfall stations appeared in the patterns (1, 2, 4, 5, 6, 8 and 10). These patterns show that whenever other stations are L then the conclusion is that St 2 and St8 are H with confidence more than (0.6) and support (0.6). The expert analyzed the patterns and verified that some patterns are new and useful. There is some interesting relationship between some rainfall stations. For example there is relation between St4, St6 and St8 even though the distance between St4 and St6 is high but the patterns show meaningful information that which says that when St4 and St6 are L then St8 is H.

Table 11 Example of the abnormal rainfall patterns in February

Patterns				Conf	Sup
St6 = L	St8 = H	==>	St4 = L	0.69	0.6
St2 = H	St5 = L	==>	St1 = L	0.67	0.6
St1 = L	St2 = H	==>	St5 = L	0.63	0.6
St5 = H	St10 = L	==>	St4 = L	0.63	0.6
St4 = L	St8 = H	==>	St6 = L	0.62	0.6

Fig. 6 Map of the three Selangor rainfall stations in April

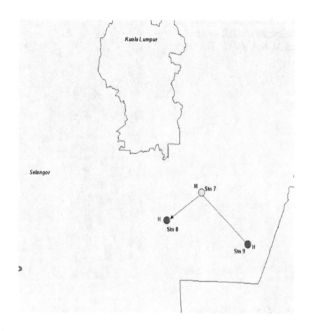

In the episodes mining phase, interesting and new patterns are discovered. In the frequent patterns phase, serial episodes are detected that explain the overall rainfall pattern over 12 months of the year. It shows that rainfall is light in 4 months, moderate in 4 months, and heavy in 6 months of the year. Some patterns can be used as rules for prediction, since those patterns show very detailed changes in the rainfall when it moves from L to M and then from M to H. This pattern shows when the rain is going to be heavy and how it will develop. The selected patterns are based on high confidence and different time periods. In the sequential patterns phase, the discovered patterns are new and useful. It indicates that sequential rules for the 10 rainfall stations can be seen in different months. These discovered patterns show meaningful and trustable rules since those patterns can be explained and corroborated by experts. Some patterns are an accurate expression of reality for certain rainfall stations in certain months. For example, one pattern with conf(0.7) and sup(0.6) shows that if St9 = H and St8 = M then St7 = H in April, as shown in Fig. 6. Hence, these are valid rules since the locations of the stations are in the same area. It should be noted that when rainfall is heavy at St9 then the rain moves to St7 and becomes moderate, then it moves to St8 where it becomes heavy rain.

One pattern that was detected in April is shown in Fig. 6. This discovered patterns show meaningful and trustable rules since this pattern can be explained and corroborated by experts, since this pattern is an accurate expression of reality for certain rainfall stations in certain months. For example, one pattern has conf(1) and supp(0.6) and says that if St9 = H and St8 = H then St7 = M in April, as shown in

Fig. 7 Map of the three Selangor rainfall stations (ST4, St8, St6) in February

Fig. 8 Map of the three Selangor rainfall stations (St2, St5, St1) in February

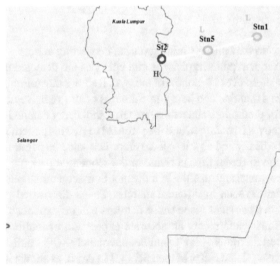

Fig. 6. Hence, this is a valid rule since the locations of the stations are in the same area. It should be noted that when rainfall is moderate at St7 then the rain moves to St8 and becomes heavy, then it moves to St9 where it becomes heavy rain.

Another pattern was detected in February as shown in Fig. 7. For certain rainfall stations in certain months, there is a unexpected relation between those 3 rainfall stations. St4 is located at the north region and St6 is located in south region. What this means is that even where the distance between those stations is high we can see an interesting pattern that says that it might be possible to have a station in the north region affecting a station in the south region. Another pattern was detected in February as shown in Fig. 8. This pattern indicates that the three stations are in the same regions in the north and the distance between them is very short. The pattern says that if St1 and St5 are L in February then St2 is H with conf(0.7) and supp(0.6). Hence experts have classified this pattern as abnormal patterns.

5 Conclusion

The pattern discovery approach based on data mining techniques specifically the episodes and sequential pattern mining algorithms were employed to extract interesting patterns for weather prediction problem. The novel attempts has shown promising discovery in this domain. This approach can be used for climate change problem such as spatial temporal mining, abrupt change detection and discovering new weather signatures for specific area. In this study, interesting new patterns of Malaysia rainfall are discovered through episodes and sequential pattern mining techniques. Series of episodes are detected that explain the overall rainfall pattern over the 12 months of the year. Some patterns can be used as rules for prediction, and the others can be used in detecting abnormal or surprising patterns. Those patterns can be used to help experts to build prediction models for all rainfall stations.

Acknowledgments This project is under the Climate Informatics Research Project with Exploratory Research Grant Scheme ERGS/1/2012/STG07/UKM/01/1 and FRGS/1/2014/ICT02/UKM/01/2 Ministry of Higher Education Malaysia.

References

1. Isa, I.S., Omar, S., Saad, Z., Noor, N.M., Osman, M.K.: Weather forecasting using photovoltaic system and neural network. In: 2nd International Conference on Computational Intelligence, Communication Systems and Networks, pp. 96–100 (2010)
2. Rahman, I.I.A., Alias, N.M.A,: Rainfall forecasting using an artificial neural network model to prevent flash floods. In: IEEE High Capaciy Optical Networks and Enabling Technologies, pp. 223–228 (2011)
3. Lee, R., Liu, J.: iJADE WeatherMAN. A weather forecasting system using intelligent multiagent-based fuzzy neuro network. IEEE Trans. Syst. Man Cybern. Part C: Appl. Rev. **34**, 369–377 (2004)

4. Zavala, V.M., Constantinescu, E.M., Anitescu, M.: Economic impacts of advance weather forecasting on energy system operations. In: IEEE PES, pp. 1–7 (2007)
5. Nguyen, V.G.N., Vo, T.T., Huynh, H.X., Drogoul, A.: On weather affecting to brown plant hopper invasion using an agent-based model. In: MEDES, pp. 150–157 (2011)
6. Saima, H., Jaafar, J., Belhaouari, S., Jillani, T.A.: Intelligent methods for weather forecasting: a review. In: National Postgraduate Conference (NPC 2011), pp. 1–6 (2011)
7. Tripathy, A.K., Mohapatra, S., Beura, S., Pradhan, G.: Weather forecasting using ANN and PSO. Int. J. Sci. Eng. Res. **2**, 1–5 (2011)
8. Gill, J., Singh, B., Singh, S.: Training back propagation neural networks with genetic algorithm for weather forecasting. In: 8th International Symposium on Intelligent Systems and Informatics, pp. 465–469 (2010)
9. Wang, C., Li, Y., Liu, X.: Research on natural disaster forecasting data processing and visualization technology. In: 4th International Congress on Image and Signal Processing, pp. 1775–1778 (2011)
10. Mannila, H., Toivonen, H., Verkamo, A.I.: Discovery of frequent episodes in event sequences. Data Mining Knowl. Discov. 259–289 (1997)
11. Katoh, T., Hirata, K., Harao, M.: Mining frequent diamond episodes from event sequences. MDAI 477–488 (2007)
12. Katoh, T., Arimura, H., Hirata, K.: Mining frequent bipartite episode from event sequences. Discov. Sci. 136–151 (2009)
13. Lin, E.K.J., Lonardi, S., Chiu, B.: A symbolic representation of time series, with implications for streaming algorithms. in: Proceedings of the 8th ACM SIGMOD workshop on Research issues in data mining and knowledge discovery, pp. 2–11. ACM, San Diego (2003)
14. Chakrabarti, K., Keogh, E., Mehrotra, S., Pazzani, M.: Locally adaptive dimensionality reduction for indexing large time series databases. ACM Trans. Database Syst. **27**, 188–228 (2002)
15. Azuraliza. A.B., Almahdi M.A., Abdul Razak H.: A harmony search algorithm with multi-pitch adjustment rate for symbolic time series data representation. Int. J. Modern Educ. Comput. Sci. (IJMECS) **6**, 58–70 (2014)
16. Höppner, F.: Learning dependencies in multivariate time series. In: Proceedings of the ECAI'02 Workshop on Knowledge Discovery, pp. 25–31 (2002)
17. Allen, J.F.: Maintaining knowledge about temporal intervals. Commun. ACM **26**(11), 832–843 (1983)

A Class-Based Strategy to User Behavior Modeling in Recommender Systems

Roberto Saia, Ludovico Boratto and Salvatore Carta

Abstract A recommender system is a tool employed to filter the huge amounts of data that companies have to deal with, and produce effective suggestions to the users. The estimation of the interest of a user toward an item, however, is usually performed at the level of a single item, i.e., for each item not evaluated by a user, canonical approaches look for the rating given by similar users for that item, or for an item with similar content. Such approach leads toward the so-called *overspecialization/serendipity* problem, in which the recommended items are trivial and users do not come across surprising items. This work first shows that user preferences are actually distributed over a small set of classes of items, leading the recommended items to be too similar to the ones already evaluated, then we propose a novel model, named *Class Path Information (CPI)*, able to represent the current and future preferences of the users in terms of a ranked set of classes of items. The proposed approach is based on a semantic analysis of the items evaluated by the users, in order to extend their ground truth and infer the future preferences. The performed experiments show that our approach, by including in the *CPI* model the same classes predicted by a state-of-the-art recommender system, is able to accurately model the user preferences in terms of classes, instead of in terms of single items, allowing to recommend non trivial items.

This work is partially funded by Regione Sardegna under project Social Glue, through PIA—Pacchetti Integrati di Agevolazione "Industria Artigianato e Servizi" (annualità 2010), and by MIUR PRIN 2010–11 under project "Security Horizons".

R. Saia (✉) · L. Boratto · S. Carta
Dipartimento di Matematica e Informatica, Università di Cagliari,
Via Ospedale 72, 09124 Cagliari, Italy
e-mail: roberto.saia@unica.it

L. Boratto
e-mail: ludovico.boratto@unica.it

S. Carta
e-mail: salvatore@unica.it

© Springer International Publishing Switzerland 2016   241
L. Chen et al. (eds.), *Emerging Trends and Advanced Technologies
for Computational Intelligence*, Studies in Computational Intelligence 647,
DOI 10.1007/978-3-319-33353-3_13

1   Introduction

The goal of a recommender system is to produce meaningful suggestions for the users, related to items or products that might interest them [24] (e.g., goods on Amazon, movies on Netflix, and so on). In this context, the literature highlights that the rating prediction represents the core task of a recommender system [4, 24]. The importance of this aspect has been further evidenced by the Netflix prize [8], and recent studies showed its effectiveness also in improving classification tasks [2, 5, 34]. However, there are widely-known problems in the recommendation process.

Overspecialization/Serendipity. Independently from the approach used to build the predictions, recommender systems usually suggest items that have a strong similarity with the user profile, consequently the user always receives recommendations for items very similar to those that she/he already considered and never receives suggestions for unexpected, surprising, and novel items. This recommender systems limit, known in the literature as overspecialization/serendipity problem, worsens the user experience and does not give the users the opportunity to explore new items and to improve their knowledge [29]. It is known that this problem affects both the most used recommendation strategies, i.e., the content-based [17] and the collaborative filtering approaches [36]. In fact, on the one hand content-based recommender systems build their predictions by calculating the similarity between the items' content, while on the other hand collaborative filtering looks for items evaluated by the users similar to the target user who has to receive the recommendations. In the literature, several researches also highlight that the serendipity of a resource can be computed by measuring its distance from the items previously considered by the target user [14, 17, 29, 35].

Preference stability. To complicate the previous scenario, there are domains like movies in which the preferences tend to be stable over time [9] (i.e., users tend to watch movies of the same genres or by the same director/actor). This is useful to maintain high-quality knowledge sources, but does not allow a system to diversify the recommendations. Preference stability also leads to the fact that when users get in touch with diverse items, diversity is not valued [20]. On the one side, users tend to access to agreeable information (a phenomenon known as *filter bubble* [22]) and this leads to the overspecialization problem, while on the other side they do not want to face diversity.

Our contributions. In this paper we want to address the following research question: *can we exploit user preferences and represent them in a broader way, in order to suggest non trivial items, but not too diverse from those the user already evaluated?* In order to face this problem, we present a representation model, named *Class Path Information (CPI)*, built as a ranking of the classes of items that each user prefers. The *CPI* model is built with a novel approach that performs a semantic analysis of the items already evaluated by a user. The goal is to extend the ground truth and infer if the terms used to describe the items evaluated by a user that belong to a class (e.g., the movies of a specific genre) also characterize other classes of items, which the user may have or may have not evaluated. By modeling user preferences in terms of classes and by predicting where the future preferences of the users will go, a recommender system can generate serendipity without recommending to the users something too far from their preferences. Moreover, by understanding the context in which recommendations should be produced in terms of classes, we avoid calculating the semantic distance between single items, which is a heavy process in terms of computational costs. Another advantage offered by this approach is that the generated models can be used to produce recommendations with any approach. Indeed, the *CPI* provides information of the classes of items the user prefers, which can be exploited by any recommendation technique.

The main contributions coming from our proposal are the following:

- we show that preference stability exists in terms of classes of items. An analysis performed on two real-world datasets shows that user preferences are distributed over a small set of classes;
- we characterize each class of items using a set of *Semantic Binary Sieves* (SBS), a novel type of filter able to weigh the relevance of each class for each user;
- we develop an algorithm able to evaluate a relevance score of each class of items for each user by using the *SBS* filters;
- we introduce the novel concept of *Class Path Information* (CPI) model, which builds a relevance score of the classes of items each user prefers, and define an algorithm to create it;
- we evaluate our approach on a real-world dataset and show that the classes available in the model have a large overlap with those of the items predicted by a state-of-the-art recommender system.

This paper extends the work presented in [26] in the following ways:

(i) we define the practical implementation of the *Semantic Binary Sieves*, by introducing the algorithm needed to build them, explaining in detail how each step works (Sect. 4.3.1);
(ii) we define the practical implementation of the *Class Path Information*, by introducing the algorithm needed to build it, explaining in detail how each step works (Sect. 4.4.1);
(iii) we formalize the concept of *CPI Valid Length*, a parameter that aims to determine the number of CPI elements (CPI length) to be taken into account, on the basis of the operating environment (Sect. 4.4.2);
(iv) we perform a series of additional experiments aimed to compare the computational load of our approach, with that of SVD, reporting the differences in terms of time needed to complete a recommendation process (Sect. 5.4.3).

Roadmap. The rest of the paper is organized as follows: Sect. 2 provides a background on the concepts handled by our proposal and the formal definition of our problem; Sect. 3 presents an analysis of preference stability on two real-world datasets; Sect. 4 describes the details of the proposed approach to model user preferences in terms of classes; Sect. 5 describes the experimental framework used to evaluate our proposal; Sect. 6 discusses related work; Sect. 7 contains conclusions and future work.

2 Preliminaries

Background. For many years the item descriptions were analyzed with a word vector space model, where all the terms of each item description are processed by TF-IDF [27] and stored in a weighted vector of terms. Due to the fact that this approach based on a simple *bag of words* is not able to perform a semantic disambiguation of the words in an item description, and motivated by the fact that exploiting a taxonomy for categorization purposes is an approach recognized in the literature [3] and by the fact that a semantic analysis is useful to improve the accuracy of a classification [6, 7], we decided to exploit the functionalities offered by the WordNet environment. Wordnet is a large lexical database of English, where *nouns*, *verbs*, *adjectives*, and *adverbs* are grouped into sets of cognitive synonyms (synsets), each expressing a distinct concept. Synsets are interlinked by means of conceptual-semantic and lexical relations.

Wordnet currently contains about 155,287 words, organized into 117,659 synsets for a total of 206,941 word-sense pairs [12]. In a short, the main relation among words in WordNet is the synonymy and the synsets are unordered sets of grouped words that denote the same concept and are interchangeable in many contexts. Each synset is linked to other synsets through a small number of *conceptual relations*. Word forms with several distinct meanings are represented in as many distinct synsets, in this way each form-meaning pair in WordNet will be unique (e.g., the *fly* noun and the *fly* verb belong to two distinct synsets). Most of the WordNet relations connect words that belong to the same part-of-speech (POS). There are four POS: *nouns*, *verbs*, *adjectives*, and *adverbs*. Both nouns and verbs are organized into precise hierarchies, defined by hypernym or *is-a* relationships. For example, the first sense of the word *radio* would have the following hypernym hierarchy, where the words at the same level are synonyms of each other: some sense of *radio* is synonymous with some other senses of *radiocommunication* or *wireless*, and so on. Each synset has a unique index and shares its properties, such as a gloss or dictionary definition. We use the synsets to perform both the definition of binary filters and the evaluation of the relevance scores of the classes in a user profile.

Notation. We are given a set of users $U = \{u_1, \ldots, u_N\}$, a set of items $I = \{i_1, \ldots, i_M\}$, and a set V of values used to express the user preferences (e.g., $V = [1, 5]$ or $V = \{like, dislike\}$). The set of all possible preferences expressed by the users is a ternary relation $P \subseteq U \times I \times V$. We denote as $P_+ \subseteq P$ the subset of preferences with a positive value (i.e., $P_+ = \{(u, i, v) \in P | v \geq \bar{v} \vee v = like\}$), where $\bar{v}$ indicates the mean value (in the previous example, $\bar{v} = 3$). Moreover, we denote as $I_+ = \{i \in I | \exists (u, i, v) \in P_+\}$ the set of items for which there is a positive preference, and as $I_u = \{i \in I | \exists (u, i, v) \in P_+ \wedge u \in U\}$ the set of items a user u likes. Let $C = \{c_1, \ldots, c_K\}$ be a set of classes used to classify the items; we denote as $C_i \subseteq C$ the set of classes used to classify an item i (e.g., C_i might be the set of genres that a movie i was classified with), and with $C_u = \{c \in C | \exists (u, i, v) \in P_+ \wedge i \in C_i\}$ the classes associated to the items that a user likes.

Let $BoW = \{t_1, \ldots, t_W\}$ be the bag of words used to describe the items in I; we denote as d_i be the binary vector used to describe each item $i \in I$ (each vector is such that $|d_i| = |BoW|$). We define as $S = \{s_1, \ldots, s_W\}$ the set of synsets associated to BoW (that is, for each term used to describe an item, we consider its associated synset), and as sd_i the semantic description of i. The set of semantic descriptions is denoted as $D = \{sd_1, \ldots, sd_M\}$ (note that we have a semantic description for each item, so $|D| = |I|$). The approach used to extract sd_i from d_i is described in detail in Sect. 4.

Problem definition. Given a set of positive preferences P_+ that characterize the items each user likes, a set of classes C used to classify the items, and a set of semantic descriptions D, our goal is to assign a relevance score $r_u(c)$ for each user u and each class c, based on the semantic descriptions D. Each relevance score will be combined into a model CPI_u, defined as follows:

$$CPI_u = (r_u(c_1), \ldots, r_u(c_K)) \tag{1}$$

Each CPI_u must respect the following properties:

- $r_u(c_1) \geq \cdots \geq r_u(c_K)$
- $CPI_u \supseteq C_u$

So, each CPI model contains a list of classes ranked by relevance score and the classes available in the model are a superset of the classes for which a user expressed a preference (i.e., we are going to predict the future preferences of the users, based on the semantic analysis of the items she/he likes).

3   Characterizing Preference Stability

In order to understand if preference stability can be characterized in terms of the classes used to classify the items, in this section we are going to present the distribution of the classes C_u related to the items a user likes. For each user $u \in U$ and each class $c \in C_u$, we consider how many positive preferences the user expressed for that class. We call this value the *popularity* of the class for that user, and define it as the percentage of items that the user likes and belong to that class:

$$popularity(u, c) = \frac{|\{(u, i, v) \in P_+ | i \in c\}|}{|\{(u, i, v) \in P_+\}|} \tag{2}$$

Then, we ordered the *popularity* values of each user in decreasing order and average all the *popularity* values at the position j in the list of each user (i.e., if $j = 1$, we calculate the average amount of preferences each user expressed for the items in the most popular class).

The study has been performed on the following real-world datasets:

Yahoo! Webscope R4.[1] The dataset contains a large amount of data related to users preferences expressed by the Yahoo! Movies community that are rated on the

---

[1] http://webscope.sandbox.yahoo.com.

Table 1 Yahoo! Webscope R4 genres

Class	Genre	Class	Genre
01	Comedy	11	Reality
02	Drama	12	Kids/Family
03	Action/Adventure	13	Crime/Gangster
04	Miscellaneous	14	Romance
05	Suspense/Horror	15	Western
06	Sci-Fi/Fantasy	16	Musical/Arts
07	Thriller	17	Documentary
08	Art/Foreign	18	Special interest
09	Animation	19	Adult audience
10	Horror	20	Features

base of two different scales, from 1 to 13 and from 1 to 5 (we have chosen to use the latter). The training data is composed by 7,642 users ($|U|$), 11,915 movies/items ($|I|$), and 211,231 ratings ($|R|$), and all users involved have rated at least 10 items and all items are rated by at least one user. The test data is composed by 2,309 users, 2,380 items, and 10,136 ratings. There are no test users/items that do not also appear in the training data. Each user in the training and test data is represented by a unique ID. As shown in Table 1, the items are classified by Yahoo in 20 different classes (movie genres), and it should be noted that each item may be classified in multiple classes.

Movielens 10M.[2] This dataset contains 10,000,054 ratings and 95,580 tags related to 10,681 movies by 71,567 users that were selected at random from MovieLens (a movie recommendation website). All the users in the dataset had rated at least 20 movies, and each user is represented by a unique ID. The ratings of the items are based on a *5-star* scale, with *half-star* increments. As shown in Table 2, in this dataset the items are classified by Movielens in 18 different classes (movie genres), and it should be noted that also in this case each item may be classified in multiple classes.

Figures 1 and 2 show the distribution of the popularities for the Yahoo! Webscope R4 and the Movielens 10M datasets. In Fig. 1, we can see that 41 % of the preferences are all in a single class (in other words, nearly half of the positive ratings given by the users are for the same genre of movies) and, by considering as characterizing only the classes with *popularity* $\geq$ 1 %, it is possible to observe that user preferences are distributed on 6 out of 20 classes. Figure 2 shows that in the Movielens dataset preference stability has a lower impact. In fact, 26 % of the ratings are in the most important class for each user, and 10 out of 18 classes are involved in the user preferences.

---

[2]http://grouplens.org/datasets/movielens/.

Table 2 Movielens 10M genres

Class	Genre	Class	Genre
01	Action	10	Film-Noir
02	Adventure	11	Horror
03	Animation	12	Musical
04	Children's	13	Mystery
05	Comedy	14	Romance
06	Crime	15	Sci-Fi
07	Documentary	16	Thriller
08	Drama	17	War
09	Fantasy	18	Western

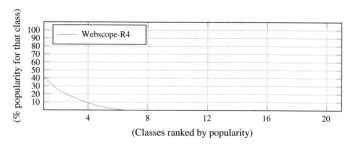

Fig. 1 Webscope-R4—involved classes in the user preferences

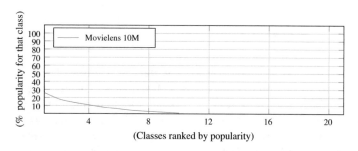

Fig. 2 Movielens 10M—involved classes in the user preferences

This analysis showed that preferences stability exists in terms of classes, and that user preferences are distributed between 30 and 55 % of the classes. Based on these results, in the next section we are going to deepen our knowledge on the user preferences in terms of classes, in order to accurately model them.

4 Our Approach

In this section we present our approach, which performs a semantic analysis of the descriptions of the items the users like, in order to build a model that infers where the future preferences will go. The goal is to understand which terms used to describe an item that a user likes characterize other classes of items. Our modeling approach performs four steps:

1. **Text Preprocessing**: processing of the textual information (description, title, etc.) present in all items, in order to remove the useless elements for the subsequent operation of synset retrieving;
2. **User Modeling**: creation of a model that contains which synsets are present in the items a user likes;
3. **Semantic Binary Sieve Construction**: creation of a *binary vector* for each class of items, and subsequent definition of the *Semantic Binary Sieves* (SBS), a series of filters that we use to estimate which synsets are relevant for that class;
4. **Class Path Information Modeling**: definition of the *Class Path Information* (CPI) model that, based on the semantic analyses performed in the previous steps, infers the user preferences in terms of classes.

Note that all steps are based on the use of WordNet synsets, which allow us to consider the semantics of the content, without performing complex operations on it. In the following, we will describe in detail how each step works.

4.1 Text Preprocessing

Before extracting the WordNet synsets from the text that describes each item, we need to follow several preprocessing tasks. The first is to detect the correct *Part-Of-Speech* (POS) for each word in the text; in order to perform this task, we used the *Stanford Log-linear Part-Of-Speech Tagger* [33]. Then, we remove punctuation marks and *stop-words*, which represent noise during the semantic analysis. Several *stop-words* lists can be found in the Internet, and in this work we used a list of 429 *stop-words* made available with the *Onix Text Retrieval Toolkit*.[3] After we have determined the lemma of each word using the Java API implementation for WordNet Searching JAWS,[4] we perform the so-called word sense disambiguation, a process where the correct sense of each word is determined. The best sense of each word in a sentence was found using the Java implementation of the adapted Lesk algorithm provided by the *Denmark Technical University* similarity application [28]. All the collected synsets form the set $S = \{s_1, \ldots, s_W\}$ defined in Sect. 2. The output of this step is the semantic disambiguation of the textual description of each item $i \in I$,

---

[3]http://www.lextek.com/manuals/onix/stopwords.html.

[4]http://lyle.smu.edu/~tspell/jaws/index.html.

which is stored in a binary vector ds_i; each element of the vector $ds_i[w]$ is 1 if the corresponding synset appears in the item description, and 0 otherwise.

4.2 User Modeling

For each user $u \in U$, this step considers the set of items I_u she/he likes, and builds a user model m_u that describes which synsets characterize the user profile (i.e., which synsets appear in the semantic description of these items). Each model m_u is a binary vector that contains an element for each synset $s_w \in S$.

In order to build the vector, we consider the semantic description ds_i of each item $i \in I_u$ for which the user expressed a positive preference. This step builds m_u, by performing the following operation on each element w:

$$m_u[w] = \begin{cases} 1, & \text{if } ds_i[w] = 1 \\ m_u[w], & \text{otherwise} \end{cases} \tag{3}$$

This means that if the semantic description of an item i contains the synset s_w, the synset becomes relevant for the user, and we set to 1 the bit at position w in the user model m_u; otherwise, its value remains unaltered. By performing this operation for all the items $i \in I_u$, we model which synsets are relevant for the user. The output of this step is a set $M = \{m_1, \ldots, m_N\}$ of user models (note that we have a model for each user, so $|M| = |U|$).

4.3 Semantic Binary Sieve Construction

For each class $c \in C$, we create a binary vector that will store which synsets are relevant for that class. These vectors, called *Semantic Binary Sieves*, will be stored in a set $B = \{b_1, \ldots, b_K\}$ (note that $|B| = |C|$, since we have a vector for each class). Each vector $b_k \in B$ contains an element for each synset $s_w \in S$ (i.e., $|b_k| = |S|$).

In order to build the vector, we consider the semantic description ds_i of each item $i \in I_+$ for which there is a positive preference, and each class c_k with whom i was classified. The binary vector b_k will store which synsets are relevant for a class c_k, by performing the following operation on each element $b_k[w]$ of the vector:

$$b_k[w] = \begin{cases} 1, & \text{if } ds_i[w] = 1 \wedge i \in c_k \\ b_k[w], & \text{otherwise} \end{cases} \tag{4}$$

In other words, if the semantic description of an item i contains the synset s_w, the synset becomes relevant for each class c_k that classifies i, and the semantic binary sieve b_k associated to c_k has the bit at position w set to 1; otherwise, its value remains unaltered. By performing this operation for all the items $i \in I_+$ that are classified with c_k, we know which synsets are relevant for the class.

4.3.1 Algorithm

Based on the above considerations, we can now define the Algorithm 1 used to create a *Semantic Binary Sieve* for each class (i.e., genre) of items. The algorithm takes as input a class $c \in C$ and the set of items I, and returns as output the SBS related with the class c.

Algorithm 1 Create SBS

Input: c=class, I= set of items
Output: SBS=Semantic Binary Sieves of c
1: $N = |GetDistinctSynsets(I)|$
2: $SBS = NewVector[N]$
3: $I_c = GetClassItems(c)$
4: **for** each item i in I_c **do**
5:     $d = GetItemDescription(i)$
6:     $S \leftarrow RetrieveAllSynsets(d)$
7:     **for** each s in S **do**
8:         $idx = GetSynsetIndex(s)$
9:         $SBS[idx] = 1$
10:    **end for**
11: **end for**
12: Return SBS

We start by storing in N the number of distinct synsets in the set I, which represents our synset ontology (step 1), then we create a SBS vector of size N (step 2). In the next step, we put in the set I_c all items that belong to a class c (step 3). For each of these items, we extract the description and the related synsets (steps 5 and 6), setting to 1 the element of the vector *SBS* that correspond with the index of these synsets (steps from 7 to 10). The algorithm ends by returning the SBS vector.

4.4 Class Path Information Modeling

This step compares the output of the two previous steps (i.e., the set B of binary vectors related to the *Semantic Binary Sieves*, and the set M of binary vectors related to the *user models*), in order to infer which classes are relevant for a user and where the future user preferences will go. The main idea is to consider which synsets are relevant for a user u (this information is stored in the user model m_u) and evaluate which classes are characterized by the synsets in m_u (this information is contained in each vector b_k, which contains the synsets that are relevant for the class c_k). The objective is to build a relevance score $r_u[k]$, which indicates the relevance of the class c_k for the user u.

The key concept behind this step is that *we do not consider the items a user evaluated anymore*. Each vector in B is used as a filter (for this reason the vectors are called *semantic binary sieves*), which allows us to estimate the relevance of each class for that user. Therefore, the relevance score of a class for a user can be used to infer where the future preferences of the users will go, since *a user might be associated to classes of items she/he never expressed a preference for, but characterized by synsets that also characterize the user model*. By ordering the relevance scores in decreasing order (from the most to the least relevant), we can build a model, named *Class Path Information (CPI)*, which can be used to generate recommendations for the users. Indeed, a recommender system might use this model to know which classes are relevant for the user, and with which score.

By considering each semantic binary sieve $b_k \in B$ associated to the class c_k and the user model m_u, we define a matching criterion $\ominus$ between each synset $m_u[w]$ in the user model, and the corresponding synset $b_k[w]$ in the semantic binary sieve, by adding 1 to the relevance score of that class for the user (element $r_u[k]$) if the synset is set to 1 both in the semantic binary sieve and in the user model, and leaving the current value as it is otherwise. The semantics of the operator is shown in Eq. (5).

$$b_k[w] \ominus m_u[w] = \begin{cases} r_u[k]++, & \text{if } m_u[w] = 1 \wedge b_k[w] = 1 \\ r_u[k], & \text{otherwise} \end{cases} \tag{5}$$

By comparing a user model m_u with each vector $b_k \in B$, we obtain a vector r_u that contains the relevance score of each class for the user (i.e., $|r_u| = |C|$). The relevance scores of each class for each user are sorted in decreasing order to build the *CPI* model for a user u (i.e., each model respects the following property: $r_u(c_1) \geq \cdots \geq r_u(c_K)$):

$$CPI_u = (r_u(c_1), \ldots, r_u(c_K)) \tag{6}$$

The output of this step is a Class Path Information model CPI_u for each user $u \in U$.

4.4.1 Algorithm

Algorithm 2 is used to create a *Class Path Information* for a user. It requires as input the set of items I, the user profile I_u (i.e., the set of items positively evaluated by the user u), and the set of classes C. It returns as output the CPI of a user u, which represents her/his preferences in terms of classes, sorted in decreasing order.

Algorithm 2 Create CPI

Input: *I*= set of items, I_u= User profile, *C*=Classes of items
Output: *CPI* = Class Path Information of *u*
1: $d = GetItemsDescription(I_u)$
2: $S \leftarrow RetrieveAllSynsets(d)$
3: $CPI = NewVector[|C|]$
4: **for** each class *c* in *C* **do**
5:    $SBS = CreateSBS(c, I)$
6:    **for** each *s* in *S* **do**
7:       $x = GetSynsetIndex(s)$
8:       $y = GetClassIndex(c)$
9:       **if** $SBS[x] == 1$ **then**
10:          $CPI[y]+ = 1$
11:       **end if**
12:    **end for**
13: **end for**
14: $CPI \leftarrow Sort(CPI, Descending)$
15: Return *CPI*

We start by appending in *d* the descriptions of the items in the user profile I_u, retrieving the related synsets (steps 1 and 2). In the step 3 we create a CPI vector of size $|C|$ (i.e., the number of classes), then for each class $c \in C$, we retrieve its SBS (step 5). In the steps from 6 to 12, we check each synset extracted from the user profile, and when the corresponding synset value in the SBS is 1, we increase by 1 unit the element of the vector *CPI* that corresponds with the class that we are processing. The algorithm ends by returning the CPI vector, sorted in decreasing order (steps 14 and 15).

4.4.2 CPI Valid Length

A possible way to determine the number of CPI elements to be taken into account, is by introducing the parameter θ, which denotes the number of CPI elements to consider. For instance, given $\theta = 2$, we take into account only the classes stored in the CPI with a *weight* > 2. We calculate the optimal value of θ as showed in Eq. 7, where $Max(C)$ denotes the maximum value of weight assumed by a class, while $|U|$ and $|C|$ denote, respectively, the number of users and classes, and the $Zeros(C)$ denotes the number of classes with a value equal to 0.

$$\theta = \left\lfloor \frac{Max(C)}{(|U| \cdot |C|) - Zeros(C)} \cdot Classes \right\rfloor \tag{7}$$

The CVL (Check Valid Length) Algorithm 3 implements this operation. It requires as input the CPI of a user *u* and the θ value, and returns the CPI valid length for the user u, i.e., the part of CPI to be taken into account. For each class stored in the CPI value (step 2), it counts how many classes have a weight higher than θ (step 3), returning this value when there is a class with a weight not greater than θ (step 5).

Algorithm 3 Get CVL

Input: CPI_u= CPI value of the user u, θ=Weight level
Output: L = Valid length of CPI
1: L=0
2: **for** each class c in CPI_u **do**
3:    **if** $Weight(c) > \theta$ **then** L++
4:    **else**
5:       Return L
6:    **end if**
7: **end for**

---

5 Experimental Framework

The experimental framework was developed by using a machine with an Intel Pentium CPU P6100 Dual Core (2 GHz × 2) and a Linux 64-bit Operating System (Debian Wheezy) with 4 GBytes of RAM. The environment for this work is based on the Java language, with the support of Java API implementation for WordNet Searching (JAWS) to perform the semantic measures, and the support of Apache Mahout[5] Java framework to implement the state-of-the-art approach that we compare our CPI modeling approach with.

This section first describes the dataset and the preprocessing performed on the data, then we describe the strategy used to perform the evaluation, the metrics, and we conclude by presenting the experimental results.

5.1 Dataset and Data Preprocessing

We performed our experiments using the Yahoo! Webscope Movie dataset (R4) described in Sect. 3. Note that we had to limit our evaluation only to one of the two datasets previously considered, since the Movielens 10M dataset does not contain any textual description of the items.

In order to create a binary sieve for each class used to build a CPI model for every user (we take in account only the 2,309 users available in the test set), we need to define an ontology of synsets based on the descriptions of the items. To perform

---

[5]https://mahout.apache.org.

this operation we considered the description and title of each movie, and since the used algorithm takes into account only the items with a rating above the average, we selected only the movies with a rating ≥ 3.

5.2 Strategy

The objective of our approach is to create a model that infers the preferences of the users in terms of classes, not only relying on the ground truth. As stated in the motivation of our work, the main domain of application that could benefit of this modeling approach are the recommender systems, which build predictions for the items not yet evaluated. Therefore, we applied a state-of-the-art recommender system to our dataset, and evaluated for each user u the set of classes C_u for which a positive value was predicted, and compared them with the CPI_u model built for that user.

The system chosen for the comparison is SVD++ [16], the Koren's version of SVD [10] that has been proved to be one of the most accurate approaches. The SVD++ approach, which we implemented through the Mahout functionalities, in addition to the training dataset requires two additional parameters: the number of target features and the number of training steps to run. After a training of the parameters, the algorithm was run with the following setting: the first parameter would be equivalent to the number of involved genres, thus we have set this value to 20; about the second parameter, considering that larger values mean longer training time, and that we have not experienced significant improvements with higher values, we have chosen the value of 4.

We required the system to produce N recommendations for each user and tested different values of the parameter (more specifically, $N = \{20, 40, ..., 100\}$). The classes involved in the recommended items were almost identical in all the settings, therefore we chose $N = 100$ to perform an evaluation in which as much information as possible was available for the comparison.

We compare the results in relation about three different aspects:

- **Evaluation of preference stability**. We perform the same analysis performed in Sect. 3, and measure the average number of classes involved in our *CPI* models and their *popularity* (i.e., how many positive preferences are associated to each class). The objective is to understand how capable our modeling approach is at reducing the effects of preference stability (which, as highlighted in the Introduction, introduces overspecialization problems), by extending the ground truth;
- **Evaluation of the classes included in the model**. In order to evaluate the significance of the produced models, we evaluate the overlap between the classes produced by the *CPI* model and the classes involved in the items recommended by SVD++, by measuring the Jaccard index between the sets C_u and CPI_u;
- **Evaluation of the computational load**. We evaluate the computational load of our approach, by comparing it with the SVD one, through a series of experiments aimed to measure the differences in terms of time needed to complete a recommendation process.

5.3 Metrics

The *popularity* metric, which allows to measure preference stability, was already introduced in Sect. 3 (Eq. 2). The evaluation of the classes included in the model has been performed by measuring the *Jaccard* index, in order to measure the overlap between the classes included in our *CPI* model (denoted as $C_u(CPI)$), and those that classify the items recommended by SVD++ (denoted as $C_u(SDV++)$):

$$J(C_u(CPI), C_u(SDV++)) = \frac{|C_u(CPI) \cap C_u(SDV++)|}{|C_u(CPI) \cup C_u(SDV++)|} \quad (8)$$

During this operation we considered the three most relevant classes identified by the two approaches.

5.4 Experimental Results

This section presents the results obtained by the three evaluations previously presented.

5.4.1 Evaluation of Preference Stability

Figure 3 shows preference stability for both approaches. The results show that the effect of preference stability is strongly reduced by our approach. In fact the number of classes involved in the model is now 10 (remember that in Sect. 3 we showed that user preferences were distributed over 6 classes). This shows an important first result, which is the capability of our approach to extend the ground truth and to be able to characterize user preferences over a larger set of classes, without considering the preferences of the other users. By enlarging the set of classes, we can also see that the *popularity* of each class (i.e., the number of preferences expressed for the items that belong to a class), is also strongly reduced; indeed, we move from 46 % of preferences

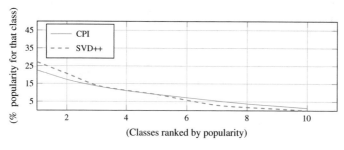

Fig. 3 Class distribution by popularity

that characterize the main class of each user to a 27.2 % value. We can also notice that the amount of classes inferred by our model that has a *popularity* $\geq$ 1 % is exactly the same available in the $C_u(SVD++)$ models. This means that our approach is able to characterize user preferences in terms of classes without producing any comparison with the preferences of the other users, thus strongly reducing the computational load necessary to infer what a user is going to like.

5.4.2   Evaluation of the Classes Included in the Model

The previous analysis showed that the number of classes involved in our *CPI* models is the same of those produced by the *SVD++* predictions. However, this result is not enough to validate our model, as the two sets of classes might be completely different. For each user we measured the Jaccard index on the sets of classes and averaged the obtained values. The average value of all results is 0.8218, which demonstrates that the most popular classes recommended through our CPI model are almost the same of the SVD++ approach. This demonstrates that in spite its simplicity, the CPI approach operates within the same range of items of a canonical approach at the state-of-the-art.

5.4.3   Evaluation of the Computational Load

In order to compare the computational load of our approach, with that of SVD, we performed a series of experiments, whose results are presented in Fig. 4. We show the differences, between our approach and SVD, in terms of time needed to complete a recommendation process. In more detail, we compare the performance by comparing the time needed to recommend N items, with $N = \{10, 20, ..., 100\}$. The value in the Y axis denotes the elapsed time needed to recommend the items, and the value on the X axis denotes the number of the recommendations. As we can observe, the results show a considerable difference in term of computational load, in favor of our approach.

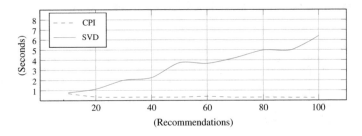

Fig. 4 Computational load comparison

6   Related Work

Likewise to other contexts, in recommender systems the preferences of the users about new choices (goods or services) tend to follow the behavior of the other users with similar tastes. This is a well-known phenomenon, called *homophily*, that in the recommender systems environment is embraced by the most common strategies used to recommend new items (e.g., *content-based* [17] and *collaborative-filtering* [11, 15] approaches). If on the one hand this approach leads toward items of likely interest to users, on the other hand it reduces the range of items of potential interest that a system could recommend, augmenting the serendipity problem, in a scenario known as the *filter bubble* [22]. The *serendipity* problem, i.e., the ability for a recommender system to suggest items of potential interest to the user that are not trivial, i.e., too similar to those in a user profile. In some works, such as [29], *serendipity* is briefly described as *a measure of how surprising the successful recommendations are*. The same work discusses the *serendipity* as the deviation from the *natural* prediction [21], and introduces the opportunity to estimate an optimal deviation value to use in order to make recommendations, underlining the risk related to an inappropriate measurement, which can lead toward a loss of user trust in regard to the recommendation system. A disadvantage that affects those recommender systems that take into account the diversity [1] is that they still operate within the classes of items for which users have expressed an explicit liking.

Another well-known problem is the so called *selective exposure*, i.e., the tendency of users to make their choices (goods or services) based only on their usual preferences, a typical way to proceed that excludes the possibility for the users to find new items that may be of interest to them [13]. The literature presents several approaches that try to reduce this problem, e.g., the *NewsCube* [23] that operates offering to the users several points of views, in order to stimulate them to make different and not usual choices.

In order to represent the user preferences and improve the effectiveness of the suggestions, in the literature we can find several approaches. For instance, users can be classified based on some explicit features (e.g., their demographic data) by extracting this information from sources such as Twitter [19], or based on other implicit features extracted through a more complex analysis of the same sources [18]. The problem of modeling semantically correlated items has been tackled in [30–32], by considering the temporal aspect. In [25], an approach to preprocess the users profiles in order to detect and remove the items that generate noise and could make the profiles not adherent to the real tastes of the users is also presented. The *serendipity* problem shows that a challenge in this area is then to identify a method able to make effective predictions, exploiting not only the information present in the user profiles. Our approach faces this problem by introducing a new modeling approach based on the class of items instead of considering the items. This is a new strategy of classification based on the classes that extends the set of possible items to recommend, taking also into account those distant from the previous choices of the users, although within their favorite classes of items. The use of a *class-based* model is also able to reduce the entity of the *selective exposure* problem, because it selects the items within a broader set of classes.

7 Conclusions and Future Work

The approach proposed in this work represents a novel way to model the user preferences, in terms of classes instead of single items, with the goal to generate non trivial recommendations. It is based on a semantic process able to create models that characterize each class and each user in terms of preferences. Such process starts by generating a set of binary filters (called *Semantic Binary Sieves*) that characterize each class of items, using them to build a class relevance score for each user. The class relevance scores are combined in a ranked list, called *Class Path Information* model, which gives us information about the user preferences in terms of classes.

Experimental results showed the capability of our approach to extend the ground truth and infer where the future preferences of the users will go. Indeed, by comparing our models with the set of classes of the items predicted by a state-of-the-art recommender system, we highlighted a strong overlap between them. This means that our approach is able to accurately infer user preferences in terms of classes, and that the generated models can be employed by a recommender system to select items within the set of classes included in the model. The advantages are both on the computational load, since the system avoids calculating the semantic distance between the items, and on the possibility to recommend items that are not too dissimilar to those she/he already positively evaluated.

A possible follow up of this work could be the implementation of our modeling technique in a recommender system. The objective is to produce the recommendations considering the classes available in the model, which are semantically related to the items a user likes, but not the same. This kind of approach would allow us to reduce the negative consequences caused by diversity, producing serendipitous but effective item recommendations.

References

1. Abbar, S., Amer-Yahia, S., Indyk, P., Mahabadi, S.: Real-time recommendation of diverse related articles. In: Schwabe, D., Almeida, V.A.F., Glaser, H., Baeza-Yates, R.A., Moon, S.B. (eds.) 22nd International World Wide Web Conference, WWW '13, pp. 1–12. International World Wide Web Conferences Steering Committee/ACM, Rio de Janeiro, Brazil, 13–17 May 2013
2. Addis, A., Armano, G., Giuliani, A., Vargiu, E.: A recommender system based on a generic contextual advertising approach. In: Proceedings of the 15th IEEE Symposium on Computers and Communications, ISCC 2010, pp. 859–861. IEEE, Riccione, Italy, 22–25 June 2010
3. Addis, A., Armano, G., Vargiu, E.: Assessing progressive filtering to perform hierarchical text categorization in presence of input imbalance. In: Fred, A.L.N., Filipe, J. (eds.) KDIR

2010—Proceedings of the International Conference on Knowledge Discovery and Information Retrieval, pp. 14–23. SciTePress, Valencia, Spain, 25–28 Oct 2010

4. Adomavicius, G., Tuzhilin, A.: Toward the next generation of recommender systems: a survey of the state-of-the-art and possible extensions. IEEE Trans. Knowl. Data Eng. **17**(6), 734–749 (2005)

5. Armano, G., Vargiu, E.: A unifying view of contextual advertising and recommender systems. In: Fred, A.L.N., Filipe, J. (eds.) KDIR 2010—Proceedings of the International Conference on Knowledge Discovery and Information Retrieval, pp. 463–466. SciTePress, Valencia, Spain, 25–28 Oct 2010

6. Armano, G., Giuliani, A., Vargiu, E.: Semantic enrichment of contextual advertising by using concepts. In: Filipe, J., Fred, A.L.N. (eds.) KDIR 2011—Proceedings of the International Conference on Knowledge Discovery and Information Retrieval, pp. 232–237. SciTePress, Paris, France, 26–29 Oct 2011

7. Armano, G., Giuliani, A., Vargiu, E.: Studying the impact of text summarization on contextual advertising. In: Morvan, F., Tjoa, A.M., Wagner, R. (eds.) 2011 Database and Expert Systems Applications, DEXA, International Workshops, pp. 172–176. IEEE Computer Society, Toulouse, France, 29 Aug–2 Sept 2011

8. Bennett, J., Elkan, C., Liu, B., Smyth, P., Tikk, D.: Kdd cup and workshop 2007. SIGKDD Explor. Newsl. **9**(2), 51–52 (2007)

9. Burke, R.D., Ramezani, M.: Matching recommendation technologies and domains. In: Ricci, F., Rokach, L., Shapira, B., Kantor, P.B. (eds.) Recommender Systems Handbook, pp. 367–386. Springer (2011)

10. Daniel Billsus, M.J.P.: Learning collaborative information filters. In: Shavlik, J.W. (ed.) Proceedings of the Fifteenth International Conference on Machine Learning (ICML 1998), pp. 46–54. Morgan Kaufmann, Madison, Wisconsin, USA, 24–27 July 1998

11. Desrosiers, C., Karypis, G.: A comprehensive survey of neighborhood-based recommendation methods. In: Ricci, F., Rokach, L., Shapira, B., Kantor, P.B. (eds.) Recommender Systems Handbook, pp. 107–144. Springer (2011)

12. Fellbaum, C.: WordNet: An Electronic Lexical Database. Bradford Books (1998)

13. Festinger, L.: A Theory of Cognitive Dissonance, vol. 2. Stanford university press (1962)

14. Iaquinta, L., de Gemmis, M., Lops, P., Semeraro, G., Filannino, M., Molino, P.: Introducing serendipity in a content-based recommender system. In: HIS, pp. 168–173. IEEE Computer Society (2008)

15. Koren, Y., Bell, R.M.: Advances in collaborative filtering. In: Ricci, F., Rokach, L., Shapira, B., Kantor, P.B. (eds.) Recommender Systems Handbook, pp. 145–186. Springer (2011)

16. Koren, Y., Volinsky, C., Bell, R.M.: Matrix factorization techniques for recommender systems. IEEE. Computer **42**(8), 30–37 (2009)

17. Lops, P., de Gemmis, M., Semeraro, G.: Content-based recommender systems: State of the art and trends. In: Ricci, F., Rokach, L., Shapira, B., Kantor, P.B. (eds.) Recommender Systems Handbook, pp. 73–105. Springer (2011)

18. Michelson, M., Macskassy, S.A.: Discovering users' topics of interest on twitter: a first look. In: Proceedings of the Fourth Workshop on Analytics for Noisy Unstructured Text Data, AND 2010 (in conjunction with CIKM 2010). ACM, Toronto, Ontario, Canada, 26th Oct 2010

19. Mislove, A., Lehmann, S., Ahn, Y., Onnela, J., Rosenquist, J.N.: Understanding the demographics of twitter users. In: Proceedings of the Fifth International Conference on Weblogs and Social Media. The AAAI Press, Barcelona, Catalonia, Spain, 17–21 July 2011

20. Munson, S.A., Resnick, P.: Presenting diverse political opinions: how and how much. In: Proceedings of the SIGCHI Conference on Human Factors in Computing Systems, CHI '10, pp. 1457–1466. ACM, New York, NY, USA, 2010

21. Murakami, T., Mori, K., Orihara, R.: Metrics for evaluating the serendipity of recommendation lists. In: Satoh, K., Inokuchi, A., Nagao, K., Kawamura, T. (eds.) New Frontiers in Artificial Intelligence, JSAI 2007 Conference and Workshops, Miyazaki, Japan, 18–22 June 2007, Revised Selected Papers, Springer (2008)

22. Pariser, E.: The Filter Bubble: What the Internet Is Hiding from You. The Penguin Group (2011)
23. Park, S., Kang, S., Chung, S., Song, J.: Newscube: delivering multiple aspects of news to mitigate media bias. In: Proceedings of the 27th International Conference on Human Factors in Computing Systems, CHI 2009. ACM, Boston, MA, USA, 4–9 April 2009
24. Ricci, F., Rokach, L., Shapira, B.: Introduction to recommender systems handbook. In: Ricci, F., Rokach, L., Shapira, B., Kantor, P.B. (eds.) Recommender Systems Handbook, pp. 1–35. Springer (2011)
25. Saia, R., Boratto, L., Carta, S.: Semantic coherence-based user profile modeling in the recommender systems context. In: Proceedings of the 6th International Conference on Knowledge Discovery and Information Retrieval, KDIR 2014, pp. 154–161. SciTePress, Rome, Italy, 21–24 Oct 2014
26. Saia, R., Boratto, L., Carta, S.: A new perspective on recommender systems: a class path information model. In: Science and Information Conference (SAI), pp. 578–585. IEEE (2015)
27. Salton, G., Buckley, C.: Term-weighting approaches in automatic text retrieval. Inf. Process. Manage. 24(5), 513–523 (1988)
28. Salton, G., Wong, A., Yang, C.S.: A vector space model for automatic indexing. Commun. ACM 18(11), 613–620 (1975)
29. Shani, G., Gunawardana, A.: Evaluating recommendation systems. In: Ricci, F., Rokach, L., Shapira, B., Kantor, P.B. (eds.) Recommender Systems Handbook, pp. 257–297. Springer (2011)
30. Stilo, G., Velardi, P.: Temporal semantics: time-varying hashtag sense clustering. In: Knowledge Engineering and Knowledge Management, volume 8876 of Lecture Notes in Computer Science, pp. 563–578. Springer International Publishing (2014)
31. Stilo, G., Velardi, P.: Time makes sense: event discovery in twitter using temporal similarity. In: Proceedings of the 2014 IEEE/WIC/ACM International Joint Conferences on Web Intelligence (WI) and Intelligent Agent Technologies (IAT), WI-IAT '14, vol. 02, pp. 186–193. IEEE Computer Society, Washington, DC, USA (2014)
32. Stilo, G., Velardi, P.: Efficient temporal mining of micro-blog texts and its application to event discovery. Data Mining and Knowledge Discovery (2015)
33. Toutanova, K., Klein, D., Manning, C.D., Singer, Y.: Feature-rich part-of-speech tagging with a cyclic dependency network. In: Proceedings of the 2003 Conference of the North American Chapter of the Association for Computational Linguistics on Human Language Technology, NAACL '03, vol. 1, pp. 173–180. Association for Computational Linguistics, Stroudsburg, PA, USA (2003)
34. Vargiu, E., Giuliani, A., Armano, G.: Improving contextual advertising by adopting collaborative filtering. ACM Trans. Web 7(3), 13:1–13:22 (2013)
35. Zhang, M., Hurley, N.: Avoiding monotony: improving the diversity of recommendation lists. In: RecSys, pp. 123–130. ACM (2008)
36. Ziegler, C.-N., McNee, S.M., Konstan, J.A., Lausen, G.: Improving recommendation lists through topic diversification. In: Ellis, A., Hagino, T. (eds.) Proceedings of the 14th international conference on World Wide Web, WWW 2005, pp. 22–32. ACM, Chiba, Japan, 10–14 May 2005

Feature Correspondence in Low Quality CCTV Videos

Craig Henderson and Ebroul Izquierdo

Abstract Closed-circuit television cameras are used extensively to monitor streets for the security of the public. Whether passively recording day-to-day life, or actively monitoring a developing situation such as public disorder, the videos recorded have proven invaluable to police forces world wide to trace suspects and victims alike. The volume of video produced from the array of camera covering even a small area is large, and growing in modern society, and post-event analysis of collected video is a time consuming problem for police forces that is increasing. Automated computer vision analysis is desirable, but current systems are unable to reliably process videos from CCTV cameras. The video quality is low, and computer vision algorithms are unable to perform sufficiently to achieve usable results. In this chapter, we describe some of the reasons for the failure of contemporary algorithms and focus on the fundamental task of feature correspondence between frames of video—a well-studied and often considered solved problem in high quality videos, but still a challenge in low quality imagery. We present solutions to some of the problems that we acknowledge, and provide a comprehensive analysis where we demonstrate feature matching using a 138-dimensional descriptor that improves the matching performance of a state-of-the-art 384-dimension colour descriptor with just 36 % of the storage requirements.

1 Closed-Circuit Television Videos

Street-scene *closed-circuit television* (CCTV) cameras record videos in natural environments without any controlled change in focus, lighting or position. As a result, images have poor colour clarity and little discriminative or representative texture definition (Fig. 1). Many variations occur over long-running continuous video, even

C. Henderson (✉) · E. Izquierdo
Multimedia and Vision Lab, School of Electronic Engineering and Computer Science,
Queen Mary University of London, Mile End Road, London E1 4NS, UK
e-mail: hello@craighenderson.co.uk

E. Izquierdo
e-mail: ebroul.izquierdo@qmul.ac.uk; e.izquierdo@qmul.ac.uk

© Springer International Publishing Switzerland 2016
L. Chen et al. (eds.), *Emerging Trends and Advanced Technologies
for Computational Intelligence*, Studies in Computational Intelligence 647,
DOI 10.1007/978-3-319-33353-3_14

Fig. 1 Typical images from a single CCTV camera, with poor lighting and long range camera views. When a subject appears in the distance, the colour and texture definition is poor and inconsistent with frames when the subject is closer

of the same scene. Weather over time causes very different images to be captured by a camera at different times; a change from sun to cloud affects the light intensity and colour definitions within the image and rain or snow can appear as noise and even occlusions in extreme conditions. At night the scene changes to artificial ambient lighting and spot lighting, particularly from vehicle headlights. Fluctuating lighting conditions caused by fire and emergency vehicle lights are commonplace in video that undergoes forensic analysis.

Mounted CCTV cameras fall into three categories. In static surveillance cameras the focal length and field of view are both fixed, and do not follow any activity. A car or a person that subsequently becomes of interest to police does not stay within shot, or even within focus. These fleeting glances can be important to an investigation but could easily be missed by reviewers scanning many hours of CCTV video. An alternative to static cameras are those which passively record following a pre-defined motion path, with the camera mounted on a bracket that automatically moves around a loop or figure-of-eight to maximise the coverage of an area with a single camera. Objects will move in and out of view regularly within a sequence of frames. Other cameras are human-operated and can record very erratic movement with dramatic changes of focus and rapid zoom as the camera operator wrestles with the controls to record action on the streets. Individual frames can therefore be very blurred. The fast movement in pan and zoom, either in the manually controlled camera or to a lesser extent in a fixed-path motion camera degrades the image quality further, and is somewhat unique to the security videos.

Video is often without metadata which may otherwise be useful, such as the video frame rate and a date/time stamp (time sequence). Time sequences would enable software to be able to synchronise video captured from multiple cameras, for example, based upon the time information associated with the video sequence. Edelman [12] reported on a system at the Netherlands Forensic Institute which uses Optical Character Recognition to read video timestamps from the video frame images. Such a technique is not reliable enough to provide sufficient metadata for steering Computer Vision algorithms, however; the Metropolitan Police in London observe that camera timestamps depend on the ongoing maintenance of the CCTV system, and their reliability varies considerably between local authority, police and private owners of surveillance systems. Standard Met Police procedure now is to record the actual current time and the presented CCTV time when a security video is seized for an investigation. This enables the police to calculate the offset of the CCTV time, but is fragile to the system clock having been altered since footage was recorded.

CCTV camera videos vary considerably in their frame rate and image resolution. Established methods of feature detection, extraction and correspondence matching perform less well on these videos than on high-definition images with sharp focus and controlled lighting conditions [17]. The frame rate of a video is a measurement of the number of frames per second, *fps*, that are recorded. With a lower frame rate, the time between frames is greater, features are further away relative to the previous frame and move greater distances relative to each other. Adjacent frames therefore have a greater visual difference than those from a high frame rate video. This difference can significantly affect the robustness of computer vision algorithms that often rely on the *a priori* knowledge that two adjacent frames in an video are very similar. As an example, a feature tracking algorithm makes the determination of whether features are related or not based on the amount of global and relative movement between frames, known as *spatial consistency*. In a low frame rate video, such determination becomes less robust as the movement threshold must be increased to compensate for the additional movement, and this can introduce noise and mis-classifications. It would be possible to configure spatial consistency algorithms using a video's metadata, for example to adapt the spatial distance threshold of related features based on the frame rate of the video. In our area of interest, surveillance videos very often have no associated metadata, and cannot therefore be used as a reliable input into algorithmic choices for spatial consistency parameters.

Low frame rates reduce the number of images that make up the video sequence and a low resolution reduces the size of each video frame. Together these two attributes can significantly reduce the amount of storage required, and therefore the cost of storing the captured video and so are often reduced by organisations who seek to minimise the overhead of their security operations. The clarity of images from different security cameras also vary considerably, and this inconsistency can cause difficulties. Images are often low resolution with poor colour definition and have little discriminative or representative texture definition, and images from these need to be matched with those from higher definition images. Quality is further reduced by varying weather conditions where the changes in light, presence of rain, snow, mist or fog, direct sunlight and shadows can all affect the image, and the ability for a feature extractor to consistently describe an image region. More information about the impurity of CCTV videos can be found in [17].

2   Characterising Video Images

An image *feature* is a low-feature attribute that contains sufficient data that it can be used to identify an aspect of the image. Features can be identified in the image spatial domain (from raw pixel values) or from a frequency domain after some processing of the image pixels. An algorithm that can find features within an image is called a *feature detector*, and some popular examples from the image domain are corner detectors [23] such as that described by Harris and Stephens [15], edge detectors such as Canny [9] or approaches based on the intensity histogram or pixel regions such as SIFT [21] and SURF [5]. These examples are more specifically defined as *keypoint detectors* where the resulting feature represents a keypoint of potential interest consisting of an x, y position in image co-ordinates. In keypoint detectors other than corner detectors, a size measured in pixels and an angle of orientation is also included in the definition of a keypoint. Measurement of position and size are in image pixel co-ordinates, but usually to sub-pixel accuracy, depending upon the detector. Such accuracy is usually achieved by performing a detection at a number of scales and combining the results from each scale to provide a solution accurate to a number of decimal points. A feature is therefore defined in real number co-ordinates rather than integer co-ordinates. Other detectors are based on global colour attributes, or find regions such as maximally stable regions (in grey-scale images [14] and colour images [13]) that discover areas with low variation in pixel values.

A *feature descriptor* is an n-dimensional vector $d = \mathbb{R}^n$ where n is defined by the descriptor algorithm. Feature descriptors are high-dimensional vectors used to *describe* the image at a keypoint or region. A full set of feature descriptors for an image can be said to describe the visual content of the image in a mathematical way such that similarities can be measured for a number of different applications. SIFT and SURF are perhaps two of the most popular feature descriptors used in contemporary algorithms. Both original papers [5, 21] describe a method of feature detection and a method of feature description, which can be used together or interchanged with each other, or other algorithms.

SIFT descriptors, which are histograms, were designed for use with Euclidean distance measures for comparison and matching [21]. However, it is well known that using Euclidean distance to compare histograms often yields inferior performance than using χ^2 [20] or Hellinger measures. Arandjelović and Zisserman made this observation and proposed *rootSIFT* [3], which transforms the SIFT descriptor such that the Euclidean distance between two descriptors is equivalent to using the Hellinger kernel, also known as Bhattacharyya's coefficient [6]. *rootSIFT* is dubbed Hellinger distance for SIFT, and yields a significantly more accurate result in calculating the distance between two descriptors used in feature descriptor matching.

Given $\mathbf{x}$ and $\mathbf{y}$ as n-vectors with unit Euclidean norm ($\|x\|_2 = 1$), the Euclidean distance $d_E(x, y)$ between them is related to their similarity (kernel) $S_E(x, y)$ as

$$d_E(\mathbf{x}, \mathbf{y})^2 = \|\mathbf{x} - \mathbf{y}\|_2^2 = \|\mathbf{x}\|_2^2 + \|\mathbf{y}\|_2^2 - 2\mathbf{x}^T\mathbf{y} = 2 - 2S_E(\mathbf{x}, \mathbf{y}) \qquad (1)$$

where $S_E(\mathbf{x}, \mathbf{y}) = \mathbf{x}^T\mathbf{y}$, and the norms $\|\mathbf{x}\|_2^2 = \|\mathbf{y}\|_2^2 = 1$. Observing that the Hellinger kernel for two L_1 normalised histograms is defined as

$$H(\mathbf{x}, \mathbf{y}) = \sum_{i=1}^{n} \sqrt{x_i y_i} \qquad (2)$$

Arandjelović and Zisserman proceed to demonstrate that SIFT vectors can be compared using a Hellinger kernel with two simple steps of algebraic manipulation: (i) L_1 normalize the SIFT vector, and (ii) square root each element of the vector. Since

$$S_E(\sqrt{\mathbf{x}}, \sqrt{\mathbf{y}}) = \sqrt{\mathbf{x}}^T \sqrt{\mathbf{y}} = H(\mathbf{x}, \mathbf{y}) \qquad (3)$$

and the resulting vectors are L_2 normalised, then

$$S_E(\sqrt{\mathbf{x}}, \sqrt{\mathbf{x}}) = \sum_{i=1}^{n} x_i = 1, \qquad (4)$$

rootSIFT is then defined as an element-wise square root of the L_1 normalised SIFT vectors [3].

There have been a number of proposals for colour descriptors that describe colour attributes of an image. These are conveniently small in dimensionality (30 to 45) and represent the colour information around a key point using a colour histogram. A detailed description of histogram based colour descriptors is provided in [37]. Colour alone is not robust for achieving good correspondences between images. Many descriptors have been proposed that use texture with various colour channel combinations, combining the texture from each channel. OpponentSIFT identifies features in opponent colour channels, red-green (RG) and yellow-blue (YB) [35] by computing SIFT descriptors in each of them. OpponentSURF uses the same technique with SURF features. In each of these, colour information is used to detect features, however, the colour detail of the image area around the feature is not encoded into the descriptor and is not used to discriminate between similar features. HueSIFT [38] describes a concatenation of a quantised Hue Histogram of 37 dimensions with the SIFT descriptor, concentrating on the effective detection of features without consideration for the descriptor encoding. Colour descriptors that use three colour channels for feature descriptions, such as OpponentSIFT and OpponentSURF, typically increase the dimensionality three times, compared with their intensity based counterparts, and the size of each descriptor becomes problematic for efficient computation and storage. A comprehensive review of colour descriptors is provide in [37].

Matching features of different images has been well researched in computer vision for many years. Empirical research algorithms typically assess descriptor matching using high quality images such as cinematic films *Groundhog Day* (Ramis 1993) and *Run Lola Run* (Tykwer 1998) used in [1, 30, 32] and subsequent comparative papers, or *Casablanca* (Curtiz 1942) in [32]. A seminal paper [30] was presented at the 9th IEEE *International Conference on Computer Vision* (ICCV 2003) that provided a breakthrough in image description. This led to a lot of subsequent research in image description and retrieval techniques in derived works [2, 25, 31–34] and by other researchers [1, 19, 29]. The paper applies text retrieval theory and practices to video searching, defined *visual words* for describing structure in images and a method to find robust visual words in video sequences. The method tracks features across *shots*, a continuous sequence of frames taken from a single camera within a scene. The number of frames and features is manageable because of the relatively short period of time covered by a shot—the average shot length in feature films was 8–11 s before 1960 and had reduced to 4–6 s by 2006 [7]. Short shots provide a natural delineation of features to process to keep the data volume a manageable size. A later processing stage can join feature tracks across shot boundaries to extend tracking of an object further, as required. Boundary shot detection is well documented as an important prerequisite step to automatic video content analysis [39] as shots are regarded as the basic unit to organise the sequenced content of video and primitives [4].

3   Bridging the Video Quality Gap

The context in which street-scene videos are recorded differ in a number of significant ways from a feature film and produce scenes that are challenging to computer vision algorithms. CCTV cameras produce a video consisting of a single shot that can last for hours, and thousands of individual frames. Without a natural delineation of shot change, contemporary methods of object tracking and mining become unmanageable as they demand large computing resource to process. Contemporary methods are not robust to the challenges of low quality images that result from these systems, which have poor colour or texture definition. Forensic analysis of security camera video sequences is a less well studied field and demands adaptation of contemporary methods to accommodate the image quality differences that exist. The quality of images from each security camera varies considerably, and this inconsistency can cause difficulties in matching features between camera images.

In the rest of this chapter, we'll take a look at two specific solutions [16] to the problems in finding features and correspondence of them between low quality images.

3.1   Blur-Sensitive Feature Detection

Detecting features in an image is sensitive to the value of individual pixels. If an image is blurred, the pixel values are different to those of a sharp image of the same scene, and consequently the features that are detected will be different in number, position, size and orientation. Finding a correspondence between images relies on matching similar features, so consecutive frames of a video that differ in the amount of blur can be a challenging problem.

If one frame of a video is blurred to a greater or lesser extent than the next, then matching features between the frames will not be very successful. We create a technique to help with this by estimating the amount of blur in each of the frames and synthetically blurring the sharpest image to more closely match the blurred. In doing this, the two images are more similar, and the feature detection is more likely to find consistent features that can then be matched.

Accurate models for calculating the motion blur of an image have been described, from estimating the parameters of a Point Spread Function [11] to using machine learning [28]. We do not need an accurate calculation of the blur parameters, but want to quickly be able to estimate the degree to which an image, or part of an image, is blurred. We observe that a blurred image will contain fewer sharp edges than a sharp image and deduce that the number of edges in an image can be used as an expression of image blurriness (or, conversely, image sharpness). Using a Canny edge detector [9] with a 3×3 Gaussian kernel and empirically chosen thresholds 175 and 225, we construct a binary edge map E from image I of size $m \times n$,

$$E_I = e_{ij} \quad i = 1, 2, \ldots, m$$
$$j = 1, 2, \ldots, n \qquad (5)$$
$$e_{ij} \in \{0, 1\}$$

The *sharpness* $S(I)$ of image I is the L_1 norm of E_I, normalised to $0 \ldots 1$.

$$S(I) = \frac{1}{m \times n} \sum_{i=1}^{m} \sum_{j=1}^{n} e_{ij} = \frac{1}{m \times n} \|E_I\|_1 \qquad (6)$$

An image is assumed to be *blurred* the sharpness is below a threshold value λ,

$$\text{image } I \text{ is blurred} = \begin{cases} true & \text{if } S(I) \leqslant \lambda \\ false & \text{otherwise} \end{cases} \qquad (7)$$

We establish a relationship map M_s to correspond the properties of a 2D Gaussian kernel G_k of size $k \times k$ with the sharpness measure of the query image I_Q after convolution with G_k. The map holds sharpness values for the query image after convolution using kernel size p.
Let

$$\Gamma \equiv \{p \in \mathbb{N} | p = 2q - 1 \wedge q \in \mathbb{N}\} \qquad (8)$$

The sharpness is calculated for each kernel size $k \in \Gamma$ where $k \leq \alpha$, and stored in a map $k \to S(I)$, thus

$$M_s(k) = S(G_k * I_Q) \quad \forall k \in \Gamma \wedge k \leq \alpha \qquad (9)$$

where

$M_s(\cdot)$	an entry in an associative map
$S(I)$	image sharpness, from Eq. (6)
G_k	Gaussian filter of kernel size k
$*$	two-dimensional convolution
α	upper bound on the Guassian kernel size

Let I_Q and I_T be two images in which to find correspondences. Assume, too, that I_T is more blurred than I_Q. Let S_a be the difference between the sharpness measured in the two images.

$$S_a = S(I_Q) - S(I_T) \tag{10}$$

$$m = \arg \max_k \left\{ S_a - S(G_k * I_Q) \right\} \qquad m \geq 0 \tag{11}$$

$$I'_Q = G_k * I_Q \tag{12}$$

We use S_a to find the corresponding Gaussian kernel size k in M_s which, when convolved with I_Q will produce an image I'_Q with sharpness that will most closely match $S(I_T)$. Features are detected in, and extracted from I'_Q, and treated as though they are from I_Q. Features can then be matched from I'_Q with those from I_T, to more robustly find correspondences between the original frames.

We evaluate the *blur sensitive feature detection* technique using intensity descriptors SIFT and SURF and colour descriptors OpponentSIFT and OpponentSURF extracted from keypoints found by seven feature detectors; the Harris corner detector, SIFT, SURF, BRISK, FAST, MSCR and MSER. We use an empirical value $\lambda := \frac{1}{32}$ in Eq. (7), so if edges are present in 3.125 % of the image or less, then the image is deemed to be blurred. Our sharpness map contains convolutions with Gaussian kernels up to 11×11, thus $\alpha := 11$ in Eq. (9).

Figure 2 shows examples of the relationship between the size of a Gaussian kernel used to blur example query images and the sharpness of the resulting image in M_s. This demonstrates the variance in the correlation between the steepness in the decline in sharpness (increase in image blur) with steadily increasing kernel sizes for different query image regions.

The method improves the matching of features in 38 of our 40 tests (Fig. 3). The percentage improvements achieved is subject to the choice of feature detector, which is expected because the synthetic blurring of the image will effect each detector differently. The two combinations that were not improved were the Harris Corner keypoints where rootSIFT and OpponentSIFT descriptors were extracted. BRISK features yielded consistently low improvements, and matching SURF features was generally more improved, with rootSIFT descriptors extracted from SURF key points being improved the most, by 92.8 %.

Fig. 2 Relationship between the size of a Gaussian kernel (*x*-axis) used to artificially blur images (*curves*) and the sharpness of the resulting image (*y*-axis)

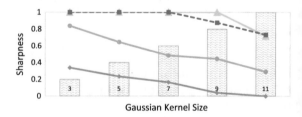

Fig. 3 Matching accuracy improvement using *blur sensitive feature detection*

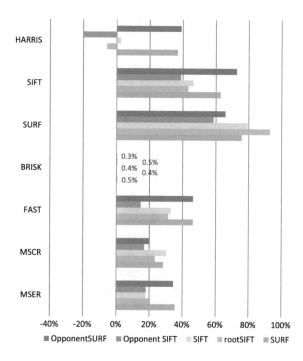

3.2 Matching Features of Texture and Colour

A composite feature descriptor is used to represent local keypoint features alongside colour information from the surrounding region. First, a local feature detector is used to find feature locations, and both a key point and a region are defined for each. In the case of a key point detector, a circular region is created with its centre at the key point co-ordinates. For region detectors, the region is approximated using an ellipse fitting algorithm through the region boundary points and a circular key point of radius $\sqrt{major\_axis^2 + minor\_axis^2}$ is defined at the centre of the ellipse. A texture feature descriptor is extracted at each key point. The colour histogram is created from the region and transformed into a feature descriptor using quantised histogram bins to form the descriptor values. The texture descriptor and colour descriptor are concatenated into a composite feature descriptor **f**, consisting the texture descriptor *t* and colour histogram *h*, thus

$$\mathbf{f} = (\mathbf{t}, h) \qquad (13)$$

The RGB colour space is known to be a poor representation for colour segmentation as there is no direct correlation between the RGB channel values and the intensity of a particular colour. The histogram is therefore constructed using the HSV colour space. The Hue (H) channel determines the colour, the Saturation (S) is the intensity of the colour and the brightness or luminance (V) can be used to find non-colour white, grey and black. Each colour in the region is quantised to its closest histogram

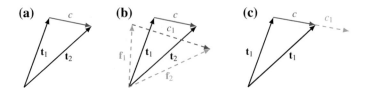

Fig. 4 Extending vectors in feature space. The principle extends to n-dimensions

bin by calculating a partial distance in HSV colour space. For colour entries in the histogram, the distance is determined by the Euclidean distance of the Hue and Saturation components, $d = \sqrt{H_i^2 + S_i^2}$. Distance to the additional three non-colour entries in the histogram—white, black and grey—are calculated using the Euclidean distance of the Saturation and Value (luminance) components, $d = \sqrt{S_i^2 + V_i^2}$. Measuring colour distances in the HSV colour space in this way maintains robustness against affine illumination changes in the image.

It is useful to think of a feature descriptor as a directional vector in high-dimensional space (Fig. 4), and refer to *feature descriptors* as *feature vectors* when visualising how two descriptors relate to each other. In designing an algorithm to extend an existing feature descriptor, consideration must made to the impact on the descriptor's position in this space. Popular feature descriptors are designed as vectors in Euclidean space, such that **a** and **b** represent high-dimensional features of size n,

$$
\begin{aligned}
\mathbf{a} &= (a_1, a_2, \ldots, a_n) \\
\mathbf{b} &= (b_1, b_2, \ldots, b_n)
\end{aligned}
\tag{14}
$$

and the distance between them is given by the length of the vector between them $\mathbf{c} = \mathbf{b} - \mathbf{a}$ (Fig. 4a), i.e. the norm $\|\mathbf{c}\|_2$, thus

$$
\|\mathbf{c}\|_2 = \sqrt{\left(\sum_{i=1}^{n} (b_i - a_i)^2 \right)}
\tag{15}
$$

Colour histogram feature space is also multi-dimensional, but the distance between points in most colour spaces are more accurately calculated using methods such as χ^2 [20], Bhattacharyya distance [6] or the Earth Mover's Distance [27]. A fast calculation of distance between two histograms of identical palettes is the *Normalised Histogram Intersection* (NHI) [36]. NHI is a light-weight calculation of similarity, and subtracted from 1 gives a dissimilarity, or a distance measure between two histograms. Given two histograms of identical palettes, u and v,

$$
H(u, v) = 1 - \frac{\sum_{j=1}^{n} \min(u_j, v_j)}{\sum_{j=1}^{n} u_j}
\tag{16}
$$

Fig. 5 Correlation of colour
histogram distance using
Euclidean distance (*left axis*)
and similarity based on
Normalised Histogram
Intersection (*right axis*)

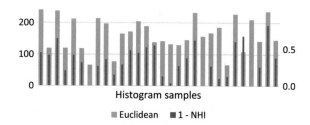

Figure 5 shows the correlation between using *Euclidean distance* and *histogram similarity* to measure the closeness of features from a typical CCTV dataset. The low positive correlation value of 0.49 shows that a Euclidean distance will generally give a reasonable indicative result but is less accurate than the histogram similarity.

Let's consider methods to calculate the distance between two composite feature descriptors $\mathbf{f}_1$ and $\mathbf{f}_2$ (Eq. 13) in such a way as to discriminate similar features of different colours. A naïve distance calculation treats the composite descriptor as a whole and calculates the distance as $\|\mathbf{f}_2 - \mathbf{f}_1\|_2$. The colour histograms become an integral part of the feature, and the unique properties of the colour histogram will be lost. The descriptors will each move in space but their relative positions will not represent the colour properties applied to the texture that we are looking for (Fig. 4b). Instead, we consider the distance between the texture $\mathbf{t}_i$ and colour histogram h_i independently and combine them mathematically to yield a representative distance between the two composite descriptors. The individual distance measures d_1 and d_2 are therefore defined,

$$d_1 = \|\mathbf{t}_2 - \mathbf{t}_1\|_2 \tag{17}$$
$$d_2 = H(h_1, h_2) \tag{18}$$

To calculate the distance D of the composite descriptor, we add a scaled colour distance d_2 to the texture distance d_1, conceptually moving the texture descriptors closer together if the colours are similar or further apart if the colours are dissimilar. Crucially, this adjustment is made in the direction of $\mathbf{c}$ (Fig. 4c) to maintain the properties of the texture descriptor.

$$D = d_1 + \lambda d_2 \tag{19}$$

Any empirically chosen constant value for λ is not robust for the variety of challenging images from surveillance video images. Instead, $\lambda := d_1$ applies the normalised colour distance as a scalar to the distance between two texture feature descriptors, increasing the normalised value of d_2 from the range $0 \ldots 1$ into $1 \ldots 2$ and up-scaling the distance of a texture feature by multiplication.

$$D = d_1 + d_1 d_2$$
$$= d_1(1 + d_2)$$
$$\therefore \quad D = \|\mathbf{t}_1 - \mathbf{t}_2\|_2 \times \left(2 - \frac{\sum_{j=1}^{n} \min(a_j, b_j)}{\sum_{j=1}^{n} a_j} \right) \tag{20}$$

Using a scalar on the texture descriptor distance ensures that attributes of the texture descriptor such as invariance to affine scale and rotation transformations are preserved. The calculation of the colour histogram in Hue and Saturation channels maintains invariance in affine illumination transformations.

To find the closest descriptor D_c to a given descriptor D_i it is customary to use an algorithm based on Euclidean distance, such as k-Nearest Neighbour. We perform a nearest descriptor calculation in two parts. First, the k-nearest neighbours of the texture descriptor $\mathbf{t}$ are found using the standard algorithm with $k = 5$, giving $\{v_1, v_2, v_3, v_4, v_5\}$. For each of the five closest descriptors, we perform the scaling multiplication of Eq. (20) and determine the descriptor with the smallest resulting distance to be the closest, D_c

$$D_c = \arg\min_i \{D_{v_i}\} \tag{21}$$

This is not guaranteed to be optimal, but in our tests increasing k to 10 does not improve the result.

4 Empirical Evaluation

The performance of the proposed descriptor is evaluated measuring the accuracy of matching features between pairs of images. The definition of a feature match depends on the matching strategy that is applied [22]. To measure the accuracy of our new composite feature descriptor and distance calculation, the results are compared with a nearest neighbour matching algorithm without any threshold filtering, such as distance ratio (Eq. 24) to discard poor matches. Seven feature detectors are used to find initial regions of interest. Five popular intensity based key point detectors; Harris Corners detector (HARRIS), SIFT, SURF, BRISK and FAST, and two region detectors; MSER on grey scale representations and *maximally stable colour regions* (MSCR) on colour images. For each of these sets of features, we compare feature matching performance of descriptors extracted using SIFT and SURF, with and without our composite descriptor, and later using OpponentSIFT and OpponentSURF 3-channel descriptors, again with and without the composite descriptor. The key point detectors HARRIS, SIFT and SURF are chosen because of their popularity and widespread adoption in many tasks including object classification and image retrieval, and BRISK and FAST for their high performance and relevance for real-time processing.

Table 1 Colour palette from [24] used in our experiments

Colour	H (°)	S (%)	V (%)
Red	0	100	100
Brown	15.1	74.5	64.7
Yellow	60	100	100
Green	120	100	100
Blue	240	100	100
Violet	300	45.4	93.3
Pink	349.5	24.7	100
White	0	0	100
Black	0	0	0
Grey	0	0	60

4.1 Evaluating the Composite Descriptor

The 10-bin palette of Park et al. [24] is used in the experiments; seven colours and three special considerations for intensities (Table 1). The descriptor extension is therefore 10 dimensions in size.

A rectangular area of an image is specified as a query region containing features that are to be matched in subsequent frames of the video sequence. In our first test the query region represents a distinctive two-colour back-pack being worn by a person (Fig. 7). This region is matched against 250 video frames, each of which has ground truth information defining the boundaries of the back-pack within it. Descriptors are created for the query image region and each image under consideration (*candidate images*) using the method described above. The positions of the features within the candidate image that match with the query region are then assessed relative to the ground truth and determined to be a true or false positive result or a negative result. A true positive result is a feature that matched with the query region (a *query match*) lies within the ground truth region. If a query match falls outside the ground truth then the region is labeled as false negative result. A feature matched between the images from outside the query region that falls within the ground truth region is counted as a false positive result. A match between the images from outside the query region to outside the ground truth region is not used directly within our analysis but are implicitly relative to other metrics.

Feature matching accuracy is typically measured using Precision-Recall [8]. *Precision* is the ability of the system to retrieve only relevant results and *Recall* is the ability of the system to retrieve all relevant results [10]. Precision is also known as *Confidence* or *True Positive Accuracy*, and Recall is also known as *Sensitivity* or *True Positive Rate* [26].

$$precision = \frac{tp}{tp+fp} \qquad recall = \frac{tp}{tp+fn} \qquad (22)$$

where *tp* is the number of true positives, *fp* is the number of false positives, and *tn* is the number of true negatives. The *F-measure* is a weighted harmonic mean of *precision* and *recall* and defined for positive real β as [18]

$$F_\beta = \frac{(1 + \beta^2) \cdot precision \cdot recall}{(\beta^2 \cdot precision) + recall} \qquad (23)$$

β can be set to vary the trade-off [8] between the two individuals metrics within the combined metric. We favour neither precision nor recall over the other, so use the common variant, F_1 where $\beta := 1$. This balances the two components without bias.

The true positive, false positive and false negative totals are tallied using manually created ground-truth data and summed across all of the images in the sequence for each of the four descriptors with and without our extension, to measure and compare the accuracy of our composite descriptor and distance measure with well-known descriptors.

4.1.1 Intensity (Grey Scale) Descriptors

It is important to compare the feature matching performance with popular intensity descriptors because these have the smallest dimensionality. In a large-scale processing system, size of descriptors is important for minimizing memory and disk storage and data processing time. Our experiments compare the matching performance of SIFT and SURF descriptors against our composite descriptor based on SIFT and SURF using our distance measure, for features detected using Harris Corners (HARRIS), SIFT, SURF, BRISK, FAST, MSCR and MSER (Fig. 6). Feature matching is determined by the nearest neighbour feature in descriptor space. The greatest improvement was achieved with SIFT descriptors extracted from MSER features where the F_1 measure increased by 163 % using our method compared to a plain SIFT descriptor on the same MSER features. Overall, the average improvement across all of the feature descriptors in this test was 95.2 %.

Figure 7 shows two examples of matching feature descriptors from a region of interest within a query image to a subsequent frame in a surveillance video, using a SURF feature detector. Figure 7a show matches of SURF descriptors extracted from

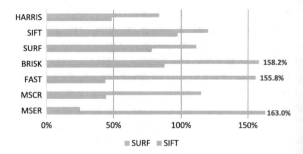

Fig. 6 Improvement of SIFT and SURF intensity descriptors using our composite descriptor and distance measure

Fig. 7 Two examples of matching SURF features on a coloured bag from a query frame (*left* in each pair) to a subsequent video frame. *Top* matches to a blurred image perform poorly using Approximate Nearest Neighbour (*blue matches*) and a significant increase in matches to the target bag using our method (*yellow matches*). *Bottom* demonstrates the significantly reduced number of false positive matches to the background using our method (*yellow*) compared with ANN matching (*blue*)

the SURF features within the region of interest in the query image (left), and a blurred frame (right). There is a notable increase in the number of features matched into the bag region in the right hand image. Figure 7b shows matches from the same query frame to a sharper subsequent frame and demonstrates the reduction in false-positive matches into background features.

The less cluttered Fig. 8 repeats the second image pairs from Fig. 7 using the *distance ratio* filter (Eq. 25) to determine if the closest match is a *good* match [21]. The distance ratio method finds the closest two features f_c and f_{c+1} and divides the nearest distance by the second closest distance,

$$distance\ ratio = \frac{\|f - f_c\|_2}{\|f - f_{c+1}\|_2} \tag{24}$$

This ratio helps to determine how reliable the match is. If the nearest feature has another feature close to it, then there is a lesser likelihood that the match is correct.

Fig. 8 *Good matches*, Eq. (25), are shown for SURF features (*top row*) and using our method (*bottom row*)

Tests in the original paper suggest that 0.8 is a reasonable threshold for this ratio, based on analysis of 40,000 key points, and that matches with a distance ratio greater than 0.8 should be considered less reliable, thus,

$$match = \begin{cases} true & \text{if } \frac{\|f-f_c\|_2}{\|f-f_{c+1}\|_2} \leqslant 0.8 \\ false & \text{otherwise} \end{cases} \tag{25}$$

In Fig. 8, the number of false positives is visibly reduced, with fewer yellow lines matched to the background in the right-hand images.

4.1.2 Colour Descriptors

We now assess our algorithm using two high-dimensional colour descriptors, OpponentSIFT (384-dimensions) and OpponentSURF (192-dimensions), with the same features from the previous section. The improvements using our method with colour descriptors are smaller than those seen with intensity texture descriptors, which is to be expected as the colour information provides a more discriminative comparison. In our tests, the best improvement was achieved using the composite OpponentSURF descriptors of FAST features, improving by 98.5 % (Fig. 9). Overall the average improvement across all of the colour feature descriptors in this test was 39.8 %.

OpponentSIFT uses colour information in the extraction of the descriptor and can be expected to out-perform those that do not use colour information in a dataset in which colour is visually distinctive. In their thorough evaluation of colour feature descriptors, van de Sande et al. conclude that OpponentSIFT is generally a better performing descriptor and is a good choice where there is no prior knowledge of the images or object/scene categories [37]. In our tests, results show that our extension method generally improves matching with this descriptor by up to 47.2 % depending on the initial feature detector.

Fig. 9 Improvement of OpponentSIFT and OpponentSURF colour descriptors using our composite descriptor and distance measure

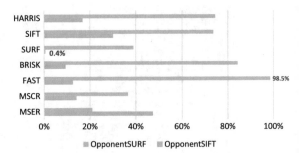

4.2 Feature Matching Results

The graphs in Fig. 10 summarize the results from our test database of 251 images. Each graph shows the F_1 measure of one of our seven selected feature detectors and all four of the feature extractors, comparing the matching performance of four methods of calculating correspondence. Each of the four methods are represented on the x-axis; the *Euclidean* correspondence using L_2 distance of unmodified texture feature descriptors is the baseline against which we measure performance improvements. *Blur sensitive* applies the blur sensitive feature detection algorithm using unmodified texture feature descriptors. *Composite* results are those achieved in using the composite texture and colour feature matching descriptor extensions and matching algorithm, and finally *Composite Blur sensitive* are results from the combined methodology described in this chapter.

The upward left-to-right trend in each of the graphs demonstrates the improvement in matching performance that is achieved with each of our method's components, and the combined methodology. The consistent closeness of the orange and yellow lines in the *Composite Blur sensitive* result is particularly striking. The performance of our method using rootSIFT descriptors (128 + 10 dimensions) closely matches the performance of the much larger OpponentSIFT 384 + 10 dimension descriptor and significantly outperforms state-of-the-art feature matching using the OpponentSIFT 128D descriptors with the Euclidean distance measure.

5 Storage Efficiency Versus Matching Performance

The effectiveness of matching features in images depends greatly on the choice of feature detector used in the initial step of the processing pipeline. Systems attempting to match features across a high volume of images are becoming increasingly common, and an important consideration for these systems is the storage requirements of the descriptors used and the trade-off between storage and accuracy. The accuracy of feature matching using contemporary techniques generally increases in line with the size of the descriptor—determined by a choice of extractor—as in the yellow bars in Fig. 11. Note the top of the bar that represents the peak performance on each descriptor is generally higher moving left-to-right. The green bars show the F_1 matching accuracy using our method, where there is a significant peak in matching accuracy at dimensionality $D = 138$ where our method using SIFT and rootSIFT descriptors outperforms all other state-of-the-art descriptor matching using Euclidean distance measures. The saving in storage using our *Composite rootSIFT* over the performance-comparable *Composite OpponentSIFT* is $394 - 138 = 256$ bytes per descriptor.

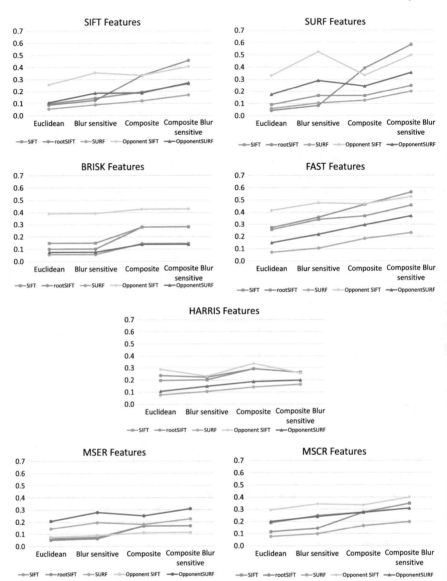

Fig. 10 Summary of the results of feature matching with each component of the method, and the combined methodology (*right-most*). Each graph shows results from a different feature detector, and compares results with each descriptors using four methods; *Euclidean*—Euclidean distance of unmodified feature descriptors is the baseline, *Blur sensitive*—blur sensitive feature detection algorithm using unmodified feature descriptors, *Composite*—composite texture and colour feature matching descriptor extensions and matching algorithm, *Composite Blur sensitive*—the combined methodology described in this chapter

Fig. 11 The correlation between descriptor size and matching accuracy. *Yellow bars* show measures for established descriptors and *Green bars* are accuracy measures using our method. Using our method with SIFT and rootSIFT 138D composite descriptors out-perform descriptors of almost 3 times the size

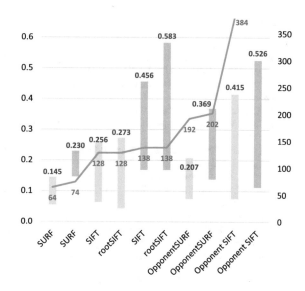

6 Conclusion

We have shown how pre-processing images to a more similar level of blurriness can improve the detection of consistent features that better establish correspondence between images. Further, by combining texture and colour attributes in a descriptor that is only slightly larger that a standard text descriptor, and using a distance calculation that understands the structure of the descriptor, greater matching accuracy can be achieved.

In a large-scale system of image features, the dimensionality of each feature quickly accumulates to a volume of data that can become unmanageable. This is especially true in complex or busy images, and those which are large in dimension. The method that we have described outperforms current state-of-the-art feature descriptors on low quality images, and does so using only 36 % of the storage requirement of its nearest performing state-of-the-art feature descriptor.

Acknowledgments This work is funded by the European Union's Seventh Framework Programme, specific topic *framework and tools for (semi-) automated exploitation of massive amounts of digital data for forensic purposes*, under grant agreement number 607480 (LASIE IP project). The authors also extend their thanks to the Metropolitan Police at Scotland Yard, London, UK, for the supply of and permission to use CCTV images.

References

1. Anjulan, A., Canagarajah, N.: A unified framework for object retrieval and mining. IEEE Trans. Circuits Syst. Video Technol. **19**(1), 63–76 (2009)
2. Arandjelović, R.: Advancing Large Scale Object Retrieval. PhD thesis, University of Oxford (2013)
3. Arandjelović, R., Zisserman, A.: Three things everyone should know to improve object retrieval. In: 2012 IEEE Conference on Computer Vision and Pattern Recognition, pp. 2911–2918. IEEE (2012)
4. Asghar, M.M.N., Hussain, F., Manton, R.: Video indexing: a survey. Int. J. Comput. Inf. Technol. **3**(1), 148–169 (2014)
5. Bay, H., Ess, A., Tuytelaars, T., Van Gool, L.: Speeded-up robust features (SURF). Comput. Vis. Image Underst. **110**(3), 346–359 (2008)
6. Bhattacharyya, A.: On a measure of divergence between two statistical populations defined by their probability distributions. Bull. Calcutta Math. Soc. **35**, 99–109 (1943)
7. Bordwell, D.: The Way Hollywood Tells It: Story and Style in Modern Movies. University of California Press, ISBN 978-0520246225 (2006)
8. Buckland, M., Gey, F.: The relationship between recall and precision. J. Am. Soc. Inf. Sci. **45**(1), 12–19 (1994)
9. Canny, J.: A computational approach to edge detection. IEEE Trans. Pattern Anal. Mach. Intell. **8**(6), 679–698 (1986)
10. Carpineto, C., Romano, G.: A survey of automatic query expansion in information retrieval. ACM Comput. Surv. **44**(1), 1–50 (2012)
11. Dash, R., Majhi, B.: Motion blur parameters estimation for image restoration. Opt. Int. J. Light Electron Opt. **125**(5), pp. 1634–1640 (2014)
12. Edelman, G., Bijhold, J.: Tracking people and cars using 3D modeling and CCTV. Forensic Sci. Int. **202**(1–3), 26–35 (2010)
13. Forssén, P.E.: Maximally stable colour regions for recognition and matching. In: Proceedings of IEEE Computer Society Conference on Computer Vision and Pattern Recognition, pp. 1–8 (2007)
14. Forssén, P.E., Lowe, D.G.: Shape descriptors for maximally stable extremal regions. In: Proceedings of IEEE International Conference on Computer Vision, pp. 1–8 (2007)
15. Harris, C., Stephens, M.: A combined corner and edge detector. In: Proceedings Alvey Vision Conference, pp. 147–151. Alvey Vision Club (1988)
16. Henderson, C., Izquierdo, E.: Robust Feature matching in the wild. In: Science and Information Conference, pp. 628–637 IEEE, London (2015)
17. Henderson, C., Blasi, S.G., Sobhani, F., Izquierdo, E.: On the impurity of street-scene video footage. In: International Conference on Imaging for Crime Detection and Prevention. IEEE, London (2015)
18. Hripcsak, G.: Agreement, the F-measure, and reliability in information retrieval. J. Am. Med. Inform. Assoc. **12**(3), 296–298 (2005)
19. Jiang, Y.G., Ngo, C.W., Yang, J.: Towards optimal bag-of-features for object categorization and semantic video retrieval. Proceedings of the 6th ACM International Conference on Image and Video Retrieval—CIVR '07, pp. 494–501 (2007)
20. Liu, H.L.H., Setiono, R.: Chi2: feature selection and discretization of numeric attributes. In: Proceedings of 7th IEEE International Conference on Tools with Artificial Intelligence, pp. 388–391 (1995)
21. Lowe, D.G.: Distinctive image features from scale-invariant keypoints. Int. J. Comput. Vis. **60**(2), 91–110 (2004)
22. Mikolajczyk, K., Schmid, C.: A performance evaluation of local descriptors. IEEE Trans. Pattern Anal. Mach. Intell. **27**(10), 1615–1630 (2005)
23. Noble, J.A.: Finding corners. Image Vis. Comput. **6**(2), 121–128 (1988)

24. Park, U., Jain, A., Kitahara, I., Kogure, K., Hagita, N.: ViSE: visual search engine using multiple networked cameras. In: 18th International Conference on Pattern Recognition, vol. 3, pp. 1204–1207. IEEE (2006)
25. Philbin, J., Chum, O., Isard, M., Sivic, J., Zisserman, A.: Object retrieval with large vocabularies and fast spatial matching. In: IEEE Conference on Computer Vision and Pattern Recognition, pp. 1–8 (2007)
26. Powers, D.: Evaluation: from precision, recall and F-measure to ROC, informedness, markedness and correlation. J. Mach. Learn. Technol. 2(1), 37–63 (2011)
27. Rubner, Y., Tomasi, C., Guibas, L.J.: Earth mover's distance as a metric for image retrieval. Int. J. Comput. Vis. 40(2), 99–121 (2000)
28. Schuler, C., Hirsch, M.: Learning to Deblur. In: NIPS 2014 Deep Learn. Represent. Learn. Workshop, Montreal (2014)
29. Shekhar, R., Jawahar, C.: Word image retrieval using bag of visual words. In: 2012 10th IAPR International Workshop on Document Analysis. Systems, pp. 297–301. IEEE (2012)
30. Sivic, J., Zisserman, A.: Video Google: a text retrieval approach to object matching in videos. In: Proceedings Ninth IEEE International Conference on Computer Vision, vol. 2, pp. 1470–1477 (2003)
31. Sivic, J., Zisserman, A.: Video data mining using configurations of viewpoint invariant regions. In: Proceedings of the 2004 IEEE Computer Society Conference on Computer Vision and Pattern Recognition, CVPR 2004, vol. 1, pp. 488–495. IEEE (2004)
32. Sivic, J., Zisserman, A.: Efficient visual search of videos cast as text retrieval. IEEE Trans. Pattern Anal. Mach. Intell. 31(4), 591–606 (2009)
33. Sivic, J., Schaffalitzky, F., Zisserman, A.: Efficient object retrieval from videos. In: Proceedings of 12th European Signal Processing Conference EUSIPCO 04, pp. 36–39, Vienna, Austria (2004)
34. Sivic, J., Schaffalitzky, F., Zisserman, A.: Object level grouping for video shots. Int. J. Comput. Vis. 67, 189–210 (2006)
35. Stokman, H., Gevers, T.: Selection and fusion of color models for feature Detection.pdf. In: IEEE Computer Society Conference on Computer Vision and Pattern Recognition, vol. 1, pp. 560–565. IEEE (2005)
36. Swain, M.J., Ballard, D.H.: Color indexing. Int. J. Comput. Vis. 7(1), 11–32 (1991)
37. Van De Sande, K.E., Gevers, T., Snoek, C.: Evaluating color descriptors for object and scene recognition. IEEE Trans. Pattern Anal. Mach. Intell. 32(9), 1582–1596 (2010)
38. van de Weijer, J., Schmid, C.: Coloring local feature extraction. In: 9th European Conference on Computer Vision, vol. 3952, pp. 334–348 (2006)
39. Yuan, J., Wang, H., Xiao, L., Zheng, W., Li, J., Lin, F., Zhang, B.: A formal study of shot boundary detection. IEEE Trans. Circuits Syst. Video Technol. 17(2), 168–186 (2007)

Automatic Detection and Severity Assessment of Crop Diseases Using Image Pattern Recognition

Liangxiu Han, Muhammad Salman Haleem and Moray Taylor

Abstract Disease diagnosis and severity assessment are necessary and critical for predicting the likely crop yield losses, evaluating the economic impact of the disease, and determining whether preventive treatments are worthwhile or particular control strategies could be taken. In this work, we propose to make advances in the field of automatic detection and diagnosis and severity assessment of crop diseases using image pattern recognition. We have developed a two-stage crop disease pattern recognition system which can automatically identify crop diseases and assess sevrity based on combination of marker-controlled watershed segmentation, superpixel based feature analysis and classification. We have conducted experimental evaluation using different feature selection and classification methods. The experimental result shows that the proposed approach can accurately detect crop diseases (i.e. Septoria and Yellow rust, which are the two most important and major types of wheat diseases in UK and across the world) and assess the disease severity with efficient processing speed.

Keywords Image processing · Machine learning/pattern recognition · Computer vision · Crop disease

L. Han (✉) · M.S. Haleem
School of Computing, Mathematics and Digital Technology,
Manchester Metropolitan University, Manchester, UK
e-mail: l.han@mmu.ac.uk

M.S. Haleem
e-mail: m.haleem@mmu.ac.uk

M. Taylor
The Food and Environment Research Agency, York, UK
e-mail: Moray.Taylor@fera.gsi.gov.uk

© Springer International Publishing Switzerland 2016   283
L. Chen et al. (eds.), *Emerging Trends and Advanced Technologies
for Computational Intelligence*, Studies in Computational Intelligence 647,
DOI 10.1007/978-3-319-33353-3_15

1 Introduction

Food is of key importance in a sustainable society. Plant diseases are a major foe to global food security, which have the potential to significantly decrease crop yields and increase postharvest losses, especially in the face of climate change. It is estimated that almost 40 % of worldwide crops are lost to diseases [1], which may cause devastating economical, social and ecological losses. For instance, yield losses from Septoria diseases of wheat, most prevalent diseases across UK, can range from 30 to 50 % [2].

Disease prevention and management are essential for the sustainability of our society. Effective control measures often rely on the symptom recognition and assessment of the severity of diseases. In the context of plants, a disease refers to any impairment of normal physiological function of a plant. Most plant diseases produce characteristic symptoms or some kind of manifestation in the visible spectrum, which are usually used for disease diagnosis and severity assessment (the proportion of the plant area affected by a disease) through visual inspection. Disease diagnosis and severity assessment are necessary and critical for predicting the likely crop yield loss using yield loss models [3], evaluating the economic impact of a disease, and determining whether preventive treatments are worthwhile or particular control strategies could be taken. There is no point in implementing a control measure if it will cost more than the economic return of the increased crop yield. Currently, the diagnosis and severity assessment are performed visually by trained surveyors, which can lead to inaccuracy and is subject to individual bias as well as being costly and time consuming. This is because the level of detail is variable and inevitably, errors arise from the subjectivity of individuals and the tedious nature of the task. The surveyors have to be regularly trained to maintain quality [4], which is costly. With large areas or fields to be inspected, the scarcity of the trained surveyors makes the monitor of disease a challenging task. Additionally, visual inspection can be destructive if samples are collected in a field for assessment later in the laboratory. Therefore, automatic, accurate and timely diagnosis and quantification of crop diseases are very important.

In this work, we have developed a novel image pattern recognition approach to automatically detect and identify crop diseases (i.e. Septoria and Yellow rust. Two types of most important and major wheat diseases in UK). Some research efforts have been made on automating plant disease identification using image processing approaches, for example, identification of banana leaf diseases [5], disease related coloration in circuit [6], detection of rice leaf disease [7] and quantification of symptoms present in unhealthy plants (e.g. pumpkin, pepper, bean) [8] using image color analysis, which tried to discriminate the diseases based on color difference. The method in [9] tried to discriminate a given disease from other pathologies of rubber tree leaves based on PCA and Neural networks, which directly applied to the RGB values of the pixels of a low resolution (15×15 pixels) and does not employ any image segmentation techniques. The system [10] was proposed to monitor the health vineyard, which mainly used thresholding to separate diseased leaves and ground from healthy leaves using both RGB and HSV color representation of the image and morphological operations. The method [11] was to detect and differentiate two diseases that affect rice crops (i.e. blast and brown spots). It first used an entropy-based thresholding and the edge detection to segment images, and then the intensity of green components was used to detect spots. The spots were resized to 80×100 pixels and the pixel values are extracted as features for final classification using self organising map (SOM).

For the purpose of automatic disease recognition from images, it is very important to extract the features and find the most discriminative features for efficient disease recognition by developing appropriate image processing and machine learning approaches.

To the best of our knowledge, there is currently no existing work for solving our problem i.e. automatically detecting crop diseases of Septoria and Yellow rust. Additionally, the existing methods are not suitable and can not be directly applied to our case. Therefore, we have developed a new approach to automatic disease detection using marker-controlled watershed segmentation, superpixel based feature analysis and classification. The proposed system can take an image as input and automatically classify it into different categories of diseases and also assess how severe that type of disease is. Our work provides added value and complement the existing research efforts. Our contributions lie in:

- Development of a new way to extract features by combining marker controlled watershed segmentation and superpixels to avoid over segmentation, reduce the complexity of subsequent image processing tasks, thus producing high quality segmentation;
- A comprehensive comparison study using different feature selection and classification algorithms;
- A prototype of two-stage crop disease pattern recognition system. The first stage determines the severity level whereas the second stage detects the disease type in a crop leaf.

The remainder of this paper is organised as follows. Section 2 describes the dataset used, Sect. 3 describes our proposed methods, which includes algorithms and methods for feature extraction, feature selection and classification. In Sect. 4, we have conducted experimental on real datasets. Section 5 concludes the proposed work and highlights the future work.

2 Ground Truth

The data were collected by domain experts in the area of food and agriculture, which were captured in real fields using the camera phone. Each image has a dimension of 3264×2448. We have two types of dataset i.e., the training set and the testing set. The training dataset is composed of 20 leaf images each diseased with 'Septoria' and 'Yellow Rust'. These images have been annotated around diseased regions. In this way, we have 40 images in our training set. For the testing dataset, we have the images from 57 leaves diseased with 'Septoria' and 60 leaves diseased with 'Yellow Rust'. The test images have been annotated with their disease severity level.

3   The Proposed Approach

A high-level overview of our classification system for disease diagnosis is shown in Fig. 1. Our case is two-stage classification framework in which the first stage determines the severity level by classification between healthy and unhealthy regions in a wheat leaf. The second stage is to determine the type of disease in a wheat leaf. During the training procedure, we need to train the classification model based on a training set of images with human annotations. At the testing procedure the performance of the classification model has to be evaluated in terms of accuracy. Finally, when the performance is satisfactory, at the deployment procedure the model has to be deployed to perform the classification of images without annotation. This work mainly focuses on the training, testing and evaluation procedures.

In most cases, crop diseases exhibit a range of visual symptoms on the leaves such as colored spots, streaks, etc. These visual symptoms continuously change their color, shape and size as disease progress. It is very important to identify features of a certain disease, extract the most discriminative features and then build a classification model using suitable image processing and machine learning approaches. The feature extraction procedure has been briefed as follows which has been explained in detail in our previous paper [12].

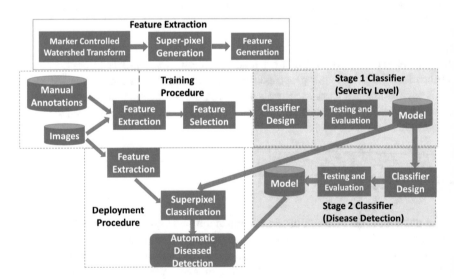

Fig. 1 A high-level overview of classification for disease diagnosis

3.1 Leaf Extraction from the Complicated Background

Our image samples are taken from real fields. Each image usually contains more than one object including foreground objects (e.g. target leaves) and background objects (e.g. non-target leaves, soil, racemes and other interference), as shown in Fig. 3a. It is critical to extract a target plant leaf first and then identify the disease patterns from that leaf using image processing approaches. Most of existing approaches for leaf extraction require manual operations or prior information and images taken in controlled environments and cannot deal with complicated background. We have first used marker-controlled watershed segmentation [13, 14] to separate the target leaf from the background. The marker-controlled watershed algorithm uses operations of mathematical morphology to place foreground markers in the blob and background markers for the areas without blobs. The details have been summarized in [12] but a block diagram representing the leaf extraction from a complicated background has been shown in Fig. 2. Figure 3b shows the result of marker-controlled watershed segmentation.

3.2 Superpixel Generation

After the target leaf is extracted, we need to further extract disease features from that target leaf. We have applied superpixels method to the target leaf for computing local image features. Superpixels [15] group pixels of an image into small meaningful regions and each region is equivalent to a pixel (A superpixel refers to a group of pixels which have similar characteristics.). The feature vector is calculated for

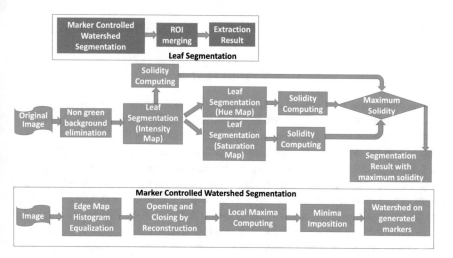

Fig. 2 Leaf extraction from complicated background

(a)   **(b)**

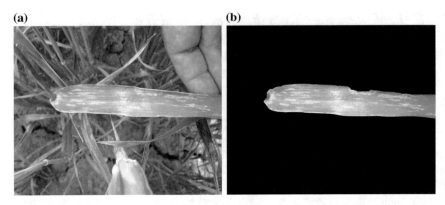

Fig. 3 a An image sample from a real crop field and **b** extraction of the target leaf from the background using marker controlled watershed transformation

each superpixel rather than pixel for high computational efficiency. It can capture redundancy in the image, reduce the complexity of subsequent image processing tasks, and produce high quality segmentations. There are various ways for superpixel grouping [16–18]. In this work, we have adopted Simple Linear Iterative Clustering (SLIC) superpixels [19]. The basic idea of SLIC superpixels is that it decomposes an image in visually homogeneous regions based on a spatially localized version of k-means clustering. Each pixel is associated to a feature vector (grayscale values in our case):

$$\Psi(x, y) = \begin{bmatrix} \lambda x \\ \lambda y \\ I(x, y) \end{bmatrix}$$

and then k-means clustering is applied. Where the coefficient λ is defined as

$$\lambda = \frac{\texttt{regularizer}}{\texttt{regionSize}}$$

where regularizer represents a nominal size of the superpixel regions and regionSize represents the strength of the spatial regularization. The image is first divided into a grid with step regionSize. The center of each grid tile is then used to initialise a corresponding k-means. The distance vector is calculated in terms of intensity values and pixel positions. After assigning each pixel to the nearest centre, an updating step adjusts the cluster centres to the mean of pixels in this group. The residual error is then calculated between new cluster centre and previous cluster centre. Such iterations continue until the error convergences. Figure 4 shows the image after using superpixels. By using superpixels, the original features from 3264×2448 can be reduced to 10,000, which dramatically increases speed and improves the quality of the results.

Fig. 4 The resulting image based on superpixels

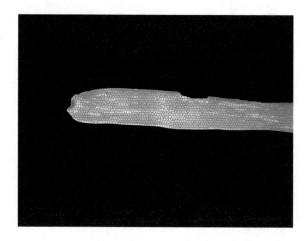

3.3 Feature Extraction

Based on the resulting image of superpixels, we have extracted textural, gradient, gabor, and biologically inspired features for each of the superpixels in the images. The details of those features are as follows:

3.3.1 Textural Features

We have focused on textural features based on Grey Level Co-occurrence Matrix (GLCM) [20] (also called Haralick features). It is used for determining how often a pixel of a grey scale value i occurs adjacent to a pixel of the value j. Four angles for observing the pixel adjacency i.e. $\theta = 0°, 45°, 90°, 135°$ are used. GLCM also needs an offset value **D** which defines pixel adjacency by certain distance. In our case, offset value is set to 1. The features, which are calculated using GLCM matrix are summarised in [12]. The mean value in each direction was taken for each Haralick feature and they were calculated from both red and green channels.

3.3.2 Gradient Features

The reason for including gradient features was illumination non-uniformity of the artefacts. In order to calculate these features, the response from Gaussian filter bank is calculated. The equation of Gaussian filter bank is given as:

$$\mathcal{N}(\sigma, i, j) = \frac{1}{2\pi\sigma^2} e^{-\frac{i^2+j^2}{2\sigma^2}}$$

The Gaussian filter bank includes Gaussian $\mathcal{N}(\sigma)$, its two first order derivatives $\mathcal{N}_x(\sigma)$ and $\mathcal{N}_y(\sigma)$ and three second order derivatives $\mathcal{N}_{xx}(\sigma)$, $\mathcal{N}_{xy}(\sigma)$ and $\mathcal{N}_{yy}(\sigma)$ in horizontal (x) and vertical (y) directions at different scales (σ). After convolving the image with the filter bank at a particular channel, the mean value is taken over of each filter response over all pixels of each superpixel.

For determining features at different smoothing scales, both red and green channels of images are convolved with the Gaussian at scales $\sigma = 1, 2, 4, 8, 16$.

3.3.3 Gabor Features

We have also applied Gabor filter [21, 22] for extracting features based on the following equation:

$$Gb(x, y, \theta, f, \sigma_x, \sigma_y) = \exp(-\frac{1}{2}(\frac{x'^2}{\sigma_x^2} + \frac{y'^2}{\sigma_y^2})) * \exp(i2\pi f x) \tag{1}$$

$$x' = x\cos\theta + y\sin\theta \tag{2}$$

$$y' = y\cos\theta - x\sin\theta \tag{3}$$

x and y are image pixel coordinates. Here we have varied $\sigma x = \sigma y = [1,2,4,8,16]$, $f = [0.333, 0.5, 1, 0.5, 0.333]$, $\theta = [0, 45, 90, 135]$.

3.3.4 Biologically Inspired Features

The Biologically Inspired Features reflect the process of the human visual cortex in image recognition tasks [23–25]. Here the crop leaf image is filtered in a number of low-level visual 'feature channels' at multiple spatial scales, for features of colour, intensity, flicker etc. It consists of 6 *Intensity Units I* and 12 *Colour Units* representing neuronal working of mammals [26]. The *Colour Units* has 4 different colour channels representing excitation as well as inhibition of colours. The excitation and inhibition of different colours make two pairs. They are formed as red-green (R-G) and blue-yellow (B-Y) channels. By using dyadic Gaussian pyramids [27] convolved on the intensity channel of an input colour image, nine spatial scales are generated with a ratio from 1:1 (scale 0) to 1:256 (scale 8). To get intensity feature maps, the centre-surround [28] operation is performed between centre levels ($c = 2, 3, 4$) and surround levels ($s = c + d$ with $d = 3, 4$ i.e., six feature maps are computed at levels of 2–5, 2–6, 3–6, 3–7, 4–7, and 4–8. Because scales are different between centre levels and surround levels, maps of surround levels are interpolated to the same size as the corresponding centre levels and are subtracted point-by-point from the corresponding centre levels to generate the relevant feature maps. For each region, we have generated the mean response of these features as:

$$I^{reg}(c, s) = \sum_i^N \frac{I(c, s, n)}{N}$$

$$RG^{reg}(c, s) = \sum_i^N \frac{RG(c, s, n)}{N} \tag{4}$$

$$BY^{reg}(c, s) = \sum_i^N \frac{BY(c, s, n)}{N}$$

where N is number of pixels in the region and,

$$I(c, s) = |I(c) - Interp_{s-c}I(s)|$$
$$RG(c, s) = |(R(c) - G(c)) - Interp_{s-c}(R(s) - G(s))| \tag{5}$$
$$BY(c, s) = |(B(c) - Y(c)) - Interp_{s-c}(B(s) - Y(s))|$$

where $Interp_{s-c}$ represent interpolation to $s - c$ level.

3.4 Feature Selection

After determination of features, we have performed feature selection on the features calculated in Sect. 3.3 so as to determine features most relevant towards classification. We have selected features for two-stage classification. For the first stage, we have selected features for classification between the diseased area and non-diseased area of the crop leaf extracted in Sect. 3.1 so as to determine severity level of the disease. For the second stage, we have performed classification between diseased region for 'Septoria' and 'Yellow Rust' so as to determine the type of disease. In our case, we have applied wrapper feature selection [29] which is an iterative procedure of feature selection until the certain task performance is maximized. For quantification of task performance, we have area under the Receiving Operating Characteristics (ROC), also known as Area Under the Curve (AUC). ROC is a graphical plot that illustrates the performance of a binary classifier system by area under it as it is created by plotting the true positive rate against the false positive rate at various threshold settings [30]. In feature selection procedure, the feature with the highest AUC is selected in the first iteration. During the next iterations, the feature which in together with previously selected features result in highest AUC. This process is repeated until there is little or no improvement. In this way there are 40 features selected for the feature set.

We have compared the performance of the features selected under the proposed criteria with other feature selection methods. i.e. Fisher Ratio [31], Quadratic Discriminant Analysis (QDA) [31], Maximum Relevancy Minimum Redundancy (mRMR) [32] and Fast Correlation-Based Filter (FCBF) [33].

Fisher Ratio is the linear discriminatory analysis which maximizes the distance between classes while minimizing the variance within each class.

Quadratic Discriminant Analysis (QDA) is same as Fisher ratio except that covariance matrix may not be identical for each class.

Maximum Relevancy Minimum Redundancy (mRMR) selects the features with highest mutual information yet, minimum redundancy among them.

Fast Correlation-Based Filter (FCBF) selects the features based on heuristics that good feature subsets contain features highly correlated with the class, yet uncorrelated with each other.

The comparison has been performed both in terms of severity level as well as disease detection procedures. The feature selection procedures such as wrapper-AUC and mRMR have been forced to stop as there has been little or no difference in their classification performance on the training set. Table 1 summarizes the total number of features selected by each feature selection procedure for both classifier stages as well as their attributes if the feature selection algorithm was converged or forced to stop. The table as well as the feature selection procedure shown in Fig. 5 shows that wrapper-AUC can perform well on least numbers of features selected compared to other feature selection procedures. The other feature selection procedure such as mRMR can perform well but their progress with number of selected features is slow. Although other feature selection procedures have been selected after improvement in their respective classification performance (e.g. Fisher Ratio in case of wrapper-LDA); their feature selection performance on Fig. 5 has been compared in terms of AUC so as to keep on one platform.

The performance has been compared with other feature selection methods in terms of ROC curves as shown in Fig. 6. We have also performed comparison with the feature set with having all features in the training set (340 features altogether).

Table 1 Number of features selected by each feature selection method for both stage 1 and stage 2 classifiers

Feature selection procedures	Stage 1 severity level		Stage 2 disease detection	
	Number of features selected	Attributes	Number of features selected	Attributes
Wrapper-AUC	40	Forced to stop	32	Forced to stop
Wrapper-LDA	23	Algorithm converged	28	Algorithm converged
Wrapper-QDA	19	Algorithm converged	17	Algorithm converged
mRMR	40	Forced to stop	40	Forced to stop
FCBF	3	Algorithm converged	4	Algorithm converged

The attributes reflect either forced stopping or the algorithm convergence of the feature selection procedure. The procedures which are forced to stop have little or no improvement while increase in feature set

Fig. 5 Comparison of feature selection procedures by increase in classification performance with number of selected features. **a** Stage-1 classifier of disease severity detection and **b** stage-2 classifier of disease detection. The classification performance have been calculated in terms of AUC

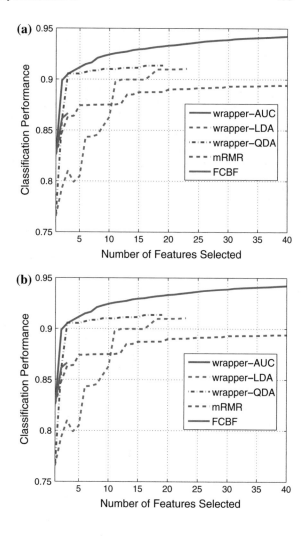

The ROC curves have been generated by 5-fold cross validation on the training set. The results show significant dominance of feature set selected by wrapper feature selection maximizing AUC. In both the cases, the 'wrapper-AUC' approach has not only outperformed its counterparts but also its performance has been near to the feature set which consists of all features. This can be quite useful in determination of classifier as classifiers with large number of features have significantly reduced computational efficiency both in terms of training as well as testing.

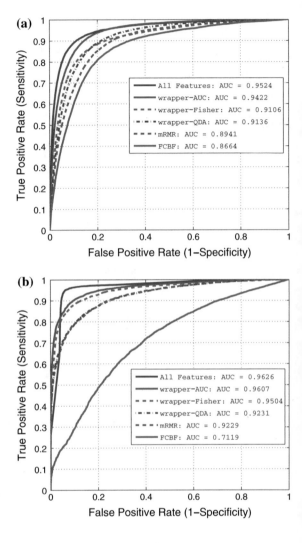

Fig. 6 Comparison of feature selection procedures in terms of ROC curves with **a** comparison of *green* area in the leaf with disease area and **b** comparison of septoria versus *yellow* rust. The result show that wrapper-AUC have performed significantly better compared to other two counterparts while performing nearest to the feature set with having all features

3.5 Classifier Design for Severity Level and Disease Diagnosis

Based on the feature set obtained from the above, we have developed the classifier for both stage 1 and stage 2 based on support vector machines (SVM) [34–36]. An SVM function can be defined as:

$$f(\mathbf{x}) = \sum_{i=1}^{N} \alpha_i y_i K(\mathbf{x}, \mathbf{x}_i) - b$$

where

- x_i is the vectors of the training set.
- N is the number of training samples used.
- α_i is a scalar, a real number that takes values between 0 and C.
- y_i is either 1 or -1, indicating the class to which the point $\mathbf{x}_i$ belongs.
- b is a scalar that shifts the output of the SVM by a constant
- $K(\mathbf{x}, \mathbf{x}_i)$ is the kernel function. A kernel is a function that takes two vectors as inputs and produces a single scalar value which is positive.

There are different types of SVM depending upon the kernel function. We decide the kernel function based upon type of data as well as with few numerical difficulties. In our case, the obvious choice was Linear SVM (LSVM) where value of kernel is set to 1. The LSVM training is based on separating the two classes by linear hyperplane. If $\mathbf{x}$ belongs to class +1, $f(x)$ is larger or equal than 1. When $\mathbf{x}$ belongs to class -1, $f(x)$ is smaller or equal than -1. The α_i values and the b value are selected to match these requirements. As for the parameters C and γ, these are meta parameters which are chosen based on past performance on a training window. We need to replicate this procedure carefully to validate the model without incurring forward-looking bias.

For the comparative study, we have different types of SVMs such as Radial Based Function (RBF)-SVM, polynomial SVM and sigmoid SVM [37]. We have also performed comparison with other classifiers such as Linear Discriminant Analysis (LDA), Quadratic Discriminant Analysis (QDA) and Artificial Neural Networks (ANN) [38]. For RBF kernel, we have chosen the kernel value equal to $k(\mathbf{x}_i, \mathbf{x}_j) = \exp(-\gamma \|\mathbf{x}_i - \mathbf{x}_j\|^2 / M)$. For ANN, we have one hidden layer with 20 neurons as there has been maximum classification accuracy compared to other ANN specifications.

The comparative results of LSVM with other classifiers have been presented in Fig. 7. For both stage 1 and stage 2, the results have been generated by performing 5-fold cross validation on the training set across different feature sets generated in Sect. 3.4 and different classifiers. The results show that feature set selected by wrapper-AUC approach has performed significantly better compared to other feature selection procedures for all of the classifiers except QDA. This shows that wrapper-AUC was an obvious choice for selecting features for both stage 1 as well as stage

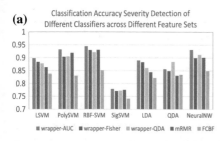
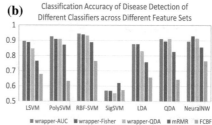

Fig. 7 Comparative results of different classifiers in stage 1 and stage 2 across different feature sets

2 classifier. Among different classifiers, all of the classifiers except sigmoid SVM have performed exceptionally well not only for severity detection (stage 1) but also for disease diagnosis (stage 2). Therefore, we omitted Sigmoid SVM for testing on our test set. As far as other classifiers are concerned, we have selected the feature set in the testing stage for each classifier which has performed the best in the respective category e.g. wrapper-AUC has performed the best for all of the classfiers except QDA, therefore, we have selected wrapper-AUC feature set for these classifiers in the testing stage. For QDA, we have selected the feature set selected by wrapper-QDA for both stage 1 and stage 2.

4 Experimental Evaluation

We have performed experimental evaluation on our test set which is composed 117 images in which 57 wheat leaf images have been taken from 'Septoria' while 60 images have been taken from 'Yellow Rust'. As stated in Sect. 3.5, we have constructed the LSVM classifier on the feature set selected by wrapper-AUC approach. For the purpose of comparison study on the test set, we have developed other classifiers on the basis of feature sets which have performed the best in the cross-validation stage in Fig. 7. We have calculated the error between the estimated severity level and annotated severity level. The severity level can be estimated by calculating the degree of overlap of the disease region with the crop leaf. We have Dice Coefficient to measure the disease severity which is the degree of overlap between the automatically determined diseased regions (D_1) and the automatically extracted leaf region (D_2) using our proposed approach. The mathematical representation of Estimated Severity (ES) is described as follows:

$$ES = \frac{2|D_1 \cap D_2|}{|D_1 + D_2|} \tag{6}$$

and therefore, the absolute difference of the particular image i can be represented as

$$E_i = |AS_i - ES_i| \tag{7}$$

We have also calculated Mean Absolute Difference (MAD) for estimating severity levels in both 'Septoria' and 'Yellow Rust' which can be represented as:

$$MAD = \frac{\sum_i E_i}{n} \tag{8}$$

where n is total number of images.

According to our collaborator i.e. Fera, the absolute difference should not be more than 10 units between the estimated severity level and the annotated severity level.

Table 2 Severity assessment across different classifiers in terms of MAD

	LSVM	RBF-SVM	PolySVM	LDA	QDA	ANN
MAD-Septoria	8.04	13.42	11.79	9.43	7.42	11.29
MAD-Yellow Rust	10.28	20.37	20.14	11.65	10.75	19.12

Table 2 represent the severity assessment across different classifiers with MAD for both 'Septoria' and 'Yellow Rust'. By observing Table 2 and Fig. 7, we come to know that our LSVM has performed significantly better compared to its other counterparts on the test set despite of being little bit inferior performance compared to RBF-SVM and PolySVM on the training set. This shows that determining the classifier model by over estimating on the training set can actually depreciate the classifier performance on the test set. The MAD for LSVM in 'Septoria' and 'Yellow Rust' is 8.04 and 10.28 which is at or below 10 units. Although the performance of QDA in severity level assessment was comparable to LSVM however, its performance with respect to disease detection was not as good as LSVM.

The performance of stage-2 classifier has been evaluated in Table 3. The stage-2 classifier has been applied on the diseased area extracted by stage-1 classifier. The disease area has been detected as either 'Septoria' or 'Yellow Rust' if the higher number of pixels have been classified as the respective class. Like stage-1, the stage-2 classifier also shows significantly better performance for LSVM compared to other classifiers.

Some visual results of automatic disease recognition are shown in Fig. 8. As shown in Fig. 8, our system can automatically detect this image which has Septoria as well as Yellow Rust diseases with severity level around comparable to the manual estimation from experts. We have also calculated computational time for our proposed approach as shown in Fig. 9. The software was run on a normal working computer (Hardware specification: CPU Intel Core i7-3630 QM, 2.4G 4 cores/8 logical processors; Memory: 8 GB (Physical)/16 GB (Virtual)). The total execution time is around 1.5 min.

Table 3 Accuracy of disease detection

	LSVM (%)	RBF-SVM (%)	PolySVM (%)	LDA (%)	QDA (%)	ANN (%)
Septoria	78.94	82.45	80.71	61.41	64.91	61.41
Yellow Rust	91.67	61.67	70	86.67	85	91.67
Overall	85.47	71.79	75.21	74.35	75.21	76.92

The second stage classifier has been applied after extracting disease area in the crop leaf

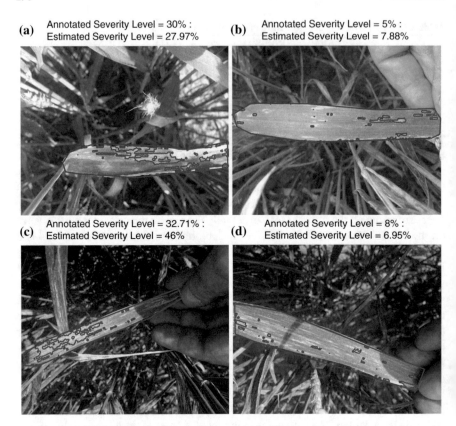

(a) Annotated Severity Level = 30% :
Estimated Severity Level = 27.97%

(b) Annotated Severity Level = 5% :
Estimated Severity Level = 7.88%

(c) Annotated Severity Level = 32.71% :
Estimated Severity Level = 46%

(d) Annotated Severity Level = 8% :
Estimated Severity Level = 6.95%

Fig. 8 Visual results of automatic septoria disease and yellow rust severity detection of crops with **a** and **b** examples from Septoria disease leaves whereas **c** and **d** are examples from Yellow Rust

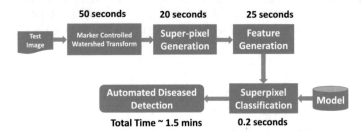

Fig. 9 The computational time for each process

5   Conclusion and Future Work

Accurate detection and identification of crop diseases plays an important role in effectively controlling and preventing diseases for sustainable agriculture and food security. Effective control measures rely on the symptom recognition and assessment of severity of diseases. In vast majority of cases, diseases exhibit a range of visual symptoms such as colored spots or streaks. Currently, the identification of diseases and their severity assessment are mainly achieved by visual observation and estimation when crop walking, which is costly and time consuming and may lead to inaccuracy and bias as well. In this paper, we have developed a novel image pattern recognition system for automatic diagnosis and severity assessment of crop diseases. There are four major contributions:

- We have developed new feature extraction methods based on combination of marker controlled watershed segmentation and superpixels methods for avoiding over segmentation, reducing the complexity of subsequent image processing tasks, and improving the quality of the results;
- We have conducted a comprehensive study on different feature selection methods (i.e. Fisher Ratio, Quadratic Discriminant Analysis (QDA), Maximum Relevancy Minimum Redundancy (mRMR) and Fast Correlation-based Filter (FCBF) and classification approaches (i.e. LSVM, RBF-SVM, PolySVM, LDA, QDA, and ANN);
- We have prototyped a two-stage crop disease pattern recognition system. The first stage determines the severity level whereas the second stage detects the disease type in a crop leaf.

The data used in the experimental evaluation were collected and captured by domain experts in the crop fields. The images are 157 in total. 40 images were used for training models and the rest of 117 images for testing. The experimental result demonstrates that our approach performs well on disease recognition and severity assessment. The preliminary result from this work is promising.

The future will be to further look into the feature extraction of diseases and collect more data from different seasons for improving robustness of algorithms. Additionally, to enable to process more images in real time, we will also develop new algorithms for parallel processing images.

Acknowledgments The work reported in this paper has formed part of the COPE project, which is funded by Sustainable Society Network+ (RCUK digital programme) and Manchester Metropolitan University in UK. This project is in collaboration with Fera (Food and Environment Research Agency). The authors acknowledge Ms. Sharon Elcock at the Fera for her support.

References

1. Strange, R.: Food security **2**(2), 111 (2010). doi:10.1007/s12571-010-0064-5
2. Basf, A.: Managing septoria in wheat (2014). http://www.agricentre.basf.co.uk/
3. Madden, I.M., Hughes, L.V.: Phytopathology **90**, 788 (2000)
4. Domiciano, G.P., Duarte, H.S.S., Moreira, E.N., Rodrigues, F.A.: Plant Pathol. **63**, 922 (2014)
5. Camargo, A., Smith, J.S.: Comput. Electr. Agric. **66**(2), 121 (2009)
6. Pydipati, R., Burks, T., Lee, W.: Comput. Electr. Agric. **52**(1–2), 49 (2006)
7. Pugoy, R.A.D.L., Mariano, V.Y.: In: Third International Conference on Digital Image Processing (ICDIP 2011), vol. 8009, pp. F1–F7 (2011)

8. Contreras-Medina, L.M., Osornio-Rios, R.A., Torres-Pacheco, I., de J. Romero-Troncoso, R., Guevara-González, R.G., Millan-Almaraz, J.R.: Sensors **12**(1), 784 (2012)
9. Abdullah, N., Rahim, A., Hashim, H., Kamal, M.: In: 2007 5th Student Conference on Research and Developmen, pp. 1–6 (2007)
10. Lloret, J., Bosch, I., Sendra, S., Serrano, A.: Sensors **11**(6), 6165 (2011)
11. Phadikar, S., Sil, J.: In: 11th International Conference on Computer and Information Technology, pp. 420–423. IEEE (2008)
12. Han, L., Haleem, M.S., Taylor, M.: In: Science and Information Conference (SAI) (2015)
13. Wang, X., Huang, D., Du, J., Xu, H., Heutte, L.: J. Appl. Math. Comput. **205**, 916 (2008)
14. Tang, X., Liu, M., Zhao, H., Tao, W.: In: Proceedings of IEEE International Congress on Image and Signal Processing, pp. 1–5 (2009)
15. Ren, X., Malik, J.: In: 9th IEEE International Conference on Computer Vision, pp. 10–17. IEEE (2003)
16. Felzenszwalb, P., Huttenlocher, D.: Int. J. Comput. Vis. **59**(2), 167 (2004)
17. Shi, J., Malik, J.: IEEE Trans. Pattern Anal. Mach. Intell. **22**(8), 888 (2000)
18. Veksler, O., Boykov, Y., Mehrani, P.: In: European Conference on Computer Vision (ECCV) (2010)
19. Achanta, R., Shaji, A., Smith, K., Lucchi, A., Fua, P., Susstrunk, S.: IEEE Trans. Pattern Anal. Mach. Intell. **34**(11), 2274 (2012)
20. Haralick, R.M., Shanmugam, K., Dinstein, I.: IEEE Trans. Syst. Man Cybern. **3** (1973)
21. Gabor, D.: J. Inst. Electr. Eng. **93**(26), 429 (1946)
22. Daugman, J.G.: IEEE Trans. ASSP **36**(7), 1169 (1988)
23. Tian, Y., Kanade, T., Cohn, J.: IEEE Trans. Pattern Anal. Mach. Intell. **23**(2), 97 (2001)
24. Siagian, C., Itti, L.: IEEE Trans. Pattern Anal. Mach. Intell. **29**(2), 300 (2007)
25. Song, D., Tao, D.: IEEE Trans. Image Process. **19**(1), 174 (2010)
26. Itti, L., Koch, C., Niebur, E.: IEEE Trans. Pattern Anal. Mach. Intell. **20**, 1254 (1998)
27. Adelson, E.H., Anderson, C.H., Bergen, J.R., Burt, P.J., Ogden, J.M.: RCA Eng. **29**, 33 (1984)
28. Sun, S.G., Kwak, D.M.: J. Multimed. **1**, 16 (2006)
29. Serrano, A.J., Soria, E., Martin, J.D., Magdalena, R., Gomez, J.: In: The International Joint Conference on Neural Networks (IJCNN), pp. 1–6 (2010)
30. Alberg, A.J., Park, J.W., Hager, B.W., Brock, M.V., Diener-West, M.: J. General Intern. Med. **19**, 460 (2004)
31. Duda, R.O., Hart, P.E., Stork, D.G. (eds.): Pattern Classification. Wiley-Interscience, New York (2000)
32. Peng, H., Long, F., Ding, C.: IEEE Trans. Pattern Anal. Mach. Intell. **27**, 1226 (2005)
33. Yu, L., Liu, H.: In: Proceedings of the Twentieth International Conference on Machine Learning, pp. 856–863 (2003)
34. Vapnik, V.N.: The nature of statistical learning theory. Springer, New York (1995). ISBN 0-387-94559-8
35. Shevade, S., Keerthi, S., Bhattacharyya, C., Murthy, K.: IEEE Trans. Neural Netw. **11**(5), 1188 (2000)
36. Muller, K., Mika, S., Ratsch, G., Tsuda, K., Scholkopf, B.: IEEE Trans. Neural Netw. **12**(2), 181 (2001)
37. Hsu, C.W., Chang, C.C., Lin, C.J.: A practical guide to support vector classifcation (2010)
38. Smola, A., Vishwanathan, S.: Introduction to Machine Learning (Cambridge University Press, 2008)

Image Complexity and Visual Working Memory Capacity

Juan Huo

Abstract This chapter presents a discussion about the relationship between the image complexity and the visual working memory capacity. In advertisement and web site design, the mismatch between the target objects and the real salient objects can represent the degree of image complexity which is an important reason of low efficiency and unpleasant reading. Many psychological experiments have also shown the effect of image complexity on short term memory. In this chapter, a method was introduced to measure this mismatch and the image complexity. The present algorithm used in this method combines the mathematic algorithm like SIFT (Scale Invariant Feature Transformation) and K-means with the cognitive science theory of visual working memory capacity. Results of the measurement method were validated by the visual working memory practical experiments. Besides, the results from EEG study of visual working memory on the same group of test images are also consistent with the value from our algorithms.

1 Introduction

As most of our knowledge and information comes from the visual system, visual working memory plays an important role in our cognitive process. However, although hundred billion ($10^{1}1$) neurons and several hundred trillion synaptic connections of human brain can process and exchange prodigious amounts of information over a distributed neural network in the matter of milliseconds, the information load we can process in a short time (within seconds) is limited [1]. It is not clear whether the limited physical resources is the major cause of visual attention, in the experiment for both human brain and primate brain, the dual-interference between simultaneous spatial attention task and spatial memory task show the competition of physical resource [2]. Even though working memory and selective attention were manipulated in two separate and unrelated tasks, the interference between them is obvious, the

J. Huo (✉)
Zhengzhou University, Zhengzhou, China
e-mail: juanhuo@126.com

© Springer International Publishing Switzerland 2016   301
L. Chen et al. (eds.), *Emerging Trends and Advanced Technologies
for Computational Intelligence*, Studies in Computational Intelligence 647,
DOI 10.1007/978-3-319-33353-3_16

distractors of visual input is crucially determined by the availability of working memory [3]. In EEG study and other kind of psychological experiments, the increased image complexity or distractors induces increased brain activity [3].

In our daily life, we have to read information from different media, either through electronic media like web site or from paper publications. Most of these information are accessed through our visual system to the brain. However, the visual information that the viewer has paid attention to may not be the useful information that the author want to show or the viewer is unpleasant with the publication due to high level of visual complexity [4]. This visual complexity caused mismatch in serious situation can even induce car accident [5], which is largely due to the dual-task interference of memory and attention [2, 6]. An algorithm which is based on SIFT and K-means algorithm has been designed to estimate this mismatch (namely the distance between the expected region of "visual conspicuity" and the real salient areas) as a metric of visual complexity [7]. As visual complexity (or image complexity called somewhere) is an important reason for this mismatch, the relationship between the saliency mismatch and the visual complexity estimator is close. Furthermore, it has been proved in cognitive experiment that the image saliency and attention priority can determine the visual working memory capacity, where the image information is stored [8]. The mismatch between the image targets and the interesting points of an image thus can be proved to have effect on our visual working memory. Until now, there have been hundreds of papers which have discussed the saliency detection of images [7]. A comprehensive survey of saliency detection algorithm is not intended here. The reason for us to employ SIFT is because of its popularity in computer vision applications and in saliency modelling [9–11]. In addition, SIFT is similar to the information process of inferior temporal cortex and has good image feature descriptor which is scale-variation free [9]. In later part of this paper, results of this SIFT & K-means algorithm are compared with two other saliency map of [12, 13], which shows similar trend of image saliency shift. Results of the SIFT & K-means is then further validated by a cognitive experiment. The cognitive experiment results show the more complex the image background is, the less objects the participants kept in their visual working memory. This is highly correlated with our new estimated value of the saliency mismatch. The algorithm of SIFT & K-means thus can be a reference to analyze the web site, publication, advertisements, movie frames and other media contents.

2 Background Knowledge

Saliency detection is considered the key for attentional mechanism. In many papers, the location of the image saliency is defined as the region where the viewers paid attention to and is also called the "conspicuity area" [7]. Information is said to be 'attended' if the information is kept in visual working memory [12]. It is believed that there is a two-component framework for the control of where the visual attention is deployed: a top down cognitive volitional control and a second faster bottom-up saliency based primitive mechanism [12]. This chapter investigates people's direct response to visual stimuli, namely the bottom-up primitive mechanism. The visual stimuli which win the competition for saliency is sent to the higher levels of brain neural networks. If the wrong stimuli or the noise is strong, the expected information can be overwhelmed and missed. Thus the mismatch between the supposed visual target and the image saliency can be a serious disturbance for viewers. The viewers can easily miss cues due to the image quality, visual complexity and the other reasons [14, 15]. In cognitive science, the limits of saliency-based information search and the shifts of visual attention is largely due to the limited visual working memory capacity [12, 16–18]. Although researchers of visual working memory have claimed different memory capacity limitation (for example in Nelson Cowan's paper, the visual working memory capacity is 4, while in Miller's research the capacity limitation is 7 ± 2) [16, 19, 20], until now the 7 [16, 20]. Thus in the following sections, the number limitation of visual working memory capacity is set to be 7, which means the maximum number of salient items that can effectively attract a viewer's attention is 7. If the number of salient items is higher than 7, some items can be possibly ignored by the viewer, which causes the so called attention competition between the expected target information and the irrelevant ones [12]. The competition results between the expected attention target and the real interest points in an image is the study issue of this paper. For the study of interesting points, SIFT is well recognized as an efficient image feature description method for image recognition [9–11]. As a bottom-up approach, the SIFT key points detection is not only popular for image feature recognition, but also widely used as a step for saliency estimation [11, 21, 22].

Until now, the improvement of saliency detection algorithm is still going on and many algorithms define different salient regions for one same image. In Sect. 3.1.5, to check the functionality of SIFT & K-means, Itti-Koch and AIM saliency map are implemented [12, 13]. The Itti-Koch method is an unsupervised method, which combines color, intensity and other texture information while the method of AIM makes use of Shannon's information measure to transfer the image feature plane into a dimension that closely corresponds to visual saliency. The saliency map generated by these two methods differs from each other for complex images but are correlated.

3 A Tentative Measurement Algorithm

3.1 Method

Based on the SIFT Density Map (SDM) described in [21], the algorithm of SIFT and K-means algorithm are implemented to calculate the locations of salient regions in this paper. Then a scale-free distance between the expected target locations and the interesting points is measured. This distance is the estimator of the mismatch. Since the computer recognized saliency differs from each other for different algorithms, it should be noted the algorithm in this paper is not to precisely locate the most possible first attention point, the locations calculated by K-means algorithm is rather a reference for viewer's possible attention.

3.1.1 SIFT and Key Points

To implement SIFT, all the images are transferred to gray scale at first. We then get the candidate key points from the scale space by finding the maxima and minimum of the convolution of image $I(x, y)$ and a variable-scale Gaussian kernel $G(x, y, \sigma)$ [9, 23]. The scale space of an image is defined as a function $L(x, y, \sigma)$.

$$G(x, y, \sigma) = \frac{1}{2\pi\sigma^2} e^{-(x^2+y^2)/2\sigma^2} \tag{1}$$

$$L(x, y, \sigma) = G(x, y, \sigma) * I(x, y) \tag{2}$$

Since the computer generated key points are closely related to the real fixation points which has been validated by [21], here the K-means algorithm is used to find the relevant locations of interesting areas in an image.

A difference-of-Gaussian function, $D(x, y, \sigma)$ is then calculated. The candidate keys are detected by the maxima and minimum of $D(x, y, \sigma)$.

$$D(x, y, \sigma) = (G(x, y, k\sigma) - G(x, y, \sigma)) * I(x, y) \tag{3}$$

$$= L(x, y, k\sigma) - L(x, y, \sigma) \tag{4}$$

where k is a constant factor.

After we got the candidate key points, the next step is to have these key points tested and the key points which have low contrast will be rejected by a threshold according to the value of $|D(\hat{x})|$. The information of locations, scale and ratio of curvatures are calculated for the selected key points.

3.1.2 Distance Parameters Q

Since the computer generated key points are closely related to the real fixation points which has been validated by [21], a K-means algorithm whose cluster center is labeled as C_i is used to calculate the possible locations of interesting points from the cluster of selected key points. The maximum number of the cluster center is $n \leq min\{7, N_T\}$, where N_T is the object number. If the nearest expected target object location is labeled as T_i, then the distance between C_i and T_i can be expressed as ΔCT_i. The scale-free parameter of ΔCT_i is represented as Q in Eq. 5, which is the reference parameter for the saliency mismatch. In Eq. 5, X is the image length and Y is the image width in pixel, and k is a constant factor.

$$Q = \frac{\sum_i^n \Delta CT_i}{n\sqrt{XY}} \tag{5}$$

where $n \leq min\{7, N_T\}$.

3.1.3 Experiments and Results

To validate this new algorithm, two experiments were carried out with a dataset of 250 images. The location of the target objects are stored in a database before the target objects are merged with different backgrounds. In our experiments, the target objects are small white balls with numerical mark on it.

3.1.4 Experiment 1

To express the measurement process, the SIFT algorithm is applied to four images shown in Fig. 1. Complexity of these images' background increases in sequence. In these images, the red diamonds represents the key points cluster center C_i. The blue arrows represents the vector of the selected key points, which indicates the key point's orientation and scale, are derived from the difference-of-Gaussian function D.

After we implement the SIFT & K-means algorithm introduced in Sect. 3.1, the mismatch value for these four images is listed in Table 1. In the first image, the background is a white plane with very low image complexity, thus the C_i is registered well with T_i. From the second image, the image background becomes more and more complex, from cloudy sky to the crowded people, the distraction from the target objects to the image background becomes more and more serious. More and more interesting points derived from the SIFT algorithm obviously shift away from the target objects.

3.1.5 Experiment 2

In the second experiment, the same images in Fig. 1 are processed by Itti-Koch and AIM algorithms to have their saliency map. Figure 2 is the result of Itti-Koch method and Fig. 3 is the result of AIM method, where the white area is the salient region. To make a comparison, the red diamond still indicates the location of C_i of Sect. 3.1.4, similarly the green star in the image represents the target objects.

The saliency map of Itti-Koch and AIM differs from each other, but nearly the same for the first image: the salient regions all registered well with the green target. Similar as the results of the SIFT & K-means method in Sect. 3.1.4, the shift from the salient points starts from the second image Fig. 2b and become most obvious for the fourth image of Fig. 2d. The distance between the green targets and the center of the

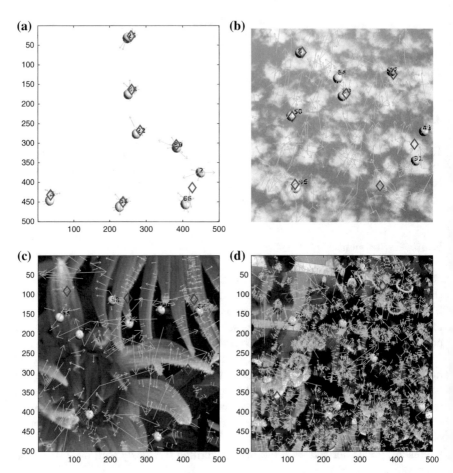

Fig. 1 Images of experiment. The *red diamond* is the key points K-means cluster center and the *blue arrows* are the vectors of key points

Table 1 Q value of SIFT-K method for Fig. 1

	(a)	(b)	(c)	(d)
Q	0.0377	0.0644	0.1334	0.1559

salient regions is then measured by Eq. 5, similar method as described in Sect. 3.1.2. The mismatch parameter Q of Itti-Koch saliency map is shown in Table 2.

The AIM method can recognize the target objects well for nearly all four images of Fig. 3. However, the shift of the attention is not obvious as the targets are still labeled as salient while the image background becomes more complex and the number of salient regions increase globally. Instead of salient region shift, the AMI method does show the increase of non-target salient information. Especially for Fig. 3c, d,

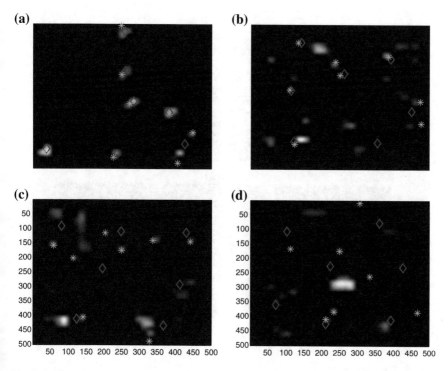

Fig. 2 Itti-Koch's saliency and target objects [12]

Table 2 Q value of Itti-Koch for Fig. 1

	(a)	(b)	(c)	(d)
Q	0.0354	0.1396	0.1423	0.2097

the number of labeled white salient regions is higher than visual attention capacity and thus the viewers can feel difficult to find the target and remember them in visual working memory.

3.1.6   The Key-Points Ratio K_{num}

Besides the distance parameter Q, the key point number K_{num} is another important parameter to evaluate the image complexity. The key numbers derived from SIFT algorithm increase with the complexity level, thus can be another important reference for the complexity level, especially when parameter Q lost its ability to distinguish the background saliency and the target memory items (there is possibility that there is overlap between the background saliency and the target memory items) (Tables 3 and 4).

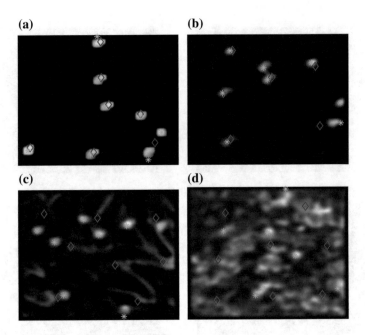

Fig. 3 AIM's saliency and target objects [13]

Table 3 K_{num} list of Fig. 1

	(a)	(b)	(c)	(d)
K_{num}	136	527	640	4004

Table 4 Five factors of input

Number of items ($Xnum$)	In each image, large $Xnum$ means higher memory load
Background complexity ($Xback$)	Higher $Xback$ increases the task difficulty
Test item's position attribute ($Xpos$)	It is a value to label the randomness of item positions

4   Human Visual Test

The stimuli are the images shown in Fig. 1. Stimuli of this experiment were presented for 70 young participants at one time with 50 % male and female at the age of 21 on the average in a large classroom. Each image was presented for 5 s and followed by 30 s memory recall time for the students to note down the numbers they have seen in each image. Four students' records were detected as outlier and rejected according to standard deviation analysis. The anova analysis of the rest 66 students' correctly recorded items is shown in Fig. 4. When the image complexity increases,

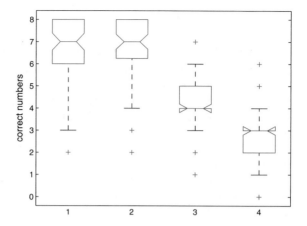

Fig. 4 Anova analysis of human visual experiment

the mean number of correctly recalled items decreases significantly. The test results were shown in Fig. 4. Generally, all three methods' complexity measurement value is inverse to students' remembered item numbers. The Spearman correlation between the tested new saliency measurement value and the mean correctly recalled number of items from the students is strong, with $|r| = 0.80$. We then perform the similar test for Itti-Koch's Q and we have $|R| = 0.78076$ while the correlation for SIFT & K-means is as high as $|R| = 0.96621$, $p < 0.05$.

Although this is not a strict cognitive experiment, this experiment does show the K-means based saliency mismatch estimation method is consistent with human's visual cognitive visual sense. Besides the above experiments, another 32 images were tested within the similar procedure of Sect. 3.1.

5 Brain Computer Interface Experiments

To further validate the above algorithm and hypothesis, a brain computer interface experiment is carried out. This task in our EEG experiments is to remember the numbers attached to white balls which are scattered in different image backgrounds as shown in Fig. 1. Each number is treated as an item for visual working memory task. The amount of items to be remembered varies from low to high. The stimuli were presented and the participants' response was recorded by EPrime software. The participants were placed 70 cm in front of a 19 in. LCD screen straight ahead of their eyes horizontally where the visual stimuli were displayed at the center of the participants' visual field. The stimuli were subtended around 6.5° visual angle. After the stimuli presentation, the participants then had another 1000ms interval of retention time before the memory recall.

6   Image Measurement

The target items to be remembered were numbers which were shown in limited digit length (<4). We have 4 participants tried 252 groups of images, namely 252 trials in total. Target items are all small white balls associated with random numbers in blue, the numbers to be remembered in every image was different from each other. These images were shown to the participants in random sequence and each participant had 50 trials on average. The four image background varies from simple to complex and classified into four levels as shown in Fig. 1. Trials for every participant include all four level image backgrounds.

To measure the image, the target item number and the background textures were labeled as a factor vector $Xnum, Xback, Xpos$. $Xnum$ is the number of items shown per image, whose value varies from 3 to 50 per image. The image complexity level of the image background is the second factor considered. Image stimuli in this research varies from a mono-color background to different texture backgrounds added in the image to increase the complexity level. The image background complexity value is labeled as ($Xback$), which is calculated according to our algorithm of SIFT & K-means. Another factor worth noting is the position distribution of the items which is referred to as the factor $Xpos$. $Xpos$ value is high if the arrangement of the items is random and is low if the arrangement of the items is in array. The memory recall process from the participants were recorded and the participants' responses were automatically calculated at the end of each trial. To make sure the participant is not aware of the factor level change and deliberately change their attention, all the factor values are arranged randomly for each image during the image generation.

6.1   Statistics of Image Measurement Results

The statistics of the total number remembered by the participants $Xnum$ was described in Fig. 6a. Although the participants were provided a glut of items in one image, the maximum number of items recalled is no more than 6 in all. Considering about that there is time related mechanism in brain, some trials were set with longer duration time. The image stimuli were presented within the same 2 s duration time. The total average time an image shown is within 2–3 s. It was also observed the short term memory limitation is also obvious even when an image's appearance repeated. In an experiment for a same participant, we have tested one image with 16 numbers which has a duration of 1 s appeared three times in less than 15 min in three separate trials. It was normally assumed that the participants got familiar with these images and correct typed in item number should increase significantly. However, participants in this research increased remembered items from 4 to 6, but no more than 6 (Fig. 5).

We have observed the main brain activity in frontal, parietal and occipital brain areas. To some extent, the increased visual information load, especially the $Xnum$ and $Xback$ not only increases the visual working memory load, but also a combination

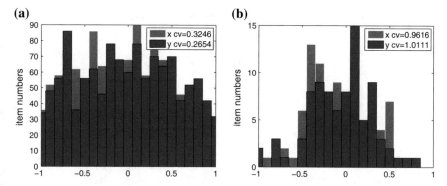

Fig. 5 **a** Shows the total tested input x, y position in the image compared to the whole image screen. **b** Shows the correctly picked number x, y positions in the screen. (0, 0) is the center of the screen while the x, y ranges from −1 to 1. In the figure, CV is the abbreviation of coefficient variation

of attention, mental and visual working memory. This is because the visual working memory and attention share some common neural substrate [24]. The attention and working memory representation of the brain cortex have been shown to be overlapped in different areas such as frontal and parietal areas [25, 26]. The increase of EEG power is the reflection of combined attention and mental load. The shared representation in brain area is also considered an important reason for the limited working memory capacity. As the attention and mental load increase, the information compete the neural resources in the same brain area.

6.2 Electrophysiological Analysis

The EEG signal was acquired from an electrode cap with 32 channels at 1000 Hz by the BrainVision recorder. The electrode impedance is kept lower than 20 KΩ. The reference signal VEOG and HEOG were also recorded. The recorded signals were then processed by a professional software BrainVision Analyzer which is a professional software for EEG signal analysis. Multiple signal processing and pattern recognition techniques were employed in this software. In this study, the signal process follows the sequence of dataset preprocessing, IIR Filters, Band Rejection, artifact rejection, frequency filtering, segment analysis and comparison. Any suspicious muscle activity-related artifact has been rejected during the process of artifact rejection and ocular correction ICA (Independent Component Analysis). Frequency higher than 70 Hz has been filtered out during band rejection transform which is used to filter out interference signals due to power supply or to poorly shielded electrical devices. The line noise or called notch frequency (50 Hz) is also filtered out in this process.

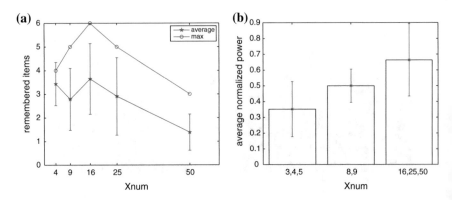

Fig. 6 **a** The average correctly recalled item numbers in each $Xnum$ group were summarized from all trials. The *error bar* is the standard deviation of each group's remembered items. **b** The item numbers are divided into 3 groups, the average EEG power from each trial in high load group ($Xnum \in \{16, 25, 50\}$) is nearly one time higher than the low load groups ($Xnum \in \{3, 4, 5\}$)

The filtered participants' EEG signal was segmented, the brain state during memory retention time (1 s duration time) is represented by its corresponding EEG power segment [27, 28]. Average EEG power was calculated from these segments based on Fast Fourier Transform (FFT) [29–32]. The average EEG power value of that segment is found to be positively correlated with the score of image complexity. Generally when the memory load is increased, the participant's brain activity also significantly increases compared to the brain's last state. However, this increase trend is not infinite. The averaged performance and the remembered item number drops after $Xnum = 16$ in Fig. 6a. The results in [33] show the similar brain power limitation for more difficult task, which indicates children and older adults have decreased alpha power with higher memory load. However, in general, the Fig. 6b shows the statistic higher EEG power level with increased target item numbers. Our results show the young adults also have on average weak EEG power when the task difficulty is far beyond their ability. The Pearson correlation test show positive correlation between the EEG power and the $Xnum$ ($p < 0.05$ Bonferroni corrected). When $Xnum$ is certain, the correlation between $Xback$ and the EEG power is also positive ($p < 0.05$). In comparison, the correlation between the $Xpos$ and the EEG power is modest, positive but not significant ($0.05 < p < 0.1$). In summary, EEG power of working memory process is closely related to the main visual complexity factor $Xnum$ and $Xback$.

7 Conclusion

In this chapter, the relationship between the visual complexity and visual working memory capacity is discussed. Although the relationship between the visual complexity and the visual attention is well known, the relationship between the visual complexity and the working memory capacity is rare to be discussed. The increased visual complexity means higher information load and caused the difficulty of correct visual attention towards the locations of memory target items, as the attention is limited by the visual working memory capacity.

Based on the relationship between the visual working memory and the visual working memory capacity, this chapter introduced a new algorithm SIFT & K-means to measure the discrepancy between the expected target object and the image salient regions. The resulting metric for mismatch and visual complexity calculated from SIFT & K-means algorithm in the first experiment is consistent with the human visual working memory experiments. Results of the second experiment show comparison between this method and two saliency detection methods show the reliability of this algorithm. The SIFT & K-means algorithm can be a reference for the measurement of image quality and image complexity. Both EEG and psychological experiments have the consistent results of visual working memory capacity as our algorithm predicted, because the EEG results clearly show the increased brain activity is needed for the correct visual attention when the visual complexity is high. Our findings from the above experiments prove the link between the visual complexity and the visual working memory capacity is close.

References

1. Marois, R., Ivanoff, J.: Capacity limits of information processing in the brain. Trends Cognit. Sci. 9(6), 296–305 (2005)
2. Watanabe, K., Funahashi, S.: Neural mechanisms of dual-task interference and cognitive capacity limitation in the prefrontal cortex. Nat. Neurosci. 17(4), 601–611 (2014)
3. de Fockert, J.W., Rees, G., Frith, C.D., Lavie, N.: The role of working memory in visual selective attention. Science 291(5509), 1803–1806 (2001)
4. Tuch, A.N., Bargas-avila, J.A., Opwis, K., Wilhelm, F.H.: Visual complexity of websites: Effects on users' experience, physiology, performance, and memory. Int. J. Hum.-Comput. Studi./Int. J. Man-Mach. Stud. 67(9), 703–715 (2009)
5. Reimer, B.: Impact of cognitive task complexity on drivers' visual tunneling. Transp. Res. Rec. 2138, 13–19 (2009)
6. Strayer, D.L., Johnston, W.A.: Driven to distraction: dual-task studies of simulated driving and conversing on a cellular telephone. Psychol. Sci. 12(6), 462–466 (2001)
7. Toet, A.: Computational versus psychophysical bottom-up image saliency: a comparative evaluation study. IEEE Trans. Pattern Anal. Mach. Intell. 33(11), 2131–2146 (2011)
8. Melcher, D., Piazza, M.: The role of attentional priority and saliency in determining capacity limits in enumeration and visual working memory. PloS One 6(12), e29296 (2011)
9. Lowe, D.G.: Distinctive image features from scale-invariant keypoints. Int. J. Comput. Vis. 60(2), 91–110 (2004)
10. Li, Z., Itti, L.: Saliency and gist features for target detection in satellite images. IEEE Trans. Image Process. 20(7), 2017–2029 (2011)
11. De Campos, T., Csurka, G., Perronnin, F.: Images as sets of locally weighted features. Comput. Vis. Image Underst. 116(1), 68–85 (2012)
12. Itti, L., Koch, C.: A saliency-based search mechanism for overt and covert shifts of visual attention. Vis. Res. 40(10–12), 1489–1506 (2000)
13. Bruce, N.D.B., Tsotsos, J.K.: Saliency, attention, and visual search: an information theoretic approach. J. Vis. 9(3) (2009)

14. Cardaci, M., Di Gesu, V., Petrou, M., Tabacchi, M.E.: A fuzzy approach to the evaluation of image complexity. Fuzzy Sets Syst. **160**(10), 1474–1484 (2009)
15. Da Silva, M.P., Courboulay, V., Estraillier, P.: Image complexity measure based on visual attention. In: 2011 18th IEEE International Conference on Image Processing (ICIP 2011), pp. 3281–3284 (2011)
16. Brzezicka, A., Wróbel, A., Kamin, J.: Neurobiology of learning and memory short-term memory capacity (7 ± 2) predicted by theta to gamma cycle length ratio. Neurobiol. Learn. Mem. **95**, 19–23 (2011)
17. Vogel, E.K., Machizawa, M.G.: Neural activity predicts individual differences in visual working memory capacity **428**, 748–751 (2004)
18. Tsubomi, H., Fukuda, K., Watanabe, K., Vogel, E.K.: Neural limits to representing objects still within view. J. Neurosci. **33**(19), 8257–8263 (2013)
19. Cowan, N.: The magical number 4 in short-term memory: a reconsideration of mental storage capacity. Behav. Brain Sci. **24**, 87–185 (2000)
20. Miller, G.A.: The magical number seven plus or minus two: some limits on our capacity for processing information. Psychol. Rev. **63**(2), 81–97 (1956)
21. Ardizzone, E., Bruno, A., Mazzola, G.: Visual saliency by keypoints distribution analysis. In: 16th International Conference on Image Analysis and Processing, ICIAP 2011, September 14, 2011–September 16, ser. Lecture Notes in Computer Science (including subseries Lecture Notes in Artificial Intelligence and Lecture Notes in Bioinformatics), vol. 6978 LNCS, pp. 691–699. Springer Verlag (2011)
22. Uchida, S., Shigeyoshi, Y., Kunishige, Y., Yaokai, F.: A keypoint-based approach toward scenery character detection. In: 11th International Conference on Document Analysis and Recognition, ICDAR 2011, September 18, 2011–September 21, ser. Proceedings of the International Conference on Document Analysis and Recognition, ICDAR, pp. 819–823. IEEE Computer Society (2011)
23. Lowe, D.G.: Object recognition from local scale-invariant features. In: Proceedings of the Seventh IEEE International Conference on Computer Vision, vol. 2, pp. 1150–1157 (1999)
24. Mayer, J.S., Bittner, R.A., Nikolic, D., Bledowski, C., Goebel, R., Linden, D.E.J.: Common neural substrates for visual working memory and attention. Neuroimage **36**(2), 441–453 (2007)
25. Buschman, T.J., Siegel, M., Roy, J.E., Miller, E.K.: Neural substrates of cognitive capacity limitations. Proc. Natl. Acad. Sci. USA **108**(27), 11252–11255 (2011)
26. Lepsien, J., Griffin, I.C., Devlin, J.T., Nobrea, A.C.: Directing spatial attention in mental representations: interactions between attentional orienting and working-memory load. Neuroimage **26**(3), 733–743 (2005)
27. Delorme, A., Makeig, S.: Eeglab: an open source toolbox for analysis of single-trial eeg dynamics including independent component analysis. J. Neurosci. Methods **134**(1), 9–21 (2004)
28. Junfeng, S., Yingying, T., Lim, K.O., Jijun, W., Shanbao, T., Hui, L., Bin, H.: Abnormal dynamics of eeg oscillations in schizophrenia patients on multiple time scales. IEEE Trans. Biomed. Eng. **61**(6), 1756–1764 (2014)
29. Ghafar, R., Hussain, A., Samad, S.A., Tahir, N.M.: Comparison of FFT and AR techniques for scalp EEG analysi. Ser. IFMBE Proceedings, vol. 21, pp. 158–161 (2008)
30. Lehmann, D., Michel, C.M.: Intracerebral dipole sources of eeg fft power maps. Brain Topogr. **2**(1–2), 155–164 (1989)
31. Michel, C.M., Lehmann, D., Henggeler, B., Brandeis, D.: Localization of the sources of eeg delta, theta, alpha and beta frequency bands using the fft dipole approximation. Electroencephalogr. Clin. Neurophysiol. **82**(1), 38–44 (1992)
32. Singh, Y., Singh, J., Sharma, R., Talwar, A.: Fft transformed quantitative eeg analysis of short term memory load. Ann. Neurosci. **22**(3), 176–179 (2015)
33. Sander, M.C., Werkle-bergner, M., Lindenberger, U.: Amplitude modulations and intertribal phase stability of alpha-oscillations differentially reflect working memory constraints across the lifespan. NeuroImage **59**, 646–654 (2012)

Immersive Brain Entrainment in Virtual Worlds: Actualizing Meditative States

Ralph Moseley

Abstract Virtual Reality with associated hardware and software advances is becoming a viable tool in neuroscience and similar fields. Technology has been harnessed to modify a user's state of mind for some time through different approaches. Combining this background with merged reality systems, it is possible to develop intelligent tools which can manipulate brain states and enhance training mechanisms.

1 Introduction

There has always been a fascination with finding ways of controlling or manipulating the human brain, probably since prehistory. As technology has developed, man has noticed that certain external situations can influence and induce a particular internal state. Possibly the first noticeable effects for our human ancestors were how the flickering of the fire would bring about a restful state, or the movement of light through trees would somehow shift the daydreamer's mind. The study of past history and current primitive cultures also reveals the use of sound, as a tool, adjust or induce trance states during shamanic or ritual. The use of light within the classical philosophical era was explored by both Apuleius (circa 125 C.E) and Ptolemy (circa 200 C.E) using and studying strobe mechanisms which appeared to alter the watcher's state of mind [1, 2].

More recently, scientists such as the 17th Century Belgian scientist Plateau, discovered how to use flicker fusion—the ability to discern whether light flashes appear to the watcher as solid unwavering light as a diagnostic tool. In healthy individuals the light would appear to be flickering at higher frequencies. Further to this, more recently still, it has been found that regular long term meditators can discern flickering at much higher frequencies than individuals who do not practice the meditation techniques [3].

R. Moseley (✉)
School of Science and Technology, Middlesex University, London, UK
e-mail: r.moseley@mdx.ac.uk; ralph@ralph-moseley.co.uk

© Springer International Publishing Switzerland 2016
L. Chen et al. (eds.), *Emerging Trends and Advanced Technologies
for Computational Intelligence*, Studies in Computational Intelligence 647,
DOI 10.1007/978-3-319-33353-3_17

Strobe lighting has also been used in this way, therapeutically. Found initially by the French psychologist Pierre Janet at the turn of the 20th Century, observed that patients at the Salpetriere Hospital in Paris became more relaxed and less affected by hysteria when exposed at particular frequencies and durations to the stroboscope [3].

2 Sound Effects

There are three distinct sound modulations which affect the human brain and need to be discussed here.

2.1 Binaural Beats

In 1839, Heinrich Wilhelm Dove, found that when tones are played in each ear but separated very slightly in frequency, a single tone is made apparent to the listener depending on the difference between the two [4]. For example, if a 300 Hz tone is played in one ear and a 310 Hz tone is played in the other through headphones, a "beat" frequency of 10 Hz is heard to be present. This tone became known as a binaural beat and is perceived by the listener as occurring totally naturally, as if without hearing the tones playing separately in each ear. For this effect to be present, the individual tones should be below 1000 Hz and the separation between the two individual tones played to each ear should be no greater than 30 Hz. If a wider band is applied than the tones become distinct and separate.

Binaural beats are formed internally by the neural output from the ears. Created inside the olivary body within the brain, it is an attempt to locate a sound source based on phase.

2.2 Monaural Beat

A monaural beat is present, for example, when two guitar strings are tuned closely but not exactly in frequency, or another example is when a binaural beat is played over loudspeakers rather than headphones and the direct separation is lost. The actual monaural beat is formed by the adding or subtracting of the two waveforms as they interact, effecting their amplitude, becoming louder and quieter in a continuing cycle. In both the binaural and monaural beats the lower tone is known as the carrier and the upper, the offset.

Interestingly, monaural beats are dependent on the amplitude of both sounds to be similar whereas the binaural tone is not, to the extent of one of the tones can be outside the hearing threshold entirely and it will still re-create the beat tone. Yet another aspect is in the use of introducing noise to such signals. In the case of the monaural beat, noise will degrade the sound, whereas in the case of the binaural beat it will become yet more prominent.

Both of these tones are rarely found in nature but are commonplace within the mechanical human world. An example of this is found where motors or engines are running at similar but slightly varying speeds and the vibrations meet in an area surrounding them.

2.3   Isochronic Tone

An isochronic tone can be defined as evenly spaced beats of a single tone which are repeated in rapid succession. They are sharp tones that quickly rise to full amplitude and fall away to nothing. This effect is again perceived most strongly in headphones.

All these tones can be embedded in music or left as they are for the listener, though this may be unpleasant, particularly in the case of isochronic tones [5].

3   Brain Entrainment with Sound and Light

The use of EEG (Electroencephalography), MRI (magnetic resonance imaging), MEG (magnetoencephalography) and NIRS (near-infrared spectroscopy) scanning devices have allowed the study of the human brain as never before and also in ways which allow for interactivity with the subject. They have allowed the mapping of human brain states which manifest in everyday life as moods or arousal of varying degrees [6].

Entrainment is an effect that has been noted whereby the human brain's prominent electrical wave frequency can be induced to "track" an external signal when the stimulus lays in the range of brainwave frequency. This frequency following response makes the brain move toward the stimulus frequency, alters the dominant waveform and therefore the overall state of the subject. Various effects are known to occur as part of the entrainment process, such as the excitement of the thalamus when exposed to the various beat tones. Other attributes of brain functionality seem to be affected by this, including spatial perception, stereo auditory recognition and also the activation of many sites in the brain.

For example, isochronic tones are widely regarded as one of the best tone-based methods for brain entrainment and elicit a strong cortical response [7]. An individual that does not respond particularly well to binaural tones often does better with the isochronic type.

The entrainment stimulus can be therefore light, sound, or a combination of both, as long as the frequencies used fall in the range that are present in the brain.

3.1 Other Methods

Besides light and sound and combinations of both there does exist other techniques for altering a subject's brain state, verifiable by scanning devices. Light and sound could be seen as objects in themselves in the meditation process. Other techniques involved silencing and witnessing (mindfulness). Another actual technique can be guided visualizations or a guided talk which directs the consciousness toward particular points in the body or even external to the body. Visualizations may involve focusing on a point or seeing something at that point such as a light. This may or may not synchronize with the breath.

A good example of spoken techniques and their usage is Headspace.com which offers guided voice for several areas toward a particular goal, for example help with:

- Stress and anxiety
- Focus
- Relationships
- Sleep
- Self esteem

Meditation, its techniques and states will be looked at in more detail.

4 Identified Brain States

Electronic scanning equipment has led to the mapping of the brain and its associated states for given activities and modes. EEG is usually non-invasive though in specific applications can be internally sited. EEG reading normally records the electrical activity along the scalp. These voltage fluctuations result from ionic current in the neurons of the brain. A recording measures the electrical activity over a given period of time via multiple electrodes placed over the surface of the scalp. Analysis and diagnosis relies on studying the spectral content of the output. Typically EEG will be used to diagnose epilepsy which causes abnormal readings but also can be used to find cases of sleep disorders, brain death, coma, encephalopathies. It can also be used as an initial diagnosis of tumours, focal brain disorders and stroke, although, this has become much less with development of other high resolution scanning methods such as MRI and CT. EEG can be limited in terms of spatial resolution and other methods are better at this aspect. However, for temporal analysis EEG can provide better data than methods that rely on heavier processing (and therefore slower).

An EEG can take electrode data from multiple points on the scalp to provide a good map of brain activity. Along with this data, sometimes other non-useful cerebral information is collected too, known as artifacts. An example of this may be the electrical activity associated with blinking and eye movements or other such ocular muscle activity. Other artifacts are also sometimes present such as cardiac, muscle activation and tongue related (glossokinetic). There are various algorithms which can attempt to clean these from the collected data.

Using the EEG (and with certain other scanning devices) it is possible to discern the most dominant waveform state in the brain (though of course other waveforms are present at different spatial locations). The useful cerebral signal falls between 1–20 Hz (though gamma can reach a lot higher) and divided up into waveform bands which signify particular overall brain states. These bandwidths are known as alpha, beta, theta, delta and gamma though may be further divided up if necessary into higher and lower versions of these components, if required.

At the lower end of the spectrum, with usually <4 Hz, are delta waves. These usually have high amplitude and show deep sleep in adults but are also present in babies. They can be related to particular pathologies such as lesions and hydrocephalus. These waveforms have been seen in some tasks involving continuous attention.

Theta waveforms appear at 4–7 Hz. These appear in young children and when adults and teens are drowsy or idling. Interestingly, they also appear when a person is trying to actively repress a response or action. The pathology that appears is similar to the delta waveform.

As the frequency increases, alpha type waves appear between 8–15 Hz. These relate to relaxed states where the person may be reflecting, for example. They are also more likely to appear with closed eyes. The amplitude of this waveform is seen to increase on the dominant side. Pathology can generally be consistent with a coma.

Mu waves, which show between 8–13 Hz, overlap with other bands. This band shows the synchronous firing of motor neurons in the rest state.

Between 16–31 Hz is the beta bandwidth which can show a range of mental activity from active calm to quite a busy state internally. However, the beta state is usually representative of active thinking, focus, highly alert or even anxious. These waveforms tend to be evenly spread, spatially but with a low amplitude waveform. The pathology here can relate to anxiety.

The gamma band is >32 Hz and shows when cross-modal activity is taking place, such as the processing of multiple sense information. It also takes place where short term memory matching is occurring of sounds, objects or tactile sensations. A decrease in gamma band activity, when combined with a similar diminution in the theta, can relate to general cognitive decline.

4.1 Targeting Meditation States

The project here focuses on the induction into what could be characterized as meditative states.

Meditation in the Eastern context is seen as a tool for spiritual development where the aim is to reach transcendental states and/or the cultivation of inner peace, focus and positive emotions. This is also seen as a way to reduce stress, agitation and negative emotions. A Western model of meditation describes a self-regulatory exercise which is used to focus the attention and through this reach a deeply restful but fully alert state. These definitions, together, actually form a mid-ground describing a model which encompasses a definition of meditation that focuses on maintaining the attention and awareness with a goal to create a better sense of well-being and serenity.

4.1.1 Positive Effects

Meditation has been studied for some time and in-depth due to its proponents' claims that it has positive effects on the human condition, general well-being and spiritual aspirations. This has been largely backed up by clinical trials and studies. There are many cases where symptoms have been improved in various disorders such as depression and anxiety [8]. Other successful areas effected include eating disorders, addictions and disorders caused by psychoactive drugs [9], stress and anxiety disorders, blood pressure and cholesterol level normalization [10]; lowering the intensity of emotional arousal [11]; enhancement of positive affect and resilience to negative affect [12]; executive functions enhancement such as attention; working memory; verbal fluency; cognitive flexibility; introspection; and better perceptual discrimination; increase compassionate behavior; improved immune function; enhanced functional and structural neuroplasticity [13] and longer life spans.

Unfortunately, there are a lot of empirical studies where there is a lack of sound methodologies applied—many stretching back to the 1970s. Some indicate biases and a lack of good control groups used. These biases, for example, may lean toward a particular technique or emphasis. Participants may self-elect to take part in a trial (such as experienced meditators) with only demographical matched controls, making the conclusions difficult to draw. Other problems have included overlapping of samples of meditation practitioners.

Despite these flaws in the methodology of some studies, there is still a large amount of evidence to suggest meditations positive effects and because of this, it is a widely used exercise which acts on the mind and body. However, to report only this is largely a one-sided view and there are studies which reveal contra-indications and negative effects to its practice.

4.1.2 Negative Effects

There are a number of both short term and long term negative effects that have been found related to meditation practice. These include, panic attacks and anxiety [14]; depersonalization syndrome [15]; high blood pressure; over-excitation of the nervous system; brain epileptization [16]. A study of twenty-seven long-term meditators [17] found 62.9 % of the group reported at least one adverse effect while 7.4 % suffered more profound negative effects. In these cases the adverse effects included relaxation-induced anxiety and panic, increases in tension, less motivation in life, depression, increased negativity, boredom, pain, impaired reality testing, confusion and disorientation, feelings of being in a "spaced out" state.

Some meditation practices include alterations in breathing and hyperventilation. These have been found to reduce cerebral blood flow by 50 %, due to central neurogenic response to hypocapnia meditated by the brain stem [18]. This leads to several negative effects in itself: ischemia; association with inducing seizures [19]; increase in cortisol and human growth hormone, which relate to depression and decreased attention and cognitive function.

Studies have also shown the occurrence of acute psychotic illness appearing in some beginner practitioners [20]. Intense practices within Yoga and meditation have given rise to the emergence of a set of symptoms labelled "Physio-Kundalini Syndrome". An individual may experience significant disruption due to the appearance of unusual phenomena labelled by some as "spiritual emergency" but could precipitate forms of mental disorders or bring forward underlying psychiatric conditions. This is recognized in the Manual of Mental Disorders (DSM) as "Qi-gong Psychotic Reaction" and is described as "an acute, time limited episode characterized by dissociative, paranoid or other non-psychotic symptoms ... Especially vulnerable are individuals who become involved in practice" [21].

Many aspects of an individual, such as their introversion versus extroversion and anxiety levels change the possible outcome of meditational practices [22] and therefore may change any positive effects into negative. It is possible, for example, that an individual suffering from depression or past trauma, who is hoping to find some relief in meditation practices, or training, may actually find they become more anxious rather than less, no matter how much they follow the instructions to concentrate on the present moment.

Positive effects of meditation can also become negative if an individual has particular constitutional neurophysiological characteristics. An example of this is the ability of long term meditators to develop high pain tolerance and circumstances where the normal reaction would be adverse, such as painful stimuli to be tolerated. This feature which is indirectly trained through the course of the meditation practice [23] could actually be undesirable in particular circumstances where a person may be instinctually motivated to pull away. This insensitivity may cause injuries or further damage to an area where it could be prevented, if felt properly. Other trained aspects in long term meditators include introspection, observing thoughts and rising emotions, as well as other objects in the world in a detached, dispassionate manner. If overexpressed in predisposed individual, this could lead to problems, leading to a psychotic world-view similar to those found in schizophrenics. This kind of psychosis is characterized by a fundamental disintegration in conscious experience and fragmentation of the self, with the inability to integrate perceptual information and thoughts [24].

The orbitofrontal cortex (OFC) is quite often enhanced in long term meditators [25]. The OFC plays a part in emotional regulation, particularly down-regulating and re-appraising negative emotional states. If overexpressed it could lead to types of mania or borderline personality disorder, leading in turn to an inability to feel empathy or compassion.

The hippocampus in meditators has been shown to have enhanced activity or larger volume [26]. This enhancement may lead to the reactivation of memories or exaggerated self-esteem. In particular individuals with particular neuropsychological characteristics could lead to serious problems involving obsessive and intrusive thoughts, dissociation and flashbacks, all relating to this type of enhancement in the hippocampus.

The default mode network (DMN) is the part of the brain that remains active when we let the attention wander. This "inattention" and the DMN's subsequent activation will come about when not attending to physical activity or engaging with the external environment, or carrying on a conversation. The DMN is active in states of daydreaming, contemplating the future, reliving the past or general mental rumination. A well balanced DMN will help the individual plan tasks, review past actions to improve future behavior and remember life detail appropriate for the individual. Some studies have revealed that long-term meditators have reduced the activity and diminished functional connectivity within the default mode network regions of the brain [27]. It is known that the DMN is linked to a sense of self and therefore this could be seen in meditators that there would be a weakening of this function. This may actually benefit some in particular traditions as the self is seen as something which should be lost but there is a risk of depersonalization and loss of personality in some where there is a neuropsychological predisposition.

It is noticeable that a lot of the negative effects show properties of the view of certain meditational practices taken to extreme—for example the diminishing of the self and the muting or dampening of particular functionality in the DMN. These seem to show how the meditation is targeting specific areas in line with the view of the technique and as we shall see in later sections of this work the main basis of meditation techniques which are in common with each other.

It is important to balance the various arguments and current studies in this area. There are large number of studies which indicate positive aspects to meditation, especially in the long-term. Though in some individuals with specific neuropsychological types there may be undesirable effects.

4.1.3 Meditation Practices and Associated States

A practice which involves the deliberate altering of the mental state with its own technology is meditation. This practice attempts to self-regulate the body and mind, affecting mental activity by utilizing specific attention mechanisms and techniques. The techniques themselves are similar to those used to induce relaxation or altered states such as trance, progressive relaxation or hypnosis. The main technique involved is the regulation of attention which is common to many methods though most meditative styles can be classified into two types, mindfulness and concentrative, depending on how this attention is focused. Mindfulness cultivates a watchfulness, or witness and usually has no object of focus and concentrative will use an object of some kind. Most meditative techniques, or styles, lie somewhere between these two.

Mindfulness will allow thoughts, sensations or feelings to arise without attaching to them or forming a judgmental response or analysis. The field of attention is kept regardless of any incursions by phenomena. Many styles lie in this area such as Vipassana, Zen and Western adaptations of mindfulness practices. Mindfulness encourages an open perceptivity and a meta-awareness of mental content.

Concentrative practices focus on some specific object, mental or sensory activity. This could be a repeated sound, a visualization or body sensation, such as the rising and falling of the breath. The Buddhist Samatha techniques, for example, use the breath as a focus. Transcendental Meditation lies in the concentration style but has some attributes belonging to mindfulness. It has the repetition of a mantra but places emphasis on lack of concentration and the development of witnessing thought-free awareness. The sound of the mantra eventually is supposed to occupy the awareness without the need for concentration to be applied. TM is therefore an example of a technique which crosses over between the two types though the goal is largely the same. Concentrative techniques do have aspects of mindfulness in that any mental phenomena that do arise are let dissipate and the object returned to, thus entering a witness state similar to mindfulness.

An example of the concentrative practices is known as Kasina (Pali; Sanskrit: kṛtsna), where a disk of a particular colour, dimensions and composition is focused on relating to a particular element or aspect. For example, the earth disk is a red-brown color formed by spreading earth or clay (or another medium producing similar colour and texture) on a screen of canvas or another backing material. These disks are concentrated on until a strong image is created in the mind. Later in the Kasina practice, only mental objects are used, taking the place of the physical disk. The idea of Kasina practice is to settle the mind and form a strong basis for training in meditation.

Practitioners of meditation have been shown to develop particular traits. These tend to vary over the length of practice. A deep sense of calm is realizable, there is a cessation of internal dialogue and development of the ability to merge the awareness with the object of meditation. One of the common experience of meditators is the shift in relationship between thoughts and feelings—there is a metacognitive distance created between arising mental phenomena and the occupying of the full attention. Long term meditators experience a deepening of the main characteristic traits: a deepened sense of calmness, heightened awareness of the sensory field and a greater depending of the shift between thoughts, feelings and alterations to the sense of self. The transcendental state of witness/observer is maintained for longer and the separation between observed and observer is perceived as growing fainter.

Various studies have been carried out to analyze the exact response by the body to particular types of meditation. The results have varied from the initial theory of relaxation response through to arousal and then more recently back to relaxation but this varies depending on the form of meditation utilized.

These techniques, above, fall into the categories defined by Lutz [28], that is, focused attention (FA), which keeps attention focused on an object and open monitoring (OM) which keeps attention involved in the monitoring process itself.

4.2   EEG and Meditation Techniques

As this project and its biofeedback technique are linked to the signature of the EEG and its programmed response within a virtual world, it is wise to look at how these appear within various techniques. Note that it is possible to break down EEG readings into separate areas: the power of the waveform (amplitude), the frequency and the coherence. EEG coherence here can be defined as the wave train becoming more rhythmic and orderly—it falls into phase and moves synchronously over large areas in the frontal regions of the brain, extending eventually toward the posterior regions. Table 1 gives a summary of states identified in normal life and below, more detail of relevant meditative states are outlined.

Table 1 Identifying predominant states and mental content

Frequency range (Hz)	Name	Related attributes and states
>40	Gamma waves	Higher mental activity, perception, problem solving, fear, and consciousness. Appears in specific meditative states, relating to Buddhist compassion meditations in the Tibetan tradition
13–39	Beta waves	The most usual state for normal everyday consciousness. Active, busy or even anxious thinking. Also appears in active concentration, arousal, cognition, and or paranoia
7–13	Alpha waves	Relaxed wakefulness, pre-sleep and pre-wake drowsiness, REM sleep, dreams and creative thought or free association. Considered as the brainwaves of meditation. These waves also appear in the relaxation process before sleep
8–12	Mu waves	Sensorimotor rhythm, Mu rhythm
4–7	Theta waves	Appears in deep meditation/relaxation, NREM sleep. Also, in hypnotic states or where some element of consciousness. A theta prominent individual may be awake but lose their sense of bodily location, for example
<4	Delta waves	Deep dreamless sleep with loss of body awareness. Does appear in the EEG of very experienced practitioners of meditation and would appear to relate to some ecstatic states. Maintaining consciousness while delta present is difficult

4.2.1 Alpha—Theta

The appearance of alpha waveforms has been seen in various types of meditation. Zazen, or Zen meditation is a kind of technique used in Zen Buddhism. A study [29] had 48 priests and disciples of Zen sects of Buddhism as subjects and their EEGs were continuously recorded before, during and after the practice itself. The results showed various stages in the technique very distinctly. In stage I, the beginning, within 50 s alpha waves appeared, notably this is whether the eyes were open or not. In stage II their amplitudes increased steadily. As the meditation technique increased over time the alpha wave train would dissipate, for stage III. Finally, at stage IV there was an appearance of a theta wave train.

These stages were identified and evaluated within the disciples' technique by a Zen master.

Yet another study [30], looked at various types of meditation—Yogic, Transcendental and Zazen. EEG was recorded in 30 normal healthy individuals practicing meditation, compared to 10 normal healthy controls not practicing. Here, alpha wave activity was prominent in meditators as compared to the controls. Meditators were said to produce strong alpha waves higher in persons performing meditation with good coherence which suggested "good homogeneity, uniformity and increased orderliness of the brain". There were 15 males practicing TM and 15 males practicing yogic meditation with 10 cases of similar age and sex who were not meditating. All meditators had been in their practice for at least 3 months. EEGs were taken before, during and after the meditation session. In the control group, EEG was taken before and during eyes closed relaxation and after relaxation. EEG was analyzed for alpha frequency, alpha voltage, alpha percentage, alpha coherence and hemisphere symmetry in both control and meditative subjects.

The study revealed no theta waves were evident in any group. Meditators showed persistence of alpha waves after eyes were opened in 28.9 % of cases, as compared to 12.4 % in control cases. In meditators good coherence was recorded during meditation. The interhemispheric time difference in alpha frequency was 12.79 ± 8.34 ms, which becomes 8.75 ± 5.65, indicating good coherence. A good hemispherical symmetry was suggested by the right and left voltage ratio, which was 0.844 before meditation and 0.876 during meditation. Cardio-respiratory readings showed that the pulse and heart rate slowed down significantly in meditator groups. Respiratory rate also become less in the group practicing meditation techniques. Blood pressure was seen to show a fall of the systolic blood pressure only, where the diastolic blood pressure remained the same. ECG did not show any abnormal signals being recorded. The alpha waves recorded here were suggestive of an increased relaxed state of mind. The alpha voltage is inversely related to mental activity; an increase in voltage is accompanied by the decrease in the frequency which occurs due to lessening in brain activity. This could be due to decreased energy metabolism of the brain [31].

An EEG pattern which characterizes that of sleep was not seen in this study. The sleep EEG is evident when there is a high voltage, slow wave pattern 12–14 Hz, sleep spindles and low voltage mixed frequency, along with rapid eye movements were not seen during the meditation. Alpha waves became present and dominating over 5–10 min of meditation. Alpha wave activity seemed to dominate where there is activation of diaphragmatic breathing, rather than thoracic breathing. A significant number of meditation techniques use breath mechanisms or breath observation as their object of awareness.

Meditation would seem to lead to the development of right hemisphere abilities [32]. Several researchers found that EEG and theta wave coherence is prominent in the right hemisphere during meditation.

The persistence of alpha waves after eye opening is an aspect which is found in the studies mentioned here and elsewhere. This follows the experience that meditators find the calm and alert state of mind continues after the actual practice.

It is interesting to note that in studies a recurring observation of results seems to take place that in deep meditation states coherence in both hemispheres of the brain appears. These two hemispheres largely split their functionality for example, speech, logical thinking, analytical thought and sense of time are thought to be resident, in the main part, in the left hemisphere. The right hemisphere is oriented toward the ability to recognize faces, comprehend maps and intuitive thinking. The right side is also focused on motor skills and spatial awareness. Meditation also seems to lead to the development of right hemisphere abilities as well as this overall synchronizing of waveforms. Some researchers [30] have found that the synchronization of both frequency and amplitude in electrical activity in all areas of the brain, this has been called "Hypersynchrony", which has been postulated as having some connection with the feeling of pure awareness or consciousness.

Alpha-theta waves have also been shown to present in Kriya yoga with some interesting research findings [33]. Following this particular meditation type, in this study, a significant rise in alpha and theta rhythms was shown in ten out of eleven subjects. Alpha waves were seen to double in certain individuals. Kriya yoga consists of a number of levels of pranayama (breathing techniques), mantra (repetition of phrases or words) and mudra (gestures).

4.2.2   Beta—Gamma

Gamma waveforms and particular types of meditation techniques in studies have been explored. For example, it appears that particular types of Tibetan meditation give rise to gamma wave trains occurring. One 2004 study [34] took eight long term meditators and placed electrodes over the head, monitoring the patterns of electrical activity as they meditated. These results were then compared by researchers to novice meditators who had been given some training and asked to meditate for an hour a day for a week preparation to the experiment itself. In a normal meditative state the two groups performed similarly, that is, the electrical activity produced and rendered by EEG was similar. However, when techniques were introduced that focused on objective compassion by the Tibetan monks the brain activity altered to show a rhythmic coherence, suggesting the neuronal structures in the brain were firing in a synchronized manner. This activity was occurring in the gamma band between 25–40 Hz. Apart from those seen in seizures, these gamma wave oscillations were the largest seen in humans. The novice meditators showed very little gamma wave production though with further experience and training the rhythmic signals from them appeared to strengthen, which possibly shows that the gamma band rhythm is trainable in itself. There would seem to be some interesting ideas that show a relationship between the tightly focused and synchronized gamma wave train and the heightened sense of consciousness, bliss and intellectual acuity subsequent to meditation.

However, some neuroscientists have disputed this gamma wave argument on the basis of two paths of reasoning; that there is a possibility of mismeasurement, for example, that the EEG could be an artifact of electromyographic activity to other artifacts such as minute eye movements. Both of these have largely been answered with careful signal separation and the use of magnetoencephalography (MEG). MEG does not have the same problem with discerning between artifacts that are associated with EEG and has identified gamma activity associated with sensory processing, mainly in the visual cortex.

4.2.3   Delta

Delta waves are a high amplitude brain wave between 0–4 Hz. They are usually associated with particular stages of sleep and have been purported to aid in the formation of declarative and explicit memory formation.

A particular type of Yogic meditation know as Yoga Nidra (literally, sleep yoga) leads to a consciousness being present while in delta-sleep. This technique leads the body to actually enter the sleep state; a series of observations are made initially by the practitioner, accustoming them to the locale they are actually in as a means of making the mind placid. A "body scan" then takes place to simply relax. The next stage is making a positive resolve or statement, known as Sankalpa in Sanskrit. A stage is then entered which is called the "rotation of consciousness". Here there is a directing of the mind to parts of the body in sequence. This sequence is usually spoken out loud via a recording or teacher. The practice continues with awareness of the body and its contact with the floor, awareness of subtle body movements, awareness of opposing sensations, a focusing on "inner space", visualizations, and movement in time (a review of the events of the day). Finally, there is the repeating of the resolve, stated earlier and a completion section, where awareness is brought back to the body, the room and external surroundings. Various studies have shown how this method leads the practitioner from normal beta activity, through the relaxation stage where alpha predominates. As the body-mind relaxes further but is still occupied in a meditative state, theta appears and finally, the individual may exhibit delta. A particular study showed that after 30 sessions of Yoga Nidra, alpha activity could be made stronger, leading eventually to introduction of theta intermixed with alpha—the signature of a deep state of relaxation.

5   Sound and Light Machines

Sound and light machines (also known as mind machines) use both visual and auditory mechanisms to deepen the overall impact of the immersive experience. This technology was developed from experiments in psychology and consciousness research and what was initially bulky machines eventually became small handheld units in the 1970s onwards. These use synchronized strobing of light and sound at known frequencies causing the frequency following response to occur and "lock on" over periods of time. The frequencies themselves can change over time visiting various states.

A sound and light machine, as available commercially, is composed of goggles or eyeglasses which are opaque, and headphones. The eye piece usually has a set of LED in front of the eyes which are controlled by the core unit, containing the processor and programs. This LED array can be single color or multiple depending on the sophistication of the model but all tend to be able to vary the intensity. More expensive sound and light machines tend to also have brighter output. Multiple colour LEDs give a range of shades and depth. It must be remembered that the eye set is worn with eyes closed and therefore give a diffuse effect on the back of the lids that you are watching. The general effect as the program runs is of mandalas and geometric shapes, spires and cracked landscapes, mesmerizing and surprisingly detailed, considering the eyes are closed.

Commercial units, such as the more basic Procyon [35], are capable of producing programs which target specific states in the progress of one session. This device produces sequences of variable light and sound pulses to specific programs which can be built-in or user-defined. The sound and light characteristics and duration are variable and programmable. It is possible to create washed out fields over the user's field of vision (this is known as a true Ganzfeld effect [36, 37], from the German meaning "complete field"), through to shimmering cascades, all with an accompanying audio sound track of pulses.

Similar to other machines, a complete program may have an overall aim such as meditation, learning stimulation or sports focus but the actual program may travel through several frequencies for varying durations to get there.

Kasina (which is actually named after the meditation technique which concentrates on an external object) is a more recent mind machine from Mindplace which has even more expanded capability [38]. This unit incorporates a more sophisticated sound synthesis engine for playing the audio pulse tones as well as the ability to play MP3 data. This allows for the playing of samples or environmental sound tracks which may have embedded frequency encoding.

A technique used by the Kasina, as well as the more rudimentary sound and light frequency pulsing, is that of music modulation. It is possible to modulate any music in such a way that a binaural or monaural frequency following response is initiated. For example, it is possible to modulate only the bass track of a piece of music using low pass filtering to select that particular band and thus avoid distorting other components that lie within the soundscape. A binaural beat could then be integrated within the frequencies present, using modulation techniques. The Kasina contains many such tracks which are musical, for example, shamanic drum patterns and Tibetan bowls, or environmental sounds such as rainforests or seashores.

Figures 1 and 2 show the output from a program for the Kasina which attempts to induce a "meditative state", that is, the brainwave "signature" exhibit properties that look like those in an adept meditator. This particular program has an initial burst to stimulate, followed by the waveform dropping in frequency steadily until the lowest

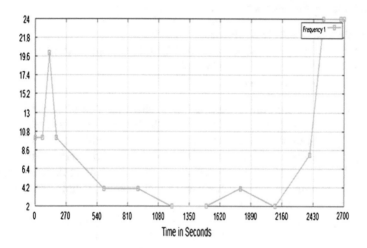

Fig. 1 Showing light frequency for a program in Kasina

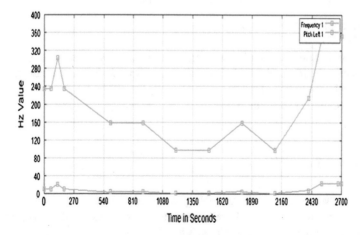

Fig. 2 Showing light frequency in relation to audio for a program in Kasina

point is reached of just 2 Hz, which is actually classed as a delta waveform state. The waveform then steadily climbs again. This entire programmed session lasts over a period of 45 min.

6 General Description of the Project

The general idea for this experimental project is the development of an environment in a 3D virtual world which reacts and adapts to a user's state and then purposely attempts to change the person's state using biofeedback mechanisms to some target EEG waveform.

There are several aspects to this:

- The collection of the EEG information
- The transferring of the data to the Virtual World in an appropriate format
- Process the stream in the Virtual World
- Adapt the virtual environment

This forms the basis of the biofeedback loop which cycles itself. The EEG headset takes the information from the scalp (actually the forehead, in this case), communicates with a program that streams this into the virtual world in a form that will be easy to process within it. This is essential as processing within the virtual world is slower and not necessarily as comprehensive in the programming language available. The virtual environment can then adapt toward the target or, simply reflect the user's state, which is a good first step.

One of the questions at this point may be: What exactly could be used within a virtual world to alter the user's brain state? For this, mind machines and the history of this field, offer plenty of scope for experimentation.

For example the following have been mentioned:

- Stroboscopic based elements
- Ganzfeld lighting effects
- Synchronized audio patterns (monaural, binaural and isochronic beats)
- Embedded audio patterns (the modulation of sound tracks or soundscapes not necessarily obvious to the user)
- Spoken or guided techniques

6.1 NeuroSky EEG

Faster processors and network technology have led to the fast real time capture and processing of brain information, efficiently and without the need for vast arrays of equipment, wiring and external computational power.

NeuroSky produce several types of EEG reader which require no wiring and are connected by fast Bluetooth interface. Mindwave Mobile [39] is a research grade EEG reader based on the TGAM (ThinkGear ASIC Module) bio-sensor chipset featuring Bluetooth interface and software to capture data. Various software is available to record or visualize the data coming from the headset. This unit measures the power output of the various bands such as alpha, beta, gamma etc. and pre-processes particular aspects of interest such as attention, meditation and eye blink. The device is made up of the headset itself, an ear clip and a sensor arm. The ear clip forms the reference and electrical ground and the EEG electrode is on a sensor armature which rests on the FP1 forehead point above the eye. The unit is powered by a single AAA battery, which lasts approximately 8 h.

The Bluetooth connection allows the EEG to be used with both desktop and mobile devices, such as phones and tablets. This flexibility has encouraged the development of software for both, with desktop software and phone apps available. These range from recording and visualization of brain EEG patterns, to some interesting BCI type control and biofeedback software. For example, one application plays movies and alters the storyline depending on the brainwave output. Another allows the user to attempt to enter a meditative state and responds to their progress by raising a ball on the screen. The height attained equaling the period spent in a characteristic meditative state. Yet another application similarly measures the attention span graphically and originally, by the screen object erupting into flame for the period that the concentration is kept focused. Software is also available on phone apps which will encourage the development of meditation using a diary, logging progress as the user visits day by day.

Other software allows the recording of EEG data into graphical format, rendering it suitable for data analysis. The data itself is collected from the FP1 point—the only electrode available on this commercial device which is somewhat limited but still useful. A good point is that data as mentioned below is to some extent pre-processed (so decomposing with FFT is largely unnecessary) and therefore persistent data which is present at the FP1 site, such as eye artifacts, can be ignored. Raw data is available too, if required.

Although there is a large range of commercial software available for the headset, it was necessary for this project to write a lower level interface to gain access to the raw and semi-processed data directly. The data stream leaves from the headset via Bluetooth and is picked up by the receiving device (in this case the desktop computer) by specific drivers and then processed by the custom software developed for this work. The data available from the headset came in the ranges (and sent as "sets") of low-alpha, high-alpha, low-beta, high-beta, low-gamma, mid-gamma, delta and theta. There are two other uniquely, pre-calculated eSense™ values named, as mentioned previously, meditation and attention or focus [40].

6.2   Virtual Worlds—Second Life and OpenSim

Both Second Life (SL) and OpenSim provide an immersive 3D world where users can create content [41]. Although it is similar to many 3D world games there is no goal, as such, or objectives. The idea being a user enjoys being there, creating content and is active in the virtual world. A viewer similar to a browser allows access to features and interaction with your avatar within this environment. The avatar acts as your character with which other users can be communicated with. This can be initiated through a built in messaging service, which can be localized to the virtual area you are in publically and also in private communications. Such virtual worlds have their own currencies and exchange rates with the outside world.

One of the main aspects of these worlds is the ability to program and create—very useful for prototyping and developing experiments. For example, it is possible to create a virtual vehicle such as a car, designing and building its appearance and then moving on to its actual functionality which is implemented by programming code stored in its constituent parts [42].

The appearance and building of items is vast in scope but has several initial basic components, for example, you can build with prims—a basic building block in various shapes. It is possible to produce gas and fluid too, for various effects. A physics environment is replicated.

The programming language within these virtual worlds is known as Linden Scripting Language (LSL) and is a fully functional programming language somewhat similar to C, Java and the JavaScript family of languages. This contains many library functions for altering and manipulating the virtual environment, controlled by the scope of various ownership permissions. For example, the above mentioned car, if built by myself, would belong to myself and unless altered, its attributes and permissions would only be modifiable by myself.

The library of functions cover all the normal aspects of programming such as string manipulation, security mechanisms, input and output as well more particular functions for the virtual world itself.

Using LSL it is possible to control all aspects of virtual object, it can, for example be made to spin on its axis or change colour or transparency or objects to create simple sounds using samples. It is also possible to reach outside of the virtual world and link up with servers and computers running other computer code, which can in turn provide or process data. So, it would be possible, for example, to have an object function as a music player with the music stored on an external server, which is streamed in when a play button in the virtual world is pressed.

This aspect of having communication with the external world allows for some interesting developments between the two.

A surface of a prim can also be made into a "media surface" which allows streamed video to be presented, which at its most simple could show a movie or video clip from the outside world. A media surface can also be linked to a website and therefore becoming an extension to YouTube or any other website. This leads to a great deal of versatility when programming the virtual world.

6.3 Advancements in Virtual Worlds

New versions of SL will eventually be available, offering greater resolution and higher frame rates. They will also be calibrated for more immersion using virtual reality headsets.

The first that should be mentioned is High Fidelity, a spin off from SL and by many of the same people, its aims lay in advancements in technology such as sensor feedback and VR. More interestingly is the fact it is open source coding [43].

The second ongoing virtual world is Project Sansar (meaning Universe in Hindi) [44], with yet more involvement with SL, the key goals here are that it works with mobile technology and has advanced expressive avatars, along with other advancements in virtual technology. The model here would seem to be different in that it is aimed at becoming the Word Press of VR, with commerce using links into the world to provide their own particular virtual areas.

6.4 Virtual Reality, SL and OpenSim

It is possible, using special adaptions to the viewer or a separate viewer built for the purpose, to use SL and OpenSim within virtual reality headsets such as Oculus Rift and Google Cardboard.

Google Cardboard is an affordable and accessible way of putting together a 3D headset. An android or Apple iPhone is used as the viewer is held at some distance from the eyes by a homemade or bought headset. The phone is useful in that it contains all of the technology required: the display, sound reproduction, tilt sensors, accelerometer, camera (if augmented reality is desired) and location sensors. Once the headset structure is acquired or made, apps exist to test the new system as a standalone device. The display provides a stereoscopic view into the 3D environment provided and the various senses detect motion and tilt, thus allowing interaction with the presentation.

To link the system up to an interactive virtual world several steps are required and software. A viewer is used with the ability to interact with a 3D headset, such as CtrlAltStudio. For Google Cardboard a server software is required on the computer to set up communication with the viewer, for example, Trinus VR and on the phone a Trinus VR client app is executed which establishes the link with the desktop. The actual connection link can be via Bluetooth, and therefore wireless, or, by USB tether.

Once the link is established, the information between the desktop and phone flows, providing 3D view to the user, along with sensor information back to the viewer app.

7   Initial Experimentation

A series of stroboscope objects were created for experimentation in SL, these ranged from simple flashing devices to entire rooms (which in itself is an object) where the textures of the prim can be controlled and changed at timed intervals. It was possible to build up complex strobe patterns by using multiple sources. It was also possible to using a media texture and stream in a stroboscope pattern from the outside world.

Similarly, fields or blankets of color which covered the field of vision were made possible along with parallax effects.

Objects were created which could generate audio frequency patterns at particular intervals and played simple synthesized audio, similar to the mind machines, as repeating samples. More complex audio could also be streamed in to receiver objects, similar to the video mentioned above, allowing for longer samples or complex synthesizer programming and modulation.

An audio track could be played from the outside world and modulated to affect the user as desired. This includes spoken word tracks which could be used for guided meditation techniques.

A problem with audio and visual elements may be the ability to synchronize them within an environment which relies on networks, servers and multitasked processing in scripts. It should be possible to have locking mechanisms and keep heavy processing in the outside world where necessary. A good example of where this can be done is where an EEG signal has come from the headset, it can be processed in the outside world and the basic elements abstracted from the data about the current waveform prior to being sent on. This keeps the stream to the minimum and also the processing in-world.

Media objects in-world can be a part of the VECSED merged reality system [45] which will be used for this project (described below) and therefore have easy access to streams and external world devices.

In all likelihood, the adaptive and programmable environment created will have a mix of these elements.

As a starting point the virtual environment which the user is situated in-world can mirror the user's current state and progress toward some pre-defined state. In this case the adaptation takes place where the colors, lights and sound in the "pod" will match this state, initially.

Meditation techniques essentially bring about their goal by two broad techniques, either based on concentration on an object, or through cultivating the witness or mindfulness state. A virtual object can be used as a concentration tool (this is similar to software available) and with this, mind machine techniques can be employed.

7.1   VECSED—Virtual Environment for Control, Simulation and Electronic Deployment

Virtual Reality systems offer excellent capabilities in terms of simulation but usually less so in terms of being able to control, access and communicate with the outside world. This project requires advanced control and communication capabilities to support the various functionalities.

This system acts as a bridge which sits between the real world and the virtual, allowing the coordination of data streams and allowing for simple or complex communication arrangements. A diagram showing the basic arrangement of the system is shown in Fig. 3.

The VECSED server/database sits between the two worlds, providing a look-up service where resources can be allocated on the basis of requirements. The main core of the VECSED server is written in PHP with a MySQL managed database. Each device (an agent) which registers itself on the system enters various information to the database on the VECSED server such as its network properties, I.P. address, location information and other such details:

- Name of the agent
- IP address and port for communication
- Type of resource
- Hostname, if relevant
- Last time communicated with (Timestamp)
- Geospatial information such as longitude and latitude

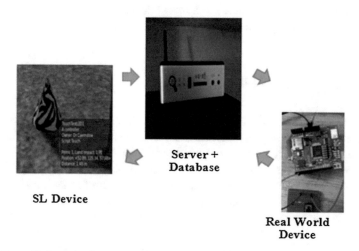

SL Device

Server + Database

Real World Device

Fig. 3 Flow of information between worlds

Once registered the device must then use secure authentication to initiate communication. When a device boots (or in the case of a virtual device, instantiates) itself, a security procedure authorizes its presence on the merged reality system.

A registered device (such as a virtual constructed object, with programmed functionality) within the virtual world can communicate with VECSED, ask for a particular real world device to link to, or any such type of device. VECSED will then respond and allow a communication link to be set up between the two. In SL any constructed object with its scripts can have an assigned URL for communication (using llRequestURL()), this is passed over to the VECSED server and logged.

A test example was developed in the form of a virtual toggle switch that controls a real world LED light on an Arduino board. Issuing the command:

ArduinoUno LED9 on now

Makes the virtual controller send the activate command via LSL and the HTTP request mechanism (llHTTPRequest()) through to the server page. The information reaches the server as HTTP POST data where it is analyzed. Any incoming commands are security checked that the source is registered first before parsing and then logging state changes etc. with the database. Finally, the information is passed on to the target device.

In this particular example, a TCP/IP socket was opened to the device and the appropriate message built from the parsed command. A C++ program was running on the WIFI equipped Arduino which listened to the port and executed commands that arrive there after looking them up. Here, a LED was activated on the board.

Similarly, a device in the real world can register itself with the VECSED server in exactly the same way, sharing its details such as resources and location. It could then establish a link with a virtual object and send data. An example in this case could be a video camera which is linked to a media device, which in turn could have its control functionality replicated within the virtual world.

7.2   EEG and VECSED Combined

An appropriate driver mechanism was written in Java which linked into the stream coming from the headset. This was necessary as the software available did not allow the degree of manipulation and analysis required for this project. The headset can provide both a raw data stream and a processed stream with strength of the waveforms in particular bands which eliminates the need for fast fourier transform (FFT) analysis. The protocol for the stream packet is detailed in Table 2.

The main task of the EEG reader was to interpret the data by parsing the data stream containing PoorSignal, EEG Raw Value, EEG Raw Value Volts, Attention Level, Meditation Level, Blink Strength, Delta (1–3 Hz), Theta (4–7 Hz), Alpha Low (8–9 Hz), Alpha High (10–12 Hz), Beta Low (13–17 Hz), Beta High (18–30 Hz), Gamma Low (31–40 Hz), and Gamma Mid (41–50 Hz). The data stream itself is composed of packets, an initial code identifying the following information in the row itself.

Table 2 Data stream protocol from the NeuroSky Mindwave Mobile

Code	Byte length	Data value meaning
0x02	1	POOR_SIGNAL quality (0–255)
0x03	1	HEART_RATE (0–255)
0x04	1	ATTENTION eSense (0–100)
0x05	1	MEDITATION eSense (0–100)
0x06	1	8BIT_RAW wave value (0–255)
0x07	1	RAW_MARKER section start
0x80	2	RAW wave value
0x81	32	EEG_POWER: eight big-endian 4 byte floating point values representing the bands (delta, theta, low-alpha etc.)
0x83	24	ASIC_EEG_POWER: eight big endian 3-byte unsigned integer values representing delta, theta etc.
0x86	2	RRINTERVAL: two byte big endian unsigned integer representing the milliseconds between two R-peaks
0x55	–	Not used (reserved for [EXCODE])
0xAA	–	Not used (reserved for [SYNC])

A parser was written in Java that extracts the information that is used in the project and passed along to any subscriber listeners to this device.

This piece of software is packaged as an agent which links into the VECSED system. On start-up it initializes communication with the EEG hardware and the VECSED server supplying identifying information and necessary handshake. It can then be communicated with and the stream read by any device in the real or virtual world which joins the VECSED network.

An experiment was performed to check the stream and its software agent could be communicated with inside the virtual world. A beacon was created in the virtual world which responded with a change of color depending on the incoming stream. This worked well, changing very quickly in response to changes in the EEG.

7.3   Head Mounted Display Immersion—Oculus Rift and Google Cardboard

To increase the sense of immersion a VR headset (or HMD, head mounted display) was used. Many headsets are being developed but two of the most popular are Oculus Rift and the somewhat home-made Google Cardboard.

Both have some support in the virtual worlds mentioned here; viewers for SL and OpenSim incorporate capabilities for both, or can be adapted. In this project Google Cardboard was used.

7.3.1 Building the Complete Headset

The headset was composed of several parts:

- The VR visual renderer
- The EEG
- Headphones

It was possible, just to situate all these on to the head at the same time, thanks to the limited coverage on the head of each. For the Google Cardboard mechanism a bought plastic version was used, to hold the phone in place securely. The NeuroSky Mobile Wave was used as the EEG reader due to its compactness and wireless capabilities. Finally, a good set of high quality headphones complete the head gear.

7.3.2 Biofeedback—and Initial Tests

The list of our inducing agents is as follows: Stroboscopic based elements, Ganzfeld lighting effects, Synchronized audio patterns (monaural, binaural and isochronic beats), Embedded audio patterns (the modulation of sound tracks or soundscapes not necessarily obvious to the user) and spoken or guided techniques. These offer interesting ideas for objects that can be developed inside a virtual world.

A simple "virtual environment pod" was created which changed its color at given program rates. When inside the pod and with the VR headset active the effect was similar to the sound and light machines and rates of pulsing could be set, which would give a Ganzfeld effect.

The EEG stream receiver was connected to the pod via object messaging, available in LSL.

A biofeedback mechanism was added whereby the pulsing would lead the incoming EEG in the appropriate direction (toward a particular state) but only change if the EEG was following and correct itself as it went.

The Ganzfeld mode of the pod was relatively easy to incorporate as all that was required was a functional control over the speed, color and type of color change over time. The software for the pod allowed user defined programs to be stored, similar to the mind machines. In effect these were simply a data store of timing values, rate of flash and duration for specific sections.

The next step was addition of synchronized basic sound pulses to match the field color changes and strobes. This was done by adding another virtual object (yet another agent in the VECSED system) which played in sync, appropriate sound pulses. It should be noted here that it may have been possible to incorporate the sound player within the pod itself but for the sake of modularity it was thought to split this functionality off to an entirely separate object. To create a sync which was in time, the pod and sound object did handshake by sending messages at intervals.

To extend this, a stream receiver was added which accepted an incoming audio stream. The idea here was to create an auditory soundscape which could be

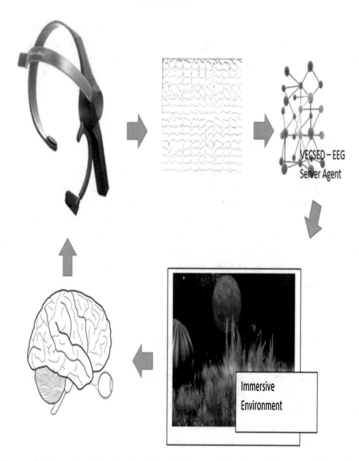

Fig. 4 System diagram showing feedback loop between adaptive environment and user

"environmental" so, rainforest or rain etc. This agent linked into VECSED and an audio stream player was linked in to cycle a long sample of sound.

The next stimulus device to be developed was a more "directed" strobe mechanism—a moving strobe that can be split to present a synchronized lightshow. Figure 4 shows the biofeedback loop occurring which encapsulates the system.

8 Final Developments of the Pod

The pod has several ways of being operated. In the first mode it can simply track the users EEG with correspondences made between particular states and colors. In the second mode, a particular program is run which will attempt to induce the user toward a specific state. Here, the concept of the "scene" was developed further as a programming tool. A scene is a specific set of visual and audio cues which induce, or at least take the user toward a particular brain state. An example may be a scene which aims at concentration and involves the user stepping on counted bricks in a sequence.

8.1 Programming Language

A programming language was developed for the objects which function within the virtual world. This programming language allowed the sequencing of events and loading of particular colors and attributes, textures and media streams within the pod. At its most simple a program could be made to load a particular scene and cycle through various control options until a specific EEG state is reached. This could be simple as activating a 3D space Ganzfeld effect, pulsing until alpha wave is predominant and then either holding for a period or changing to a new pattern. This allows for an adaptive environment.

8.2 Initial Tests with the Full System

Once the full system was operating, more meaningful tests were created toward the primary goal—exploration of the meditative state using virtual reality. A ball was created from the particle type within SL, this expanded in size as one of the EEG attributes increased using red for attention and blue for meditation.

Yet another experiment involved the creation of two 3D animations, a complex "mechanical" set of objects which moved as a progression, the other, being more liquid and surreal. These could be interacted with using the avatar, or simply viewed. Again the ball was used to indicate attention and meditation states and it could be seen to be dominated by the attention waveform for the mechanical progression or more meditative for the surreal experiment.

8.3 Reflecting the User

The developed, adaptive pod, with corresponding repertoire for all the states possible was set up initially to reflect the user's state. When initialized into SL the pod forms itself and all tools that it requires to function, such as the virtual process devices which communicate via VECSED. A process known as the palette is initialized which can change the color of the environment, media output to walls, audio and other necessary functionality.

With the avatar standing in the pod and the user in the real world with all the headset items (EEG, headphones and display goggles). The EEG signal is fed into the VECSED system and processed. The pod's processes access the palette and depending on the program running selects an environment which reflects the user's incoming data-stream relating to the brain-wave pattern. The program could reflect this in:

- Environment (the pod) colors
- Textures within the environment
- Images and video
- Audio

8.4 Inducing States

With the reflection mechanism working, further experiments were necessary to actually manipulate the user toward specific states.

The reflection exercises utilized palettes which activated when a state was reached; here they are used within scenes to reach particular goal or way-point. As well as these palettes, more complex objects were developed which were task oriented:

- Meditation object (abstract, soothing, tending toward "hypnotic")
- Concentration object (tasks involving focus such as stepping, counting or basic visual analysis)

Both of these objects are instantiated within the virtual world on demand and a selection made to utilize any of their sub-tasks. The concentration object included stepping stones, target focus and calculation. The meditation object included elements which were more abstract in type, utilizing textures, sounds, video and images.

Refinements were made in the objects and programmed environment to integrate the known inducing agents mentioned so far:

- Ganzfeld lighting
- Stroboscopic elements
- Synchronized audio patterns (monaural, binaural and isochronic beats)
- Embedding of audio (and the various modulation possibilities)

These were relatively easy to implement within the virtual worlds of SL and OpenSim due to their completeness, flexibility and power of expression with the programming language (LSL). The virtual environmental pod itself was made so elements could be built-in such as the Ganzfeld capability. It could also strobe.

Sound in the virtual worlds used could be sample-based or incoming streams. An object was created which could handle either of these possibilities and therefore act as an audio interface. The capacity to play samples is limited to some extent due to the size, although for most purposes it is simple cycling of tones used in any case (for example, square wave audio trains). More complex sounds, such as environmental backgrounds or speech track can be streamed into the world. A server database of stored sound was connected through and could be called up as required. The server could store various assets such as video, sound or other media in the database which allows for a complex library. The environmental pod therefore could send a message to the interface object to call up an appropriate resource, when required, for rendering or playback.

Other media objects allowed their surfaces to present particular images or video, again streamed from the server or held locally.

With these capabilities, the virtual environmental pod becomes a fully functioning 3D light and sound synthesizer, capable of either reflecting the user's EEG or taking part in the biofeedback loop and reacting to it.

8.5   Programming for Scenes

More advanced techniques were required for scenes, where the user is taken on a journey through various brain states, watching the EEG stabilize in a target range before gently moving to the next way-point in the scenes mapping.

A meta-programming language was developed to control and re-create scenes. This language was parsed by the main system within the environmental pod which then sent messages out to the listening instantiated objects. A simple program might be the creation of a scene with appropriate palette and cycle until a given state is reached, an example of which is shown in Fig. 5.

Only a small sample of individuals were tested—the focus for this project was to develop the technical capability and provide a suitable environment for experimentation and further research. The results show that the system, as it is, can induce individuals fairly easily into meditative states (along with others, such as attention/focus). A noticeable aspect was not just the reaching of a state but the stabilization once achieved which was something the system was particularly good at compared to other means, such as mind machines. Figure 6 shows a typical induction session after a relatively short period of 3 min.

Mind machines, as sold commercially, do not react to the user they simply execute a light and sound show for the user, who may drift around the EEG spectrum somewhat. The virtual environment captures the individual and reacts, tailoring the output intelligently to the user's current state, which it can attempt to manipulate.

In terms of the practical usage of the system there were only a few technical problems. The NeuroSky EEG, for example, can send an invalid data stream at times. This was found to be due to lack of a complete contact with the ear of the user via the clip and therefore a lack of a reference earth from a "floating connection". A solution to this seemed to be to dampen the lobe a little to help the connection. Aside from this the eSense^{TM} values were very stable and the fact that a lot of the number crunching could be done on the ASIC chip helped matters. The values seemed to be very accurate at deducing the focused and meditative states, though it would be interesting to see the means by which they derive what constitutes a meditation

```
// instantiate scene "focus ball" and initiate animation/sound
// sequence
load scene "focus ball";
repeat
  // set initial ball state
  scene element["ball".message["blue"]];


  // wait until the incoming EEG stream predominantly what
  // is required
  wait until eegstream(WAVE) == BETA;


  // State achieved set ball to red
  scene element["ball"].message["red"];


  wait until eegstream(WAVE) != BETA;
until FINISH;
```

Fig. 5 Simple initial feedback loop with system

Fig. 6 Showing initial induction into "meditative" state after 3 min

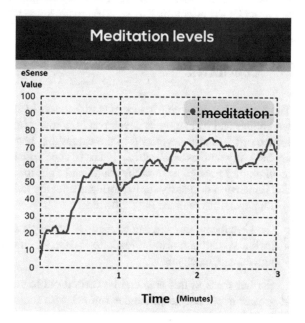

state (for as has been described, there are many types). This is not a real problem as other values can be used, as they were here, in the various bands to determine a particular state, rather than relying on the two predefined variables for attention and meditation.

9 Future Scope

The system presented here is a merged reality system operating in both the real world and the virtual. It is an intriguing area with many possibilities for development and experimentation. The biofeedback mechanism within the virtual is novel and stable, presenting good initial studies of how an intelligent system can be built in synergy with the human brain.

The use of Google Cardboard to add depth of immersion with the virtual world appears to deepen the experience of the user. Other headsets could be used such as Oculus Rift and the Microsoft HoloLens. More work could be done to see how performed tasks in the virtual environment by the user affects the EEG, while attempting meditative or concentrative tasks.

It may be possible to integrate Augmented Reality aspects into this system without too much alteration of the hardware and software set-up.

Other virtual environments, such as those already mentioned, High Fidelity and Project Sansar, should offer even greater frame rates and other interactivity to extend the experience. The system could easily be adapted to work in these.

These kind of systems could be used in the training of individuals, or for treatment of particular disorders, such as post-traumatic stress disorder [46].

10 Conclusion

The greater the immersion in virtual worlds the easier it seems to become to utilize the techniques and technologies developed here toward specific brain states. Hardware technology has also developed to the extent that headsets can be developed which are basically wireless and extremely fast to cope with the processing required. These technologies now go beyond the sound and light machines and create the isolation and immersion necessary for exploring artificially induced states.

This work has realized the following:

- The virtual environment can be made to be responsive to a user's brain states
- Such a system can also be made to induce specific target states with intelligent biofeedback mechanisms

Further work in this area can be carried out to study the effects on individuals at greater depth and technologies refined with the information gained, as data is collected.

References

1. Mayer, l., Bhikha, R.: The Historical Significance of Colour, Tibb Institute, online article http://www.tibb.co.za/articles/Part-2-Historical-significance-of-colour.pdf (2014). Accessed 20 Nov 2014
2. Budzynski, T.: The clinical guide to sound and light. In: Proceedings of Brainwave Entrainment Symposium (2006)
3. Hutchison, M.: Time flashes: a short history of sound and light technology, online article https://www.mindmods.com/resources/General-history.pdf. Accessed 4 Dec 2014
4. Filimon, R.C.: Beneficial subliminal music: binaural beats, hemi-sync and metamusic. In: Proceedings of Recent Advances in Acoustics and Music (2010)
5. Doherty, C.: A comparison of alpha brainwave entrainment, with and without musical accompaniment. BA Psychology Thesis, DBS School of Arts, Dublin (2014)
6. Cahn, B., Polich, J.: Meditation states and traits: EEG, ERP, and neuroimaging studies. Psychol. Bull. **132**(2), 180–211 (2006)
7. Huang, T.L., Charyton, C.: A comprehensive review of the psychological effects of brainwave entrainment, alternative therapies in health and medicine. Res. Libr. **14**(5), 38 (2008)
8. Farb, N.A.S., Anderson, A.K., Segal, Z.V.: The mindful brain and emotion regulation in mood disorders. Can. J. Psychiatry (2012)
9. Ospina, M.B., Bond, K., et al.: Meditation practices for health: state of the research. Evid. Report/Technol. Assess. J. (2007)
10. Anderson, J.W., Liu, C., Kryscio, R.J.: Blood pressure response to transcendental meditation: a meta-analysis. Am. J. Hypertension (2008)
11. Nielsen, L., Kaszniak, A.W.: Awareness of subtle emotional feelings: a comparison of long-term meditators and non-meditators. Emotion **6** (2006)
12. Chambers, R., Gullone, E., Allen, N.B.: Mindful emotion regulation: an integrative review. Clin. Psychol. Rev. (2009)
13. Davidson, R.J., Lutz, A.: Buddha's brain: neuroplasticity and meditation. IEEE Signal Process. Mag. **25** (2008)
14. Lazarus, A.A., Mayne, T.J.: Relaxation: some limitations, side effects and proposed solutions. Psychotherapy **27** (1990)
15. Lukoff, D.: From spiritual emergency to spiritual problem: the transpersonal roots of the new DSM-IV category. J. Hum. Psychol. **38** (1998)
16. Lansky, E.P., St Louis, E.K.: Transcendental Meditation: a double-edged sword in epilepsy? Epilepsy Behav. **9** (2006)
17. Shapiro, Jr., D.H.: Adverse effects of meditation a preliminary investigation of long-term meditators. Int. J. Psycho. Med. **39** (1992)
18. Patel, V.M., Maulsby, R.L.: How hyperventilation alters the electroencephalogram: a review of controversial viewpoints emphasizing neurophysiological mechanisms. J. Clin. Neurophysiol. **4** (1987)
19. Prevett, M.C., Duncan, J.S., et al.: Demonstration of thalamic activation during typical absence seizures using $H_2(15)O$ and PET. Neurology **45** (1995)
20. Bharadwaj, B.: Proof-of-concept studies in yoga and mental health. Int. J. Yoga **5** (2012)
21. American Psychiatric Association: Diagnostic and Statistical Manual of Mental Disorders: DSM-IV. Appendix I: Glossary of Culture-Bound Syndromes, American Psychiatric Association (1994)
22. Murata, T., Takahashi, T., et al.: Individual trait anxiety levels characterizing the properties of Zen meditation. Neuropsychobiology **50** (2004)
23. Grant, J.A., Rainville, P.: Pain sensitivity and analgesic effects of mindful states in Zen meditators: a cross-sectional study. Psycho. Med. **71** (2009)
24. Uhlhaas, P.J., Mishara, A.I.: Perceptual anomalies in schizophrenia: integrating phenomenology and cognitive neuroscience. Schizophrenia Bull. **33** (2007)
25. Westbrook, C., Creswell, J.D.: Mindful attention reduces neural and self-reported cue-induced craving in smokers. Social Cogn. Affect. Neurosci. **8** (2013)

26. Engström, M., Pihlsgard, J., Lundberg, P., Soderfelt, B.: Functional magnetic resonance imaging of hippocampal activation during silent mantra meditation. J. Altern. Complem. Med. **16** (2010)
27. Fell, J.: I think, therefore I am (unhappy). Front. Hum. Neurosci. **6** (2012)
28. Lutz, A, Slagter, H.A., Dunne, J.D., Davidson, R.J.: Attention regulation and monitoring in meditation. Trends Cogn. Sci. (2008)
29. Kasamatsu, A., Hirai, T.: An electroencephalographic study on the Zen Meditation (Zazen). Psychiatry Clin. Neurosci. **20**(4) (1966)
30. Khare, K.C.: A study of electroencephalogram in meditators. Indian J. Physiol. Pharmacol. (2000)
31. Desiraju, T., Meti, B.L., Kanchan, B.R.: Neurophysiological correlates of yogic practices of meditation and parnayama unpublished paper presented at INDO-USSR Symposium on Neurophysiology, Armenia (1983)
32. Earle Johnathan, B.: Cerebral laterality and meditation: a review of the literature. J. Transp. Psychol. (1981)
33. Hoffman, E.: Mapping the brains activity after kriya yoga. Bindu Mag. **12**
34. Lutz, A., Greischar, L.L., Rawlings, N.B., Ricard, M., Davidson, R.J.: Long-term meditators self-induce high-amplitude gamma synchrony during mental practice. Proc. Natl. Acad. Sci. USA (2004)
35. MindPlace, Procyon, http://www.mindplace.com/Mindplace-Procyon-System-Meditation-Machine/dp/B000X2BSJM. Accessed 2 Dec 2014
36. Metzger, W.: Optische Untersuchungen am Ganzfeld. Psychologische Forschung **13**, 6–29 (1930) (the first psychophysiological study with regard to Ganzfelds)
37. Dunning, A., Woodrow, P.: ColourBlind: machine imagination, closed eye hallucination and the Ganzfeld effect. Web. 3 Dec 2013 (2010)
38. MindPlace, http://www.mindplace.com/Mindplace-Kasina-Mind-Media-System/dp/B00GU1W8AS. Accessed 7 Dec 2014
39. NeuroSky, Biosensors, http://neurosky.com/products-markets/eeg-biosensors/hardware/. Accessed 5 Dec 2014
40. NeuroSky, eSense, http://developer.neurosky.com/docs/doku.php?id=esenses_tm. Accessed 6 Dec 2014
41. Linden Lab., Second Life, http://secondlife.com/. Accessed 20 Feb 2014
42. Wiki Second Life, LSL, http://wiki.secondlife.com/wiki/Getting_started_with_LSL. Accessed 7 Dec 2014
43. High Fidelity, https://highfidelity.io/. Accessed 6 Dec 2014
44. Project Sansar, http://variety.com/2015/digital/news/second-life-maker-linden-lab-wants-to-build-the-wordpress-of-virtual-reality-1201546110/. Accessed 3 Sept 2015
45. Moseley, R.: Merged reality systems: bringing together automation and tracking through immersive geospatial connected environments. In: Proceedings of Science and Information Conference (SAI) (2014)
46. Fragedakis, T.M., Toriello, P.: The development and experience of combat-related PTSD: a demand for neurofeedback as an effective form of treatment. J. Counsel. Dev. **92**(4) (2014)

A Real-Time Stereo Vision Based Obstacle Detection

Nadia Baha and Mouslim Tolba

Abstract This work aims at defining a new approach for real-time obstacle detection based on sparse disparity map. To achieve fast time execution, the image processing is performed on a subset of points. The proposed method begins by extracting the pixel primitives from a pair images. Then, a detection of the ground is made to limit obstacles area using color information. Subsequently, an extraction of subset of points is performed using the features extracted in the previous step. The use of the extracted features as well as the subset of points will enable us to calculate a sparse disparity map which is used to calculate the v-disparity image. The use of v-disparity image allows us to extract the obstacle profile in order to eliminate the ground pixels. To eliminate the obstacles belonging to the background, we use the ground limits to remove all object points outside these limits. Finally, a segmentation based on disparity value of the remaining points is applied to extract obstacles. To end, validation step is used to remove false positives. The experimental results obtained are interesting with real time execution. This allows early detection to prevent collisions.

1 Introduction

In the last decade, obstacles detection has been a major research topic in computer vision. However, it is still a hard problem to solve when automation, speed and precision are required and/or the objects present complex shapes. Indeed, obstacle avoidance is a fundamental requirement for autonomous mobile robots and vehicles, and numerous vision-based obstacle detection methods have been proposed. The obstacle detection systems generally consist of: sensor, obstacle detection, an avoidance system and an IHM. The obstacle detection is the heart of these systems. The

N. Baha (✉) · M. Tolba
Computer Science Department, University of Science and Technology USTHB,
Algiers, Algeria
e-mail: nbahatouzene@usthb.dz; bahatouzene@yahoo.fr

M. Tolba
e-mail: tolmouss@yahoo.fr; moustolba@yahoo.fr

© Springer International Publishing Switzerland 2016
L. Chen et al. (eds.), *Emerging Trends and Advanced Technologies
for Computational Intelligence*, Studies in Computational Intelligence 647,
DOI 10.1007/978-3-319-33353-3_18

development of an efficient vision-based method for mobile robot navigation is still an active research topic. The first and important step is to ensure that any collisions will be avoided through vision. Therefore, autonomous robot navigation requires almost real-time frame rates from the responsible algorithms. For this, several systems have been proposed in the literature [1–19]. In this paper, a novel method for real-time obstacle detection is proposed that makes use of color and gradient for robust obstacle detection. The method detects obstacles based on sparse disparity map.

The paper is organized as follows: Sect. 2 presents a survey of the related work. Section 3 presents the steps involved in the proposed method. Section 4 presents and discusses the experimental results. Finally, we conclude this work in Sect. 5.

2 Related Work

The majority of methods reported in the literature follow two basic steps: (1) Hypothesis Generation (HG) where the locations of potential vehicles in an image are hypothesized, and (2) Hypothesis Verification (HV) where tests are performed to verify the presence of obstacles in an image. The objective of the HG step is to find candidate obstacle locations in an image quickly for further exploration. HG approaches can be classified into one of the following three categories: knowledge-based, stereo vision based, and motion-based. We introduce in this paper a brief discussion of obstacle detection methods. This work also analyzes the strengths and weaknesses of each category.

Knowledge-based methods use a prior knowledge to determine where obstacle is located in an image [1–6]. For this, some methods use: symmetry, corners, horizontal/vertical edges, texture, color, shadow, lights. Motion-based methods use the assumption that each moving object can be considered as obstacle [7–10]. For this, we need to compute the optic flow. The mainly problem of this method is located in computing and interpreting the optic flow in real times.

Stereo-based methods use stereo information to detect obstacle. There are two types of methods using stereo information for obstacle detection. One uses disparity map [11–14] while the other uses an anti-perspective transformation (i.e., Inverse Perspective Mapping (IPM)).

Disparity map, presents a difference between pixel in left image and its corresponding pixel in right image. Many authors use stereo to detect obstacle, someone use Euclidean plan [14, 15], Homography [16] other use segmentation of disparity, v-disparity, uv-disparity [17–19]. IPM (Inverse Perspective Mapping) [20–25] is a particular use of the concept of inverse perspective transformation. IPM techniques eliminate the effect of perspective lines that are parallel in the real world and converge to a vanishing pixel in the image. Each pixel of the image is re-mapped, thereby creating a new pixel matrix wherein the perspective lines are converted into straight lines and the deformed object to object. It represents a view of the original image, as the projection of the scene on a flat surface.

The second step is the verification and validation. Validation is an important step in the obstacle detection process it allows to verify the existence of an obstacle. In this step we try to match the regions generated with a validated model to or not belonging to a class of obstacle.

3 Proposed Method

The different systems for detecting obstacles cited in the literature are based either on object detection, in this case they must be able to identify any object and classify them as obstacle and avoid them, either on the ground detection, thus any object not belonging to the ground is considered as obstacle.

The aim of our work is to propose a method for obstacle detection from a pair of images to enable a mobile robot to navigate in a scene safely in real time.

The proposed obstacle detection method uses stereoscopic pair images. Our method starts by locating the ground and computing the feature of all pixels. Then, to allow rapid execution time, we use these characteristics to extract a subset of pixels from left and right images which are used to compute a sparse disparity map. Then the v-disparity image is computed using the sparse disparity map. The v-disparity image will allow us to extract the obstacle profile in order to remove ground pixels. Then, we use a v-disparity and some rules to remove background pixels. To eliminate the indirect obstacles: those belonging to the background. We use the ground limits to remove all object points outside these limits. Finally, we apply a simple segmentation method based on the disparity value to segment the new disparity map. Figure 1 shows the different steps followed by our obstacle detection method. We propose in the next section, the steps allowing the obstacle detection.

We describe in this section the steps allowing the obstacle detection.

3.1 Features Extraction

3.1.1 Color Extraction

Color is one of the prominent image features. The characteristic color allows distinguishing regions. In this work, we used two color spaces: HSI (Hue, Saturation, Intensity) and CIELAB to compute the characteristics of all image pixels.

3.1.2 Gradient

In the literature, most operators use the same principle of applying the convolution masks along the two axes x and y. The difference between these operators consists in the difference between the value of these and how to determine them. These masks

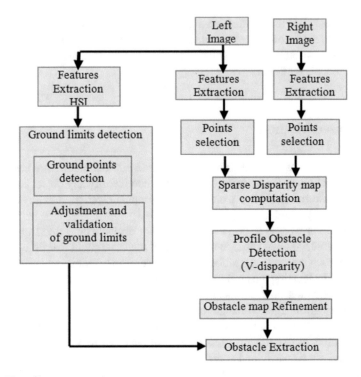

Fig. 1 Obstacle system overview

$F_x(0,0)$	$F_x(0,1)$	$F_x(0,2)$
0	0	0
$F_x(2,0)$	$F_x(2,1)$	$F_x(2,2)$

$F_y(0,0)$	0	$F_y(0,2)$
$F_y(1,0)$	0	$F_y(1,2)$
$F_y(2,0)$	0	$F_y(2,2)$

Gx, Gy dense mask

$F_x(0,0)$	0	$F_x(0,2)$
0	0	0
$F_x(2,0)$	0	$F_x(2,2)$

$F_y(0,0)$	0	$F_y(0,2)$
0	0	0
$F_y(2,0)$	0	$F_y(2,2)$

Gx ,Gy No-dense mask

Fig. 2 Gradient masks

have been designed to extract the horizontal and vertical edges (gradient in x and y). Among the most popular operators, there is the Gauss operator. To extract the features of a pixel, we use horizontal and vertical gradient. For this we apply the 3×3 Gauss operator.

In order to reduce the processing time, we introduce a no-dense gradient mask. A no-dense gradient is a gradient using a no-dense mask. This means that we use only half parameters as shown in Fig. 2. For example, for a 3×3 mask, instead using 9 parameters to compute the gradient, we use only 4 as shown in the following example.

After obtaining the horizontal gradient (Gx) and vertical (Gy) of the image points. We will use these values to generate a subset of points, to estimate the sparse disparity map.

3.2 Ground Detection

The purpose of this step is to detect the obstacle region (ground). In this step, the color characteristics are used to detect the ground. The used method follows two steps, one for the selection of the ground points and the second to confirm and adjust the detected ground.

3.2.1 Ground Points Selection

The selection of the ground points allows to retrieve the ground in the image. For this we use an area confidence as a sample of the image that will allow us the extraction of ground points. In order to achieve a good detection of the ground, we tested our method based on two distances (cylindrical and Euclidean) for each of the HSI and CIELAB spaces.

The idea consists in measuring the cylindrical distance between a ground sample and the image points. For this, first we divide the image into areas of fixed size and calculate the Hue, saturation and intensity in each region. This allows eliminating noise and generating a lower resolution image.

$$distance_{HSI} = \sqrt{S^2 + \hat{S}^2 + Sx\hat{S}^2xcos(H - \hat{H}) + (I - \hat{I})^2} \qquad (1)$$

with

S is the average saturation of the tested point
$\hat{S}$ is the average saturation of the sample
H is the average hue of the of the tested point
$\hat{H}$ is the average hue of the sample
I is the average intensity of the tested point
$\hat{I}$ is the average intensity of the sample

At the end, we apply a segmentation algorithm which uses Euclidian distance to test the belonging of the area to the ground. The same algorithm is applied with the CILAB space using Euclidian distance.

3.2.2 Confirm and Adjust the Detected Ground

This step is used to improve quality of the detected ground limits. For this we use a variation of limit of ground to correct them. So when variation is larger than a given threshold, we replace this variation by a linear variation.

3.3 Selection of a Pixels Subset

After computing the horizontal and vertical gradient (Gx, Gy) of the image points. We will use these values to generate a subset of points and to estimate the disparity map. In this step, we extract a subset of pixels in order to keep only the most significant pixels.

In order to get a good performance (real time and accuracy), we tested two algorithms:

3.3.1 Based on a Local and Global Threshold

In this method, we select only the points that have a greater gradient than the average gradient of the image and that of the average gradient in a defined window size.

3.3.2 Based on the Pixels Density

In this method, we construct a histogram where the abscissa represents the number of points having a gradient greater than the threshold of a window of defined size. Then we will proceed to the extraction of n windows which have the highest number of points.

Once the pixels are extracted, the next step is to estimate the disparity map by searching the matched pixels in both images using the criteria of classical correlation.

3.4 Sparce Disparity Map Calculation

It is known in the literature that the calculation of the disparity map has become a major challenge in the field of Machine Vision. It concerns the features matching between a pair of images of the same scene. When the stereo images are rectified, the matching points will be searched on corresponding horizontal lines and the disparity is calculated as the difference between the abscissas of matched points. The disparity values for all the image points define the disparity map. Once the stereo correspondence problem is solved the depth of the scene can be estimated. This issue is of great interest in the context of 3D reconstruction, Virtual reality, and robot navigation. Several studies have been conducted and different approaches have been proposed to solve the problem of matching that is characterized by the multiplicity of candidate points. The real-time requirements of most vehicles applications complicate the realization of such vision systems. The key to success in realizing a reliable embedded real-time-capable stereo vision system is the careful design of the core algorithm. The trade-off between execution time and quality of the matching is a difficult task, and must be handled with care.

However, for extracting dense and reliable 3D information from the observed scene, stereo matching algorithms are computationally intensive. For this we have proposed in this work a method of obstacle detection using a sparce disparity map.

3.4.1 Matching Score Evaluation

In our work, to compute the matching score between two pixels pl(xl, yl) and pr(xr, yr) in the left and right images, two pixel properties are used in the matching measure: these are intensity, the horizontal and the vertical gradient. Thus the matching score of a pixel pair is defined as the sum of absolute difference between the corresponding feature values of the pixels for each value of the disparity $d, d \in [0, dmax]$. This choice has been made because that gives more robustness to the matching measure.

$$cost(x, y, d) = |G_l x(x, y) - G_r x(x - d, y)| + |G_l y(x, y) - G_r y(x - d, y)| + |I_l(x, y) - I_r y(x - d, y)|$$
(2)

3.4.2 Aggregation Score

Aggregation is performed by summing the calculated matching costs over a squared aggregation window wxw with constant disparity d. The aggregation Sumcost (x, y, d) is defined as:

$$Sumcost(x, y, d) = \sum Cost(x', y', d)$$
(3)

with $(x', y') \in$ window $w \times w$.

In order to get the best performance (computation time and accuracy), the summing of the aggregation cost was achieved using 3 types of window:

- Dense window: all pixels in the window are used.
- Semi-dense window: the half of pixels is used.
- No dense window: the half of the contour pixels of the window is used (see Fig. 3).

We have remarked that, the use of the no-dense window gives a best performance. Considering all the aggregation values obtained varying the disparity in the range [0.. dmax], we obtain the following vector:

$$Cost(x, y, d) = [Sumcost(x', y', 0), Sumcost(x', y', 1), ...Sumcost(x', y', dmax)]$$
(4)

The size of the vector is defined by a set of interest extracted pixels. The disparity at the pixel (x, y) is chosen as the disparity d with the minimal cost (x, y) among the various costs of neighboring pixels to pl in the window W.

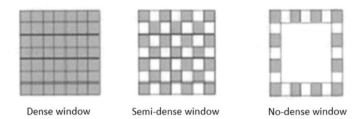

Dense window   Semi-dense window   No-dense window

Fig. 3 Different window types

3.5 Detection of Pixels Belonging to An Obstacle

In this step, we try to remove the pixels not belonging to obstacles. For this, we use a v-disparity image [17] and we take as hypothesis that the disparity of pixels belonging to one obstacle is the same. In order to improve the quality and get a better execution time we used the v-disparity to detect obstacle profile instead detecting ground profile as in [17].

This method can detect obstacles even if there is no ground in addition it is more robust because it detect obstacles without taking into account the ground. Method based on the V-disparity.

3.5.1 Method Based on the V-disparity

The v-disparity image is the projection of the image disparity along the vertical axis, where the ground appears as an inclined line (or curve) and obstacles as vertical straight line, see Fig. 4.

In our work, we are looking for vertical lines in order to detect areas of obstacle. Thus, we only keep the points belong to these lines. We begin by pre-filtering the v-disparity image, for this we check a density of disparity in vertical neighbors this allows to improve the quality of detection line. Detecting lines in the v-disparity is determined by a minimum length and a maximum number of holes. First, we test

Fig. 4 **a** Disparity image.
b V-disparity image

(a)

Disparity Image

(b)   V-disparity

the number of consecutive points while taking into account the maximum number of holes. If the maximum number of holes is reached then we analyze the built segment size, if it is greater than a given threshold then the line is accepted otherwise it will be rejected. In order to make the detection of vertical lines more effective, we use a variable threshold value depending of the disparity value. This is due to the fact that near objects has a larger size.

3.5.2 Removing Pixels Not Belonging to Obstacles

After detecting vertical lines in the v-disparity image, we use the limit of straight lines to mark pixels belonging to obstacles. Thus, only pixels having the same disparity will mark. Once the pixels belonging to the ground are removed, we use the ground limit to remove the obstacles situated outside the limits.

In order to study the efficiency of the method using the V-disparity to extract the obstacle profile, an other approach was implemented: in this approach, we use the disparity map and UV-disparity image to frame the obstacles areas. Using the v-disparity image (respectively u-disparity) permit to detect the areas likely to be obstacles, this results in vertical lines (respectively horizontal). After detecting the straight lines, a grouping based on the disparity value is done in order to frame the obstacles areas. The results obtained are less accurate with a relatively high computation time.

3.6 Refinement of the New Disparity Map

In this step, we try to eliminate erroneous disparities: disparities which are not representative enough (their number is low). For each point i of the image, we look for in a neighborhood $w' \times w$ with $w' > w$ the number of points having the same disparity that the point i. If the number of points is below a given threshold, the point is deleted. The refinement improves the quality of the segmentation by removing small areas.

3.7 Segmentation of the Disparity Map

After removing the pixels belonging to the ground, a segmentation algorithm based on the disparity value is applied on the new disparity map. This step is used to extract obstacle by merging the pixels belonging to the same obstacle.

$$F(x)=\begin{cases}1 & \textit{if } \text{Width}(x)\geq \textit{Threshold1}(d)\\ 0 & \textit{else}\end{cases}$$

$$G(x)=\begin{cases}1 & \textit{if } \text{Height}(x)\geq \textit{Threshold2}(d)\\ 0 & \textit{else}\end{cases}$$

$$H(x)=\begin{cases}1 & \textit{if } \text{Number of pixel}(x)\geq \textit{Threshold3}(d)\\ 0 & \textit{else}\end{cases}$$

$$T(x)=\begin{cases}1 & \textit{if } \text{Local UV-disparity}(x)\geq \textit{Threshold4}(d)\\ 0 & \textit{else}\end{cases}$$

With F,G,H,T a low classifiers
With d disparity of obstacle

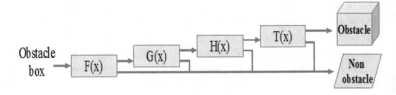

Fig. 5 Obstacle detection process

3.8 *Validation*

The validation step allows removing the fault positive. For this we use multiple low classifier based on size, number of point and local uv-disparity. Local uv-disparity is a uv-disparity applied in a box of obstacle detected. When we remove the fault positive some left result will be merged using color. Figure 5 shows the process used to determine if a detected object is obstacle or no obstacle for this we use some low classifier to accelerate choice.

4 Experimental Results

In this section, we describe the experiments conducted to evaluate the performance of the proposed method. Our aim is to obtain an accurate obstacle detection map and a fast runtime which is the requirement of any obstacle detection system of autonomous mobile robot navigation. The proposed approach was tested with some a stereo images used by the web community extracted from some university sites in Pc Core i5 2.9 GHZ. Our method showed a reasonable accuracy and real-time performance. In the following, we present the different results obtained at each step of the proposed method.

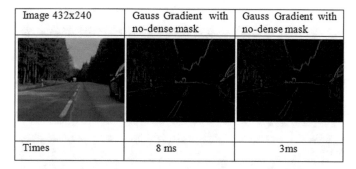

Image 432x240	Gauss Gradient with no-dense mask	Gauss Gradient with no-dense mask
Times	8 ms	3ms

Fig. 6 Processing time obtained for Gauss gradient with dense and no- dense masks

4.1 Features Extraction

This step is very important: it allows us to extract regions and to compute the sparse disparity map. The obtained results using Gauss gradient with a dense mask and Gauss gradient with no-dense mask have showed the same result, but less time when we use a Gauss gradient with no dense mask. Figure 6 shows an example of the processing time obtained for 423 × 240 image.

4.2 Ground Detection

The purpose of this step is to reduce obstacles to those belonging to the road. For this, we use some computed characteristics as color, local minimum intensity and max intensity color and some rules to detect ground.

Figure 7 illustrates the processing time of the two methods (based on cylindrical distance for HSI space and euclidian distance for CIELAB space) applied on some images. The ground detection using HSI space is faster compared to CIELAB space.

Now, we present the results of each step that allows reconstructing the ground. In Fig. 8, the first column shows test images. The second column shows the conversion of the image to 16 color classes. The third column shows the result of the segmentation using the 16 color classes and the fourth column shows the result of the rectification on the detected ground image. The presented results show that we have obtained good ground detection with interesting processing times.

4.3 Selection of a Pixels Subset

The quality of a pixels subset allows to increase the quality of the disparity map, obstacle extraction and to decrease computation time. The next figure shows an image   CIELAB   HSI

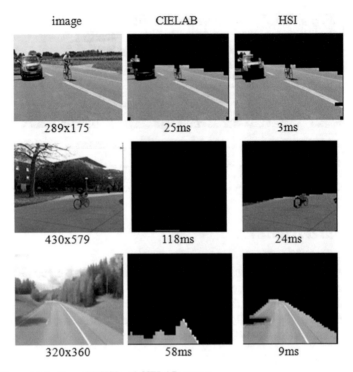

289x175   25ms   3ms

430x579   118ms   24ms

320x360   58ms   9ms

Fig. 7 Ground detection with HSI and CIELAB spaces

example of selecting a subset of pixels and the corresponding processing time. The results are similar with a less time execution for the method based on local and global levels (see Fig. 9).

4.4 Sparse Disparity Map Calculation

In this section we present the results obtained by applying different window sizes (7 × 7, 3 × 7, 1 × 7) and different type of windows (dense, semi-dense and no dense) to compute a sparse disparity. Figure 10 shows the disparity maps obtained, The same quality result are obtained for dense, semi-dense and no-dense windows with least time for a no dense window.

4.5 Detection of Pixels Belonging to An Obstacle

Figure 11 shows in the first column: the reference image, the second column: the sparse disparity map obtained, the third column: the v-disparity and the fourth column: obstacles map. At this level the pixels of the ground are removed.

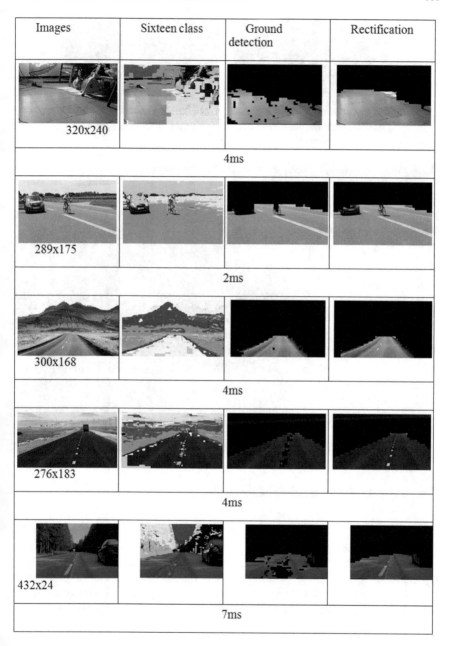

Images	Sixteen class	Ground detection	Rectification
320x240			
4ms			
289x175			
2ms			
300x168			
4ms			
276x183			
4ms			
432x24			
7ms			

Fig. 8 Processing times obtained by each step of the ground detection

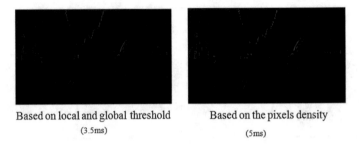

Based on local and global threshold	Based on the pixels density
(3.5ms)	(5ms)

Fig. 9 Processing time obtained by the pixels subset selection step

Fig. 10 Sparse disparity map obtained with different window sizes

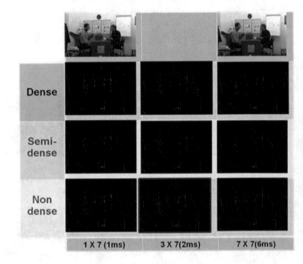

4.6 Removal of No Direct Obstacles

Figure 12 shows the results obtained after the removal of obstacles belonging to the back ground using ground limits.

Experiments are conducted in order to study the influence of the disparity map type(dense or sparse) on the performance of our method, Fig. 13 shows some results with dense disparity map. Thus our method can be applied on sparse and dense disparity map. In order to get real-time performance, we choose the sparse disparity map.

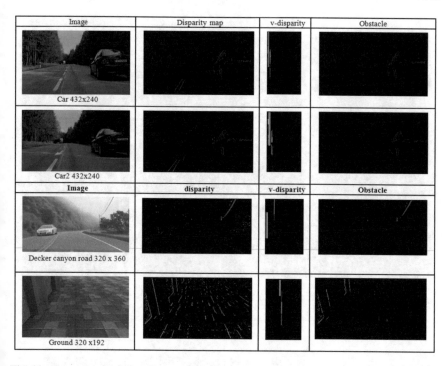

Fig. 11 *First column* reference image, the *second column* the sparse disparity map obtained, the *third column* the v-disparity and the *fourth column* obstacles map

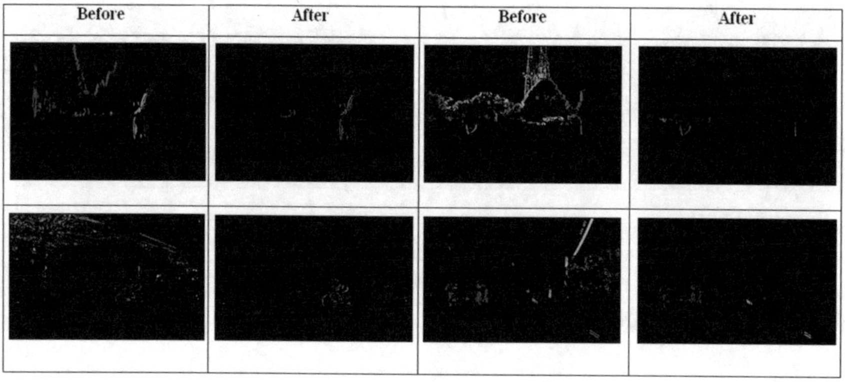

Fig. 12 Removal of obstacles belonging to the back ground using ground limits

| Image | Obstacle image | Image | Obstacle image |

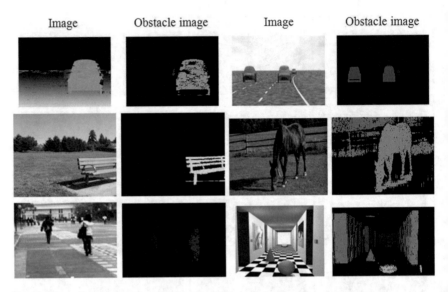

Fig. 13 Some examples of obstacle detection using a dense disparity map

| Before | After | Before | After |

Fig. 14 Validation step on some images

4.7 Validation Step

This step allows removing a fault positive. Figure 14 illustrates the result of the validation step for some examples. Table 1 shows the computation time (ms) obtained by each step of our obstacle detection system.

Table 1 Processing time obtained for each step of obstacle detection system

Images	Ground detection	Gradient	Pixels selection	Disparity calculation	Obstacle	Segmentation	Validation	Total (ms)
432 × 240	7	6	5	6	2	2	2	29
320 × 360	8	8	6	6	1	1	2	32
430 × 579	21	17	18	9	2	1	2	70
430 × 579	19	18	19	8	2	1	1	68
640 × 480	25	21	22	12	2	1	1	84
222 × 177	3	2	2	1	1	1	1	11
227 × 170	3	2	2	2	1	1	1	12

5 Conclusion

In this paper, we proposed an obstacle detection method that makes use of color information and v-disparity method. The v-disparity is used to detect the obstacle profile instead ground profile. The proposed method allows to detect obstacles with least time for own example 30ms then achieving real-time performance. To improve the result, we need to use adaptive threshold, for this we use width of ground and sample disparity ground. The first for segmentation and the second for line detection in v-disparity. Test results show that our algorithm performs better than available color and v-disparity based obstacle detection approaches.

References

1. Fletcheret, L. et al.: Vision in and out of vehicles. IEEE Intell. Syst. **18**(3), 12–17 (2003)
2. Tada, N., Murata, K., Saitoh, T., Osaki, T., Konishi, R.: Monocular Vision based Indoor Mobile Robot. In: The 23rd International Technical Conference on Circuits/Systems, Computers and Communications. Conference Publications (2008)
3. Chaudhry, R.: Histograms of oriented optical flow and Binet-Cauchy kernels on nonlinear dynamical systems for the recognition of human actions. In: Computer Vision and Pattern Recognition, Conference Publications. pp. 1932–1939 (2009)
4. Mittal, A., Sofat, S.: A novel color coherence vector based obstacle detection algorithm for textured environment. In: IEEE International Conference on Machine Vision, Conference Publications (2010)
5. Metzler, S., Nieuwenhuisen, M., Behnke, S.: Robot Soccer World Cup XV: learning visual obstacle detection using color histogram features. In: RoboCup: Robot Soccer World Cup XV. Computer Science, Conference Publications, vol. 7416, pp. 149–161 (2011)
6. Teoh, S.S., Brunl, T.: Symmetry-based monocular vehicle detection system. In: Machine Vision and Applications, Conference Publications, pp. 831–842 (2012)
7. Chirkhodaie, A., Amrani, R.: Visual terrain mapping for traversable path planning of mobile robots. In: Proceedings of SPIE, Conference Publications, vol. 5608, pp. 118–127 (2004)
8. Salmane, H., Ruichek, Y., Khoudour, L.: Object tracking using Harris corner points based optical flow propagation and Kalman filter. In: Intelligent Transportation Systems (ITSC), Conference Publications, pp. 67–73 (2011)

9. Wang, Z., Zhao, J.: Optical flow based plane detection for mobile robot navigation. In: Proceedings of the 8th World Congress on Intelligent Control and Automation, pp. 1156–1160 (2011)

10. Molineros, J., Cheng, S.Y., Owechko, Y., Levi, D., Zhang, W.: Monocular rear-view obstacle detection using residual flow. In: Computer Vision ECCV, Conference Publications, vol. 7584, pp. 504–514 (2012)

11. Han, D.: Real-Time Object Segmentation of the Disparity Map Using Projection-Based Region Merging. Vision and Image Processing Lab. Sejong University 98 Kunja-dong, Kwagjin-gu, Seoul Korea (2007)

12. Lee, C.H., Su, Y.C., Chen, L.G.: An intelligent depth-based obstacle detection system for visually-impaired aid applications. In: Image Analysis for Multimedia Interactive Services, Conference Publications, pp. 1–4 (2012)

13. Vincze, M., Wohlkinger, W., Olufs, S., Einramhof, P., Schwarz, R., Varadarajan, K.: Object detection and classification for domestic robots. Commun. Comput. Inform. Sci. 106–120 (2012)

14. Nedevschi, S., Danescu, R., Frentiu, D., Graf, T., Schmidt, R.: High accuracy stereovision approach for obstacle detection on non-planar roads. In: Proceedings of the IEEE Inteligent Engineering Systems, pp. 211–216. Cluj Napoca, Romania (2004)

15. Sappa, A., Geronimo, D., Dornaika, F., Lopez, A.: On-board camera extrinsic parameter estimation. Electron. Lett. 42, 745–747 (2006)

16. Baha, N., Larabi, S.: Obstacle detection with stereo vision based on the homogtaph. In: International Arab Conference on Information Technology, ACIT 11, Saudi Arabia, Dec 2012

17. Labayrade, R., Aubert, D.: A single framework for vehicle roll, pitch, yaw esti-mation and obstacles detection by stereovision. In: Proceedings of IEEE Intelligent Vehicles Symposium, pp. 31–36 (2003)

18. Gao, Y., Ai, X., Wang, Y., Rarity, J., Dahnoun, N.: U-V-Disparity based Obstacle Detection with 3D Camera and steerable filter Intelligent Vehicles Symposium (IV), Department of Electrical and Electronic Engineering, Bristol University, pp. 957–962. Bristol, UK, June 2011

19. Soquet, N., Aubert, D., Hautiere, N.: Road segmentation supervised by an extended V-disparity algorithm for autonomous navigation. In: Intelligent Vehicles, pp. 160–165, June 2007

20. Nieto, M., Laborda, J.A., Salgado, L.: Road environment modeling using robust perspective analysis and recursive Bayesian segmentation. Mach. Vis. Appl. 927–945 (2011)

21. Hattori, H., Maki, A.: Stereo without depth search and metric calibration. In: Proceedings of the IEEE Computer Society Conference on Computer Vision and Pattern Recognition, pp. 177–184 (2000)

22. Shu, Y., Tan, Z.: Vision-based lane detection in autonomous vehicle. In: Proceedings of the Congress on Intelligent Control and Automation, pp. 5258–5260. China, June 2004

23. Simond, N., et Parent, M.: Obstacle detection from ipm and super-homography. In: Proceedings of IEEE International Conference on Intelligent Robots and Systems, pp. 4283–4288. California, Sant Diego, USA, Nov 2007

24. Nieto, M., Laborda, J.A., et Salgado, L.: Road environment modeling using robust perspective analysis and recursive Bayesian segmentation. Mach. Vis. Appl. 927–945 (2011)

25. Nieto, M., Salgado, L., Jaureguizar, F., et Cabrera, J.: Stabilization of Inverse Perspective Mapping Images based on Robust Vanishing Point Estimation. Grupo de Tratamiento de imgenes E.T.S. Ing. Telecomunicacin, Universidad Politcnica de Madrid, Spain (2007)

Depth and Thermal Image Fusion for Human Detection with Occlusion Handling Under Poor Illumination from Mobile Robot

Saipol Hadi Hasim, Rosbi Mamat, Usman Ullah Sheikh
and Shamsuddin Hj. Mohd. Amin

Abstract In this paper we present a vision-based approach to detect multiple persons with occlusion handling from a mobile robot in real-world scenarios under two lighting conditions, good illumination (lighted) and poor illumination (dark). We use depth and thermal information that are fused for occlusion handling. First, a classifier is trained using thermal images of the human upper-body. This classifier is used to obtain the bounding box coordinates of human. The depth image is later fused with the region of interest obtained from the thermal image. Using the initial bounding box, occlusion handling is performed to determine the final position of human in the image. The proposed method significantly improves human detection even in crowded scene and poor illumination.

Keywords Robot vision · Thermal image · Depth image · Image fusion

S.H. Hasim (✉) · R. Mamat (✉) · S.H.M. Amin (✉)
Control and Mechatronics Engineering Department,
Universiti Teknologi Malaysia, 81310 Johore, Malaysia
e-mail: shshadi2@live.utm.my

R. Mamat
e-mail: rosbi@fke.utm.my

S.H.M. Amin
e-mail: sham@fke.utm.my

U.U. Sheikh (✉)
Electronics and Computer Engineering Department,
Universiti Teknologi Malaysia, 81310 Johore, Malaysia
e-mail: usman@fke.utm.my

© Springer International Publishing Switzerland 2016   365
L. Chen et al. (eds.), *Emerging Trends and Advanced Technologies
for Computational Intelligence*, Studies in Computational Intelligence 647,
DOI 10.1007/978-3-319-33353-3_19

1   Introduction

1.1   Background

The field of computer vision is concerned with problems that involve interfacing computers with their surrounding environment through visual means. One such problem, object detection, involves detecting the presence of a known object in an image, given some knowledge of what that object should look like. As humans, we take this ability for granted, as our brains are extraordinarily proficient at both learning new objects and recognizing them later. For example, for humans the colour and point of view of a car are not relevant in deciding whether it is a car or not. Similarly, in human detection, we as humans can locate a person regardless of the colour or type of clothing worn, the pose, appearance, under partial occlusion, in different lighting or even in cluttered backgrounds. However, in computer vision, this has proven to be one of the most difficult and computationally intensive problem. Computer vision systems are far behind human in performing such analysis and drawing meaningful conclusions.

1.2   Robots

Traditional surveillance systems based on non-mobile platform are mostly passive, in which the system can only detect events and trigger alarms, while active role based systems (mobile platform) can be used to interact with the environment, human and with another robots (significantly expanding the potential of these systems). Such mobile robots are able to patrol the environment such as at the airport, warehouse, bank, etc., to keep watch on valuable items, recognize people, and discriminate intruders.

Starting in the 70s, many researches were conducted in the field of robotic vision so that robots become visually literate. Various practical problems can be solved if a robot is able to see just like humans. One useful application of robotic vision is for the detection and tracking of human which can enable robots to be used for search and rescue missions involving human victims, robots as security guards and in commercial applications such as in supermarkets. In order to realize all these, the robot must have the knowledge, including abilities to detect and recognize the presence of humans, understand human activities, navigate safely in public, help human to make way for them, and also interact with the humans.

1.3   Human Detection

In the context of surveillance systems, detecting human presence is a key requirement. However detecting humans is not an easy task because humans have a non-rigid appearance. Another factor that makes this task even more challenging is the varying illumination, possible occlusions, viewpoint changes as well as background clutter that may occur in a scene. The main key to the robust system is the system can work in bright and poor illumination. A further point that makes the task of surveillance to become difficult is the crowded atmosphere in which occlusion occurs. The question is on how to propose a method that can detect human in occluded situations with unpredictable human motion. There are several limitations with existing methods in occlusion handling, among which are, that they solve the occlusion but limited to only two or three humans [1–3]. Works by [4] tried to solve the occlusion of five humans, but N-human occlusion problem has yet to be addressed. In addition, if the occlusion is too severe, most of these methods will fail as shown in [5, 6].

With so many challenges, using a single sensor usually does not produce acceptable accuracy in detection. One approach is to use sensor fusion to produce multi-modal features in order to build a robust detection. In this paper, we consider a sub-problem of human recognition: human detection, in which we are interested in detecting humans based on the characteristic patterns of information obtained from thermal camera and distance information obtained from raw depth. This approach differs from other techniques for human detection, such as those that detect humans based on face, histogram, colour, texture, or surface features. We focus on human detection with occlusion handling for surveillance system from a mobile robot by incorporating information from distinct sensors. In the proposed work, we develop a visual perception algorithm where the system is able to detect human based visual features through robust distinguishing operations using fusion of information between depth and thermal images. We follow a standard pattern recognition approach based on three main steps: (i) Pre-processing based on Small Vision System (SVS) to achieve colour constancy, (ii) ROI Generation and (iii) Classification Generation. Our proposed method is further explained in Sect. 3.

2 Related Works

2.1 Sensor

The most frequently used sensor for surveillance is the visible-light colour camera, however it is limited to strict illumination conditions. The infrared spectrum can simplify the task of detection and can be utilized in a wide range of illumination condition. Infrared cameras are very sensitive devices and measure temperature differences as small as 0.08 °C. Ideally the camera is set up at places with long queues such as passport or customs control. To achieve the correct temperature, the camera should focus on the most reliable temperature spot on the body—the corner of the eyes. The camera produces infrared images or heat pictures of a person's face and detects whether the body temperature exceeds a certain value or not. The infrared cameras are very easy to use and have proven themselves as tools that can be operated by non-specialists after a few hours of training. Infrared cameras can be used in any environment where large numbers of human are passing or staying, such as airports, train stations, the underground or building foyers. An infrared camera enables a quick and accurate identification of human [7].

In real-world, it is a key requirement for robots to interact in real time with dynamic people, so there a significant challenge in 3D environment. The emergence of commercial depth cameras allow the formation of a system that uses the dimension of depth, thus enabling robots to observe their surroundings and see in 3D space [8]. Although there are many approaches that have been proposed to detect humans using RGB cameras, the depth information is often ignored. Depth can be used together with other vision-based sensors to develop a more reliable and robust system. Multi-sensor-based systems have advantages over single sensor-based systems due to the following [9]:

- **Improved system reliability and robustness**. Multi-sensor systems have an inherent redundancy. Due to the availability of data from multiple sensors, system performance is improved. When one or more sensors fail or are unable to operate due to interference such as jamming, the system can continue to operate at a reduced performance level (graceful degradation).
- **Extended coverage**. Use of multiple sensors allows an increase in coverage, both spatial and temporal. Multiple sensors can observe a region larger than the one observable by a single sensor.
- **Increased confidence**. Sensors can confirm each other's inferences, thereby increasing confidence in the final system inference. Also, some inferences can be ruled out to generate a reduced set of feasible options, thereby reducing the effort required to search for the best solution.
- **Shorter response time**. Since more data is collected by multiple sensors, a prescribed level of performance can be attained in a shorter time.
- **Improved resolution**. Use of sensors with different resolutions can result in an inference with better resolution than any single sensor.

2.2 Humans Detection System

To explain the process of the human detection system, we use the model architecture proposed by [10].

Preprocessing

The first stage is to deal with the data derived from the sensors. In the pre-processing stage, raw sensor data are manipulated to make the data meaningful and suitable for computation in later stages. Depending on the sensors used, as well as data format, the pre-processing stage typically includes sensor data filtering, normalization and format conversion.

Table 1 Comparison of segmentation techniques

Technique	Advantage	Disadvantage
Seed region growing	Reliable	Time consuming and over-segmentation
Hierarchical clustering	Reliable	Time consuming
Watershed	Reliable	Over-segmentation
K-means	Fast	Fragmented shape

ROI Generation: Object Segmentation

They are three consecutive phases to achieve segmentation. The first phase, the depth image is scanned to obtain the possible candidates, and then the candidates will be filtered to remove any excess regions in the second phase. Finally, depth imaged is clustered and labelled on the base of the candidate depth measurement, and each cluster is stored as a single object. The work by Xia's [1] examined each region using a 3D head model. Using the depth information for authentication and extracting parameters of the head from the depth, the parameters are then used to construct the 3D head model. They then matched the 3D model against the entire detected region to make the final approximation. To achieve maximum accuracy, they proposed a segmentation-based method to segment the body and the objects that are connected to it. Table 1 shows the comparison between different segmentation techniques [11].

ROI Generation: Candidate Selection

To ease the human recognition task in vision-based systems, a candidate selection mechanism is usually applied. The choice of candidates can be obtained by performing object segmentation in either a 3D scene or a 2D image plane. Monocular daylight or NIR cameras are often cheaper than stereo vision and they hold better signal-to-noise ratio and resolution. Although they do not have the depth information, there are still several efficient approaches for candidate selection available for these cameras [12].

Classification Generation: Feature Extraction

Human detection from depth information is feasible, and it has some unique advantages compared to just using an intensity image [13]. We believe that the human appearance can be better captured if the information of the edge/local shape and the texture are combined with depth information.

A depth image is a kind of image with a particular form. It complies with the basal image format, which is an array of pixels organized into rows and columns. The main difference between depth images and grey images is the meaning of the pixel. For the grey image, the pixel value represents grey scale or photosensitive strength. For depth images, the pixel value represents the distance to the reference plane. A depth image directly reflects the three-dimensional geometric information of the scene surface and has advantages such as high accuracy and is not affected by illumination, shadow or other factors. Therefore, visual systems based on depth image information are increasingly an area of interest.

Works by Zhang's [14] introduced the Average Depth Difference (ADD) feature. Recently, Zhang's [15] proposed the Maximal Depth Difference (MDD) feature which refers to the maximum of the depth differences between any pixel and its eight neighbors in a depth image. The advantage of the MDD feature is that it not only considers the relationship between the depth difference and self-occlusion but also strengthens the effect of the depth difference feature on detection of a result.

Classification Generation: Classifier

A promising technique for human detection was proposed which relies on Haar wavelets [16] used in conjunction with an support vector machine (SVM) [17, 18]. This Wavelet is used to build a wavelet template that captures the structural relationship between instances of classes at various scales. The template used as a feature vector for SVM training. This approach has a detection rate of 70 % with a false positive rate of 1:15,000 [19] based on images of pedestrians in cluttered outdoor environments.

Inspired by SIFT descriptor, histograms of oriented gradients have been developed to detect accurately humans in arbitrary scenes [20]. This method divides an image into dense overlapping cells, with each cell consisting of an HOG (Histogram of Oriented intensity gradient). These features are then used to train SVM. The performance of this method is significantly better than wavelet-based approaches; however processing each frame is computationally expensive.

2.3   Handling Occlusion

Several researchers overcome the occlusion problem in detecting multiple objects by combining some camera inputs [21–23] and by using the mask and the appearance model to deal with the shape change and large occlusions [24]. Work by [2] developed a Bayesian segmentation approach that combines the region-based background subtraction and human shape model for detecting a person under occlusion and in [25], the author proposes a dynamic Bayesian network. Some methods use the motion model to implement fast detection that can engage with some instances of occlusion. These methods require accurate modelling of motion and fail at the non-linear motion of interacting objects.

In [26], region detector is used to solve for occlusion. In [27] an appearance model is used for detecting objects under occlusion. A template matching algorithm using appearance features and Kalman filter was proposed by [28]. Whereas, [29] presented dynamic background layer model and model each moving object as a foreground layer, together with the foreground ordering. Approach for detecting occlusion in stereo camera Augmented Reality (AR) systems have also been investigated, some of which solve the problem of occlusion using depth information delivered through stereo matching [30].

3 Methodology

We present an approach using a thermal image, depth image, and two different measurement models together with Median Filter and Connected Component Filter. With this method, a person can be detected under any illumination and independent to skin color. Our complete system for tracking people by mobile robot is shown in Fig. 1.

To build the measurement model based on the characteristics of the gray value, we use the algorithm proposed by [31]. They proposed a very fast approach in detecting objects in grey level images.

First, we perform image registration on the input depth and thermal images. This process is carried out offline to obtain camera calibration parameters so that the two images (depth and thermal) are properly registered for further processing. The process is performed only once, manually by an operator by selecting similar control points in both images. This method of registration is suitable for images that have distinct features. The rest of the algorithm consists of three stages, namely (a) preprocessing, (b) region-of-interest (ROI) generation and (c) object classification and occlusion detection. In the next subsections, we explain each stage briefly.

3.1 Pre-processing Module

The first process in the preprocessing stage is to normalize the depth information between the value of 0–255. The input depth information, I_D has a resolution of 13-bits, and is normalized using $I'_D = \frac{I_D}{2^{13}} * 255$. Next is to register the thermal image, I_T

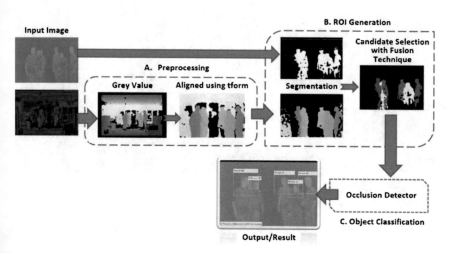

Fig. 1 Overview of the human detection system with occlusion handling

with the depth image, I'_D using the camera parameters obtained from the registration process.

We use the cascade object detector proposed by Viola-Jones to detect human based on the upper body structure. This detector is applied to the thermal image, I_T to obtain the bounding box coordinates of persons and these coordinates are stored in a matrix which is called BBC (bounding box coordinates). The cascade object detector is trained using thermal images of human body. To train the dataset, we use Cascade Training GUI proposed by [32]. Cascade classifier training requires a set of positive samples and a set of negative images. We provide a set of positive images with object of interest specified to be used as positive samples (human consisting of upper body). We also provide a set of negative images from which the function generates negative samples automatically.

3.2 ROI Generation

The purpose of ROI generation is to extract regions of interest from the thermal images in the presence of background clutter. We maintain the depth information in the range of 3–8 m. Depth range of above 8 m is inaccurate and below 3 m the thermal image is not suitable due to the focal length of the thermal camera which is 18 mm with a horizontal field of view of 29.9°. The thermal camera (Model: Raven 384) is capable of capturing radiation emitted by humans and objects in the range of

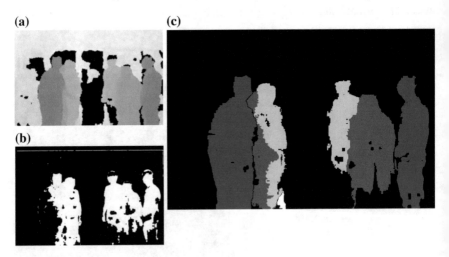

Fig. 2 Fusion of depth and thermal images. **a** A depth image after registration. **b** A thermal image after segmentation. **c** Fusion result

0–50°C. We select the region of interest with the existence of people in the range of 18–37°C (Fig. 2).

Whereas the approach of Candidate Selection with Fusion Technique is to generate Depth of Interest (DOI) of depth images accurately, we use the ROI of thermal images as a template. In this work, we are interested in using fused information, in which the I'_D image is fused with ROI of the thermal image in the spatial domain. In the first stage, we apply a median filter to filter out the noise. The second stage we use connected component labelling to measure a set of properties for each connected component (object) in the binary image and use these properties to filter unwanted objects.

3.3 Object Classification

Person Prediction—We introduce a new approach for occlusion handling, using coordinates of BBC detected in the pre-processing stage to find a new person based on the distance in each x-axis and y-axis of the bounding box. A new matrix is created to collect coordinates and distance of new persons and is called as NPC. To do this, we use the distance of persons in bounding box detected in pre-processing stage as a reference. This procedure is presented in Algorithm 1.

Verification/Refinement—Compare NPC against BBC to eliminate the coordinates of the same person. Finally, the real matrix of the bounding box is produced, and we name this matrix as RBB. This procedure is presented in Algorithm 2.

Algorithm 1: Person Prediction

Initialize the total of BBC

- BBCj = [xj, yj, hj, wj], j=1,........., N is total of number of bounding boxes

For each BBC

- j is no. of current BBC

- Compute the height and weight of current bounding box based on min x and min y.

- Initialize the reference distance (Zr).

- Compare distance in every pixel on bounding box against Zr : if distance is 0 and the same Zr, ignore and go to next pixel, and if distance is not equal to Zr, go to storage array

Matrix of new person: NPC = [distance, x, y, j]

- Store coordinate and distance of new person

Algorithm 2: Verification/Refinement

1. Initialize the total number of NPC and BBC

2. Occlusion Handling: Eliminate coordinate of the same person

For each NPC:

    - k is no. of current NPC

    - j is no. of BBC in NPC

  For each BBC:

    - d is no. of current BBC.

    - if j equal to d, ignore and go to next d.

    - if j not equal to d, compare the value of x in current NPC to the value of x-min and x-max in current BBC.

    - $X_{NPC} \geq Xmin_{BBC}$ && $X_{NPC} \leq Xmax_{BBC}$ is true, then all values of current NPC are removed.

    - $X_{NPC} \geq Xmin_{BBC}$ && $X_{NPC} \leq Xmax_{BBC}$ is false, then all values of current NPC are retained.

    - go to next k

Matrix of real person:

- update real matrix of bounding box, RBB = [BBC; NPC]

4 Experimental Results

4.1 Dataset

A new dataset was developed to assess the ability to handle occlusions of human under no/poor illumination (e.g. during night time). This dataset consists of thermal images and depth data, both taken by Raven 384 thermal camera and Microsoft Kinect respectively in a laboratory environment. Microsoft Kinect and Thermal camera are placed on a mobile platform as shown in Fig. 3 and it can take up to 30 frames per second [33].

The dataset was collected in a real-world scenario under two lighting conditions; (1) illuminated (lighted) and (2) no ilumination (dark). From the original data, 200 frame images with occlusions were selected manually to create the final dataset for testing human detection. 100 frame images are captured in good illumination, and 100 frame images is a dark condition. 1400 people in the dataset are manually labelled with bounding box. Ground truth can be used to evaluate the performance of the proposed method. The following experimental results are reported based on the proposed dataset. Complex occlusions in the dataset can be seen in Fig. 4.

Fig. 3 UTMbot Robot (*right*). An example of images captured from the Kinect camera (*top-left*) and thermal camera (*bottom-left*)

4.2 Ground Truth

We evaluate our proposed method based on our collected samples. First, we manually construct the ground truth (GT) from our dataset. The only bounding box around the upper-body is accepted as correct detection. Bounding boxes obtained from the GT are referred to as targets (N_{GT}). To indicate the quality of the detection, we compute the precision, recall, and accuracy. We define precision (P) to indicate the level of false positives (*FP*), recall (R) to indicate true positives (*TP*), and accuracy (A) to indicate the precision of the overall system.

(a) Precision (P):

$$P = \frac{TP}{TP + FP} \tag{1}$$

(b) Recall (R):

$$R = \frac{TP}{TP + FN} \tag{2}$$

(c) Accuracy (A):

$$A = \frac{TP + N_{GT}}{TP + N_{GT} + FP + FN} \tag{3}$$

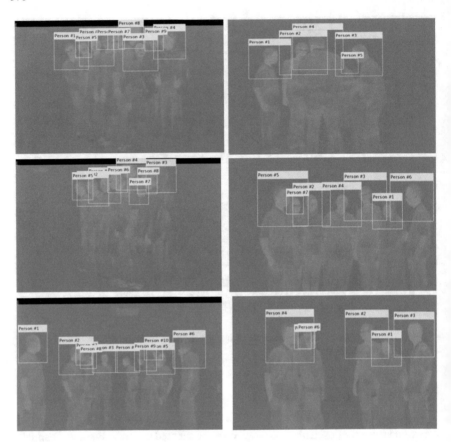

Fig. 4 Results of *Dark* dataset (*left column*) and *Lighted* dataset (*right column*)

The Root Mean Square Error (**RMSE**) is a very commonly used measure of the difference between the value predicted by the model and the actual observed values of the environment that is modeled. The RMSE serves to aggregate the values into a single measure of predictive power. The RMSE can be used to distinguish the performance of the model in the calibration and validation period, and also to compare the performance of individual models with other predictive models [34]. The RMSE of a model prediction with respect to the estimated variable H_{GT} is defined as the square root of the mean squared error:

$$RMSE = \sqrt{\frac{\sum_{i=1}^{n} (H_{GT,i} - H_{AD,i})^2}{n}} \tag{4}$$

where H_{AD} is the observed value and H_{GT} is the modelled value at time/place i.

Table 2 Detection performance of the proposed method

Condition	Lighted	Dark
Total frames	388	365
Selected frames	100	100
Max person	10	10
Total ground truth	709	712
Pre-detection (Bbox)	594 TP | 1 FP | 5 FN	588 TP | 1 FP | 7 FN
GT occlusion	110	117
Occlusion detection	96 TP | 5 FP | 14 FN	108 TP | 4 FP | 9 FN
Actual detection	690	696
Total detection	690 TP | 6 FP | 19 FN	696 TP | 5 FP | 16 FN
Total time	898 s	912 s
Average time	8.98 s	9.12 s

Table 3 Accuracy of the proposed method and *RMSE* for human detection

Evaluation	Percentage	
	Lighted	Dark
Precision, *P*	99.14 %	99.29 %
Recall, *R*	97.32 %	97.75 %
Accuracy, *A*	98.24 %	98.53 %
RMSE	1.9	1.6

4.3 Results

Table 2 shows the ground truth of which 709 persons of whom 110 persons were occluded. The results (*Lighted*) obtained from our proposed (method and dataset) of which 690 true positives were detected and 6 false positives (non person detected) and 19 false negatives (person not detected). Whereas the results of *Dark* dataset is that 696 true positives were detected, 5 and 16 false positives and false negatives respectively.

The accuracy of our method is high for both conditions (97.32 % for *Lighted*, and 97.75 % for *Dark*) and the RMSE is also lower than 2. These indicate that the proposed system performs well and successful in reducing miss rate. Table 3 shows the Precision, Recall, Accuracy and RMSE of the proposed method.

5 Conclusion

In this paper, we presented a human detection method based on the fusion of thermal and depth images for occlusion handling. In the proposed method, the segmentation of depth images is not conducted because many of the existing techniques

involve significant mathematical complexity and computational burden, thus leading to problems for real-time implementation. Here we introduced a new technique
to get the depth of interest (DOI) using image fusion technique. The proposed *Occlusion Detector* process can detect occluded persons by using the BBC to find a new
person based on the distance in each x-axis and y-axis of the bounding box. Then,
a new matrix is generated containing the coordinates of the new person (NPC). The
NPC is compared to the BBC to remove coordinates of the same person. Finally, the
real matrix of the bounding box is produced. The advantages of our method are that
it is computationally inexpensive and performs well even under crowded scene and
poor illumination as it depends only on depth and thermal information. This work can
further be extended by including human tracking as well as real-time implementation
on robot.

Acknowledgments The authors would like to thank Faculty of Electrical Engineering of Universiti
Teknologi Malaysia for providing technical support for this research work. The authors are also
grateful to the financial aid from Centre for Artificial Intelligence and Robotics (CAIRO) and
Computer Vision, Video and Image Processing Research Lab (CvviP) of Universiti Teknologi
Malaysia.

References

1. Xia, L., Chen, C., Aggarwal, J.K.: Human detection using depth information by kinect. In: Computer Society Conference on Computer Vision and Pattern Recognition Workshops (CVPRW),
   2011 IEEE, pp. 15–22 (2011)
2. Eng, H., Wang, J., Kam, A.H., Yau, W.A.: Bayesian framework for robust human detection
   and occlusion handling using human shape model. In: Proceedings of the 17th International
   Conference on ICPR 2004, vol. 2, pp. 257–260 (2004)
3. Kamijo, S., Matsushita, Y., Ikeuchi, K., Sakauchi, M.: Occlusion robust tracking utilizing
   spatio-temporal Markov random field model. In: Proceedings of 15th International Conference
   on Pattern Recognition ICPR-2000 [Internet], vol. 1, pp. 140–144. IEEE Computer Society
   (200). http://ieeexplore.ieee.org/lpdocs/epic03/wrapper.htm?arnumber=905292
4. Otsuka, K., Mukawa, N.: Multiview occlusion analysis for tracking densely populated objects
   based on 2-D visual angles. In: Proceedings of 2004 IEEE Computer Society Conference on
   Computer Vision Pattern Recognition, 2004 CVPR 2004 [Internet], vol. 1, pp. 90–97. IEEE
   (2004). http://ieeexplore.ieee.org/lpdocs/epic03/wrapper.htm?arnumber=1315018
5. Ikemura, S., Fujiyoshi, H.: Real-time human detection using relational depth similarity features.
   Computer Vision—ACCV 2010. Lecture Notes in Computer Science, vol. 6495, pp. 25–38
   (2011)
6. Camps, O., Sznaier, M.: Segmentation for robust tracking in the presence of severe occlusion.
   In: Proceedings of the 2001 IEEE Computer Society Conference on Computer Vision and
   Pattern Recognition, 2001 CVPR 2001. pp. 483–489 (2001)
7. Almerfors, A.: Infrared Camera [Internet] (2009). http://www.applegate.co.uk/b2b-news-
   articles/more-detail-regarding-infrared-camera-0020792.htm?view=NEWS_105027
8. Zhang, H., Reardon, C., Parker, L.E.: Real-time multiple human perception with color-depth
   cameras on a mobile robot. IEEE Trans Cybern. **43**(5), 1429–41 (2013)
9. Varshney, P.K.: Multisensor data fusion. Electron Commun. Eng. J. 245–253 (1997)
10. Hadi, H.S., Rosbi, M., Sheikh, U.U.: A review of infrared spectrum in human detection for
    surveillance systems. Int. J. Interact Digit Med. **1**(3), 13–20 (2013)

11. Ding, J., Kuo, C., Hong, W.: An efficient image segmentation technique by Graduate Institute of Communication Engineering, National Taiwan University. In: The 21th IPPR Conference on Computer Vision, Graphics and Image Processing (2009). http://djj.ee.ntu.edu.tw/FastS
12. Alonso, I.P., Llorca, D.F., Sotelo, M.Á., Bergasa, L.M., Member, A., De, Toro P.R., et al.: Combination of feature extraction methods for SVM pedestrian detection. IEEE Trans. Intell. Transp. Syst. **8**(2), 292–307 (2007)
13. Wu, S., Yu, S., Chen, W.: An attempt to pedestrian detection. In: Third Chinese Conference on Intelligent Visual Surveillance (IVS), pp. 9–12 (2011)
14. Zhang, S., Liu, J.: A self-occlusion detection approach based on range image of vision object 2041–2046 (2011)
15. Zhang, S., Liu, J.: A self-occlusion detection approach based on depth image using SVM. Int. J. Adv. Robot Syst. **9**, 230 (2012). http://www.intechopen.com/journals/international_journal_of_advanced_robotic_systems/a-self-occlusion-detection-approach-based-on-depth-image-using-svm
16. Mallat, S.G.: A theory for multiresolution signal decomposition? The wavelet representation. IEEE Trans. Pattern Anal. Mach. Intell. **II**(7), 674–693 (1989)
17. Oren, M., Papageorgiou, C., Sinha, P., Osuna, E., Poggio, T.: Pedestrian detection using wavelet templates. In: Proceedings of the IEEE Computer Society Conference on Computer Vision and Pattern Recognition, pp. 193–199. IEEE Computer Society (1997)
18. Shimizu, H., Poggio, T.: Direction estimation of pedestrian from multiple § tidl images 596–600 (2004)
19. Koenig, N.: Toward real-time human detection and tracking in diverse environments. In: 2007 ICDL 2007 IEEE 6th International Conference on Development and Learning, pp. 94–98 (2007)
20. Dalal, N., Triggs, B., Schmid, C.: Human detection using oriented histograms of flow and appearance. In: European Conference on Computer Vision, pp. 428–441 (2006)
21. Chang, T.-H., Gong, S., Ong, E.-J.: Tracking multiple people under occlusion using multiple cameras. In: Proceedings of 11th British Machine Vision Conference (2000)
22. Dockstader, S.L., Tekalp, A.M.: Multiple camera fusion for multi-object tracking. In: Proceedings of IEEE Workshop on Multi-Object Tracking. IEEE Computer Society, pp. 95–102 (2001)
23. Dockstader, S.L., Tekalp, A.M.: Multiple camera tracking of interacting and occluded human motion. Proc. IEEE **89**(10), 1441–1455 (2001)
24. Cucchiara, R., Grana, C., Tardini, G., Vezzani, R., Emilia, R.: Probabilistic people tracking for occlusion handling. In: Proceedings of the 17th International Conference on ICPR 2004, pp. 132–135 (2004)
25. Wu, Y., Yu, T., Hua, G.: Tracking appearances with occlusions. In: 2003 Proceedings of the IEEE Computer Society Conference on Computer Vision and Pattern Recognition, pp. I-789–I-795. IEEE Computer Society (2003)
26. Siebel, N.T., Maybank, S.: Fusion of multiple tracking algorithms for robust people tracking. In: 7th European Conference on Computer Vision, Denmark, vol. IV, pp. 373–387 (2002)
27. Senior, A., Hampapur, A., Tian, Y.-L., Brown, L., Pankanti, S., Bolle, R.: Appearance models for occlusion handling. Image Vis. Comput. **24**(11), 1233–1243 (2006)
28. Nguyen, H.T., Smeulders, A.W.M.: Fast occluded object tracking by a robust appearance filter. IEEE Trans. Pattern Anal. Mach. Intell. **26**(8), 1099–1104 (2004)
29. Tao, H., Sawhney, H.S., Kumar, R., Rd, W.: Dynamic layer representation with applications to tracking (2000)
30. Wloka, M,M., Anderson, B.G.: Resolving occlusion in augmented reality. In: Proceedings of the 1995 Symposium on Interactive 3D Graphics, pp. 5–12 (1995)
31. Viola, P., Jones, M.: Rapid object detection using a boosted cascade of simple features. In: Proceedings of 2001 IEEE Computer Society Conference on Computer Vision Pattern Recognition CVPR 2001, vol. 1, pp. I-511–I-518. IEEE Computer Society (2001)
32. Shoelson, B.: Cascade trainer: specify ground truth, train a detector [Internet] (2013). http://www.mathworks.com/matlabcentral/fileexchange/39627-cascade-trainer--specify-ground-truth--train-a-detector

33. Hadi, H.S., Rosbi, M., Sheikh, U.U., Amin, S.H.M.: Improved occlusion handling for human detection from mobile robot. In: Science and Information Conference (SAI), IEEE, pp. 694–698 (2015)
34. Chai, T., Draxler, R.R.: Root mean square error (RMSE) or mean absolute error (MAE)? Arguments against avoiding RMSE in the literature. Geosci. Model Dev. **7**(3), 1247–1250 (2014)

Exploiting the Retinal Vascular Geometry in Identifying the Progression to Diabetic Retinopathy Using Penalized Logistic Regression and Random Forests

Georgios Leontidis, Bashir Al-Diri and Andrew Hunter

Abstract Many studies have been conducted, investigating the effects that diabetes has to the retinal vasculature. Identifying and quantifying the retinal vascular changes remains a very challenging task, due to the heterogeneity of the retina. Monitoring the progression requires follow-up studies of progressed patients, since human retina naturally adapts to many different stimuli, making it hard to associate any changes with a disease. In this novel study, data from twenty five diabetic patients, who progressed to diabetic retinopathy, were used. The progression was evaluated using multiple geometric features, like vessels widths and angles, tortuosity, central retinal artery and vein equivalent, fractal dimension, lacunarity, in addition to the corresponding descriptive statistics of them. A statistical mixed model design was used to evaluate the significance of the changes between two periods: 3 years before the onset of diabetic retinopathy and the first year of diabetic retinopathy. Moreover, the discriminative power of these features was evaluated using a random forests classifier and also a penalized logistic regression. The area under the ROC curve after running a ten-fold cross validation was 0.7925 and 0.785 respectively.

Keywords Diabetic retinopathy · Diabetes · Penalized · Logistic regression · Random forests · Mixed model

1 Introduction

Diabetic retinopathy (DR) is a major disease, affecting the lives of millions of people around the world, leading to blindness, if left untreated or not diagnosed early [3, 17]. It constitutes a complication of diabetes mellitus, although it is not uncommon

G. Leontidis (✉) · B. Al-Diri · A. Hunter
University of Lincoln, Brayford Pool Campus, LN67TS Lincoln, UK
e-mail: gleontidis@lincoln.ac.uk

B. Al-Diri
e-mail: baldiri@lincoln.ac.uk

A. Hunter
e-mail: ahunter@lincoln.ac.uk

© Springer International Publishing Switzerland 2016
L. Chen et al. (eds.), *Emerging Trends and Advanced Technologies for Computational Intelligence*, Studies in Computational Intelligence 647,
DOI 10.1007/978-3-319-33353-3_20

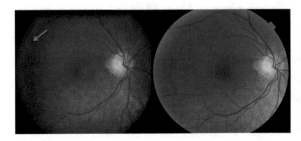

non-diabetic people to develop background retinopathy. In Fig. 1, two images can be seen, from the same patient, one during diabetes and one after the first lesions (micro-aneurysm) have appeared in the retina. It is worth pointing out that a normal/non-diabetic image does not seem to have any difference from a diabetic retinal image, since at this stage, the changes occur only to the vascular geometry, which cannot be easily identified.

Retina is a dynamic tissue and a very important, non-invasive window to the blood vessels. Retina processes light through a layer of photoreceptors. The absorbed light is converted into neural signals, in order to be forwarded through the optic nerve head directly to the brain for visual recognition [17]. Each person's retina is unique just like the fingerprints, making it very difficult to compare different retinas, since changes will inevitably and naturally exist. Therefore it is crucial, if someone wants to study the effects that a disease cause to the retinal vasculature, to look at specific segments and regions within the same subjects at different intervals. More details addressing the importance of this approach will be given in the next sections.

The underlying mechanisms that provoke diabetes are more or less known, however it still remains unclear how this sequence of events affects the retina, both structurally and functionally, leading to the development of DR. Diagnosing DR early or identifying diabetic patients with higher risk, can have a big impact on our society and possibly help clinicians deal with the disease earlier and delay the progression, by monitoring the patients more intensively [3].

For the present study, fifty high resolution (3216-by-2316 pixels) fundus images were used, taken from twenty five patients who progressed from diabetes to DR. Our aim is to understand to what extend has the retinal vascular geometry been affected by the progression and proliferation of diabetes, until the moment that the first lesions appear. To accommodate this, two groups were created; one for the period 3 years before DR and one for the very first year that DR appeared. Therefore we hypothesize that the retina is already adapting to the new underlying conditions, and that especially during the advanced stages of diabetes (few years before DR), these changes can be reliably identified and characterized. The images come from a diabetic screening database in England and all of the ethical guidelines have been followed. It is worth pointing out that in United Kingdom, all the people that are diagnosed with diabetes are entering automatically into the diabetic screening program for annual inspection of their retina. Therefore all the images are labeled and identified by the year they were captured, defining clearly the periods of diabetes, and also the initial appearance of DR.

The chapter is organized in three main sections. In the first section, all the methods, methodologies and tools will be described and analyzed, giving some essential background information of the investigated geometric features and their importance, as well as all the necessary image preprocessing. In the second part, the techniques for the statistical analysis, feature selection process and the classification approaches will be thoroughly addressed. At the final section the results will be presented, together with the inferences and the implications of the present study, including discussion, limitations, future approaches and conclusions.

2 Related Work

Retina includes both very small and very large vessels, which can range from very few μm to more than 100 μm. It can be easily inferred that it is very difficult to compare the retina of different people and include representative and balanced amount of small and large vessels, which will in any case be different among people. During progression of diabetes and also during DR the retinal geometry changes [25].

Most of the studies in the past, investigating either hemodynamic or geometric features, have been focused on the analysis of different groups of people. For instance the oxygen saturation was investigated in different groups of people ranging from normal subjects to proliferative retinopathy, finding significant differences among them [15]. In another study they evaluated the differences between patients with diabetes and DR, using as features only the vessels' widths and angles [12].

Using different subjects, when investigating the human retina, makes it hard to associate any identified changes to diabetes/DR, and not instead to the normal changes that occur to the retina during aging, or between genders, or simply because different retinas, and more importantly different areas of the retina, might also vary [3, 27]. A few follow-up studies have been conducted, studying similar periods of diabetes, without though including in any classification system or evaluating features like central retinal vein/artery equivalent or tortuosity, which is the purpose of this study [4, 18–21].

3 Methods

As mentioned previously, fifty images in total were analyzed, making sure that all the features can be measured in an equally reliable manner in all of them and thus ensure that the changes can be attributed to the progression of diabetes. All the methods and tools were carefully chosen, having always as first priority the reliability and accuracy of the measurements, rather than using the fastest or with the fewest human interventions methods. For the image preprocessing, extraction of all the features and for the mixed model design, the software Matlab 2015b was utilized. On the contrary, for the regularized random forests (RRF) and the penalized logistic regression, the open source software "R" was used.

3.1   Features and Tools

A number of features were investigated in this study, which are representative of the whole retinal vasculature. Measuring these geometric features means that many different methods and tools have to be used in all the stages. The main investigated features are the following: (a) Vessels' widths, (b) Vessels' angles, (c) Tortuosity, (d) Fractal dimension, (e) Lacunarity and (f) Central retinal artery and vein equivalents for calculating the arteriovenous ratio as well.

3.1.1   Widths and Angles

Using the tool that was implemented and described in details in a previous study [1], 1200 vessels widths (600 arteries and 600 veins in total for both groups in pixels) and 400 branching angles (in degrees) in the corresponding junctions (200 for arteries and 200 for veins in total for both groups) were measured. Although many state-of-the-art automated tools have been proposed in literature, utilizing many different methods e.g. wavelets and edge location refinement both to segment and measure retinal vessels using image profiles, computed across a spline fit of each detected centerline [5], an infinite active contour model, using an infinite perimeter regularizer and multiple region information [29] or using neighbourhood estimator before filling filter [2], still they cannot be used in large studies for evaluating the progression of the disease. Their consistency and accuracy/precision as well as the measurement errors across datasets with different image quality, do not allow us to find these subtle changes that occur inside the vasculature over time, and which we are trying to identify in the same retinas. Both widths and angles were measured twice by the same observer, yielding an intra-rater reliability of over 90% for the absolute agreement. Therefore both groups of measurements were kept by taking their average.

Empirically, the changes that we are trying to identify as a consequence of the proliferation of diabetes can be as small as 1% of change pre- to post- DR, and in the most extreme cases they can reach up to 7–10%. Therefore the semi-automated approaches are still preferred, because they let us measure the same junctions over time and be consistent to the accuracy of our measurements. From each junction's vessels' widths, the branching coefficient (BC) is derived and calculated by Eq. (1),

$$Branch.Coef. = \frac{W_1^2 + W_2^2}{W_0^2}, \tag{1}$$

Fig. 2 Segmented images from the same patients before (*left*) and after diabetic retinopathy (*right*), used for the evaluation of tortuosity. Vessels edges and centerlines are highlighted

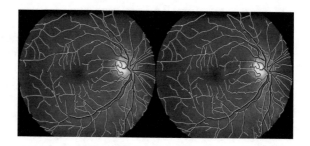

where w1, w2, w0 are the widths of the larger child vessel, smaller child vessel and parent vessel respectively. Furthermore another derivative feature was introduced, as the ratio between the junction angle and the corresponding BC Eq. (2).

$$(Angle/BC)_i = \frac{Angle_i}{BC_i},$$ (2)

3.1.2 Tortuosity

In addition to these, tortuosity of the vessels, which is a property of a curve being tortuous i.e. twisted, was also included and calculated by the method proposed in [13]. For this purpose the images were segmented, using an algorithm described in [14], and the coordinates of each segment were also extracted Fig. 2, in order to calculate the local tortuosity. The global, image-level tortuosity was then derived by using the mean, median, standard deviation and the third quartile, in a similar way like in a previous study [22].

3.1.3 Fractal Dimension and Lacunarity

Fractal dimension (FD) and lacunarity are another two important features that are included in this study. The former can give us a measure of complexity of a structure, as long as it can be considered a fractal. The latter is a measure of heterogeneity of a fractal structure.

Fractality

Fractals present various degrees of self-similarity in different scales. Human retina has been found to almost be a self-similar structure, thus being possible to be analyzed as such, giving us a measure of complexity, letting us also investigate, whether it changes during different periods [9]. Its discriminatory power was evaluated within the classification system in conjunction with the other features. Higher values of FD indicate more complex structure.

Lacunarity

Complimentary to the FD, lacunarity was also evaluated, which is a counterpart of FD, describing the gappiness between the structures, or alternatively how the fractals fill the space.

For FD, the well established method of box-counting algorithm (Minkowski–Bouligand dimension) was used [24], based on Eq. (3). For this purpose, all the images were segmented [14], obtaining the binary vascular trees, in order to apply the box-counting and gliding box methods. Each image of the same patient was processed, in order to include the same vessels, making sure that any identified differences are due to the proliferation of diabetes and not an error from the algorithm.

$$Fractal\,Dim. = \lim_{r \to 0} \frac{Log\,N(r)}{Log\,1/r}, \tag{3}$$

in which N(r) refers to the number of boxes of side length r that has to be used to cover a given area in the Euclidean n-space, by using a sequential number of descending size boxes. This occurs in multiple orientations. The final dimension in the 2D space is between 1 and 2 ($1 \le D \le 2$) [23].

Lacunarity was estimated using the gliding-box algorithm, for different grid orientations [28]. A unit box of size r is chosen randomly and the number of set points p are counted i.e. the mass. The procedure is repeated with the box centered consecutively for each point within the set, creating a distribution of masses B(p, r). Finally, we get the probability, by converting the distribution into probability distribution $Q(p, r)$, dividing by the total number of boxes (B) of size r Eq. (4).

$$Q_{p,r} = \frac{B_{(p,r)}}{B_{(p)}}, \tag{4}$$

Finally, after several transformations, the gliding box equation can be written in terms of the accumulated sum of the mean and the second moments of all boxes Eq. (5).

$$L_{GB}(r) = \frac{B_{(r)} \sum_{i=1}^{B_{(r)}} p(i, r)^2}{[\sum_{i=1}^{B_{(r)}} p(i, r)]^2}, \tag{5}$$

where the denominator is the square of the total number of elements in the data set [28].

3.1.4 Arterio-Venous Ratio

Central retinal vein (CRV) and artery (CRA) are the two major vessels of the retina. CRV leaves the optic nerve head 10 mm from the eyeball, draining the blood from the capillaries into the superior ophthalmic vein or to the cavernous sinus directly, depending on the individual [7]. On the other side, the CRA branches off the

Fig. 3 On the *left*, the mask as created by our algorithm is shown, after defining the optic disc diameter, and on the *right* the region of interest, with the veins and arteries labeled, from which the CRVE, CRAE and AVR are calculated

ophthalmic artery, crossing inferior to the optic nerve head within its dural sheath to the eyeball. Since these two vessels cannot be seen in the retinal fundus images, it has been proposed, initially by Parr [26] and then revised by Knudtson [16], a method to estimate the central retinal vein and artery equivalent, CRVE and CRAE respectively, based on the Eqs. 6 and 7, derived partly by the branching coefficient that they estimated in normotensive subjects. The region of interest is defined as shown in Fig. 3, and includes the region where the edges of the vessels course through at 0.5–1.0 disc diameters from the optic disc margin. The region between this area and the optic disc is excluded, as not having the vessels attained their status inside the retina yet. Within this area, the six largest veins and the six largest arteries are measured, following an iterative procedure of pairing up the largest vessels with the smallest ones, until a final single number is obtained. All the values are entered in Eqs. 6 and 7 for arterioles and venules respectively.

The final value for the vein is termed central retinal vein equivalent (CRVE) and the respective final value for the artery is termed central retinal artery equivalent (CRAE). The ratio CRAE/CRVE is known as arterio-venous ratio.

$$Arterioles : \hat{W} = 0.88 * \sqrt{(W_1^2 + W_2^2)} \tag{6}$$

$$Veins : \hat{W} = 0.95 * \sqrt{(W_1^2 + W_2^2)} \tag{7}$$

where $\hat{w}$ is the estimate of the parent trunk arteriole or venule and w_1, w_2 are the two branches (children).

3.2 Design and Analysis

All of the above features were evaluated separately, using a mixed model design filter, as described in the next subsection [20]. Based on this design, repeated measures analysis of variance (ANOVA) was used, in order to calculate the F-statistic and finally the p-value for each feature. In that way, we try to evaluate whether any observed differences between the two groups, for each feature, are just random observations, or whether they can be attributed to the disease's proliferation. This is also a way of defining the importance of these features and thus make an initial feature selection. It is worth mentioning that, when dealing with features that have a biological meaning, it has to more deeply be investigated, whether they should be included in a classification system, regardless of the result of the statistical analysis. The mixed model based on the repeated measures nature of the analysis, increases the statistical power, requiring fewer subjects to be analyzed [11]. Including matched junctions and the same groups of patients, could lead to the decrease of both the statistical error (difference from the unobserved population mean) and the residuals (difference from the sample mean). In order to make sure that this parametric test is the correct one for the analysis of our data, normality and sphericity tests were run for each feature. For the former, the Shapiro-Wilk test was used, and the null hypothesis that the data are normally distributed was not rejected, regardless of the feature under investigation (p-values ranging from 0.30 to 0.56). Similarly for the sphericity, the Mauchly's test was used, which again failed to reject the null hypothesis that the assumption of sphericity is met (p-values ranged from 0.16 to 0.39).

Although ANOVA is robust in marginal violations of normality, it still suffers from sphericity, which if present, causes the test to become unstable i.e. leads to an increase of Type I error; that is, the likelihood of detecting a statistically significant result when there is not one.

3.2.1 Mixed Model Filter

As mentioned above, in order to account for the different way that the features are measured, a mixed model factorial/nested design has been developed in MATLAB 2015b version, in which all the local measurements are used in the statistical analysis. As can be seen in Fig. 4, in the case of widths and angles, we have multiple measurements within each subject, in a nested formation. That means that all these observations are not independent, and thus that needs to be taken into account. Using

Fig. 4 Mixed model design filter used for the statistical analysis of each feature and for the initial feature selection

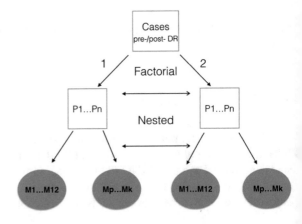

this design, each measurement in $P1jM1k$, where 1 is the first case, e.g. pre-DR group, j the corresponding patient and k the specific measurement, is related only to the corresponding measurement at the same exact junction in $P2jM2k$. This logic is applied in this model, which is then analyzed by ANOVA.

3.3 Classifiers

In order to test the discriminative power of these features, two different approaches were followed. Firstly, a regularized random forests classifier was used, slightly adjusted for the feature selection process, as proposed in [8]. Secondly, a logistic regression model was developed, using both Least Absolute Shrinkage and Selection Operator (Lasso) and ridge regression, as a hybrid penalty for the coefficients of the features (L1- and L2-norms), which is called elastic net regularization described in [10].

3.3.1 Regularized Random Forests

Random forests is a well-established supervised classifier and very popular in machine learning. It was proposed by Breiman as an improvement to the decision trees' bagging method [6]. It consists of multiple decision trees, each of which is grown on a bootstrap sample, taken from the original training data. The Gini index $(Gini(u))$ at node u, is defined as

$$Gini(u) = \sum_{c=1}^{c} \hat{p}_c^u(1 - \hat{p}_c^u) \tag{8}$$

where $\hat{p}_c^u$, is the proportion of class-c observation at node u. Subsequently, the Gini information gain of Xi for splitting node u, is the difference between the impurity at node u and the weighted average of impurities at each child node of u. This can be seen in Eq. (9) [8].

$$Gain(X_i, u) = Gini(X_i, u) - w_L Gini(X_i, u^L) - w_R Gini(X_i, u^R) \tag{9}$$

where u^L and u^R are the left and right children nodes of u respectively. Similarly w_L and w_R are the proportions of instances assigned to the left and right children nodes. The most important part of random forests is the mtry function, in which a random set of features out of P is evaluated. The feature with the highest $Gain(X_i, u)$ is used for splitting the node u. The importance score for variable X_i is then calculated,

$$Importance_i = \frac{1}{ntree} \sum_{u \in S_{X_i}} Gain(X_i, u) \tag{10}$$

where S_{Xi} refers to the set of nodes split by X_i in random forests with ntree number of tree. In short, the regularized version of random forests (RRF) can select a compact feature subset, by including an additional penalty coefficient, creating a regularized information gain Eq. (11) [8]

$$Gain_R(X_i, u) = \left\{ \begin{array}{ll} \lambda \cdot \text{Gain}(X_i, u) & i \notin F \\ \text{Gain}(X_i, u) & i \in F \end{array} \right\} \tag{11}$$

in which F refers to the set of indices of features used for splitting in the previous nodes. The parameter $\lambda \in (0, 1]$ is the penalty coefficient. When $i \notin F$ the coefficient penalizes the ith feature for splitting node u. Smaller leads to a larger penalty. Regularized random forests uses $Gain_R(X_i, u)$ at each node, and adds the index of a new feature to F. For instance a RRF with $\lambda = 1$, has the minimum regularization, however a new feature has to be more informative at a given node than the features that have already been included to the feature subset. The feature subset selected by RRF ($\lambda = 1$) is termed the least regularized subset, as it offers minimum regularization. Apart from the feature selection process, the rest of the algorithm is exactly the same as the initially proposed random forests classifier [8].

For the evaluation of the performance of RRF, the Out of Bag error (OOB) was used, which is the internal way of validating the performance of random forests classifier [6]. In addition, ten-fold cross validation was utilized.

3.3.2 Logistic Regression with Elastic Net Penalty

In this study, where the response variable is binary, a regularized logistic regression model is used [10]. The difference with the ordinary logistic regression has to do with the penalty parameter applied to the coefficients. In the case of ridge regression, the coefficients of correlated predictors are shrunk towards each other, allowing them to work together. From a Bayesian point of view, the ridge regression works better, if there are many predictors and all have non-zero coefficients.

On the other side the least absolute shrinkage selector operator (Lasso) is to some extend indifferent to very correlated predictors, tending to pick one and discard the rest. The Lasso penalty corresponds to a Laplace prior, which expects many coefficients to be zero or close to zero and a small subset of non-zero coefficients. In the middle of this, elastic net with a $= 1 - \varepsilon$ for small $\varepsilon > 0$, performs similarly to Lasso, removing however any extreme behavior caused by highly correlated predictors. The general formula Pa of elastic net, as seen in Eq. (13), introduces a compromise between ridge and Lasso. As α increases from 0 to 1 for a specific value of parameter λ, the sparsity of the solution in Eq. (15) (referring to the coefficients equal to zero), increases monotonically from 0 to the sparsity of the Lasso solution. More specifically, assuming that the response variable G $= 1, 2$, then the logistic regression model represents the class-conditional probabilities, through a linear function of the predictors, which in the logarithmic form is given by Eq. (12) [10].

$$log\frac{Pr(G = 1|x)}{Pr(G = 2|x)} = \beta_0 + x^T\beta \qquad (12)$$

where in this case the model is fit by regularized maximum binomial likelihood.

$$P_\alpha(\beta) = \sum_{j=1}^{p}[\frac{1}{2}(1 - \alpha)\beta_j^2 + \alpha|\beta_j|] \qquad (13)$$

Let $p(x_i) = Pr(G = 1|x)$ be the probability according to Eq. (14).

$$Pr(G = 1|x) = \frac{1}{1 + e^{-(\beta_0 + x^T\beta)]}} \qquad (14)$$

For an observation i at specific values for the parameters (β_0, β), the penalized log likelihood is maximized Eq. (15).

$$\max_{(\beta_0,\beta)\in\mathbf{R}^{(p+1)}}\left[\frac{1}{N}\left[\sum_{i=1}^{N}\{I(g_i = 1)logp(x_i) + I(g_i = 2)log(1 - p(x_i))\} - \lambda P_\alpha(\beta)\right]\right]$$
$$(15)$$

Replacing, the log-likelihood part of Eq. (15) takes the form,

$$l(\beta_0, \beta) = \frac{1}{N}\sum_{i=1}^{N}y_i \cdot (\beta_0 + x_i^T\beta) - log(1 + e^{(\beta_0 + x_i^T\beta)}) \qquad (16)$$

a concave function of the parameters. In this approach, for every value of λ, an outer loop is created for the computation of the quadratic approximation l_Q of Eq. (16) about the current parameters $(\tilde{\beta}_0, \tilde{\beta})$.

$$l_Q(\beta_0, \beta) = -\frac{1}{2N}\sum_{i=1}^{N}w_i(z_i - \beta_0 - x_i^T\beta)^2 + C(\tilde{\beta}_0, \tilde{\beta})^2 \qquad (17)$$

where

$$z_i = \tilde{\beta}_0 + x_i^T\tilde{\beta} + \frac{y_i - \tilde{p}(x_i)}{\tilde{p}(x_i)(1 - \tilde{p}(x_i))} \qquad (18)$$

$$w_i = \tilde{p}(x_i)(1 - \tilde{p}(x_i)), (weights) \qquad (19)$$

Finally, the penalized weighted least-squares problem can be solved by Eq. (20), using the coordinate descent approach [10].

$$\min_{(\beta_0,\beta)\in\mathbf{R}^{(p+1)}}\left[-l_Q(\beta_0, \beta) + \lambda P_\alpha(\beta)\right]. \qquad (20)$$

A number of sequential nested loops are created :

- Outer loop: Decrement λ.
- Middle loop: New quadratic approximation l_Q for the current parameters $(\tilde{\beta}_0, \tilde{\beta})$.
- Inner loop: Execute the coordinate descent algorithm on the penalized weighted least-squares problem Eq. (20).

Further information of the above method is given by Friedman et al. [10].

In the same way as RRF, ten-fold cross-validation was used to evaluate the classifier.

4   Results

This section will present the results of the three different approaches that were previously addressed.

MMF   In which the results of the analysis of every feature are presented, together with some more information about the data.

RRF   In the first part the results of the feature selection process, according to their importance will be shown, followed by the classification results based on the feature subset.

LOG   Similarly to the RRF, in the elastic net logistic regression the first part will be devoted to the selection of α and λ parameters and subsequently the feature subset, and then at the last part, the results of the classification will be shown.

All of the features were scaled (normalized), by centering the data. This was done by subtracting the mean and normalizing it dividing by the standard deviation. Especially with the gradient descent algorithms, like logistic regression, this can be beneficial, as we can achieve better numerical stability and quicker convergence.

The open source software "R" was used both for the RRF and Elastic net logistic regression classifiers, as well as for all the evaluation steps and feature selections.

4.1   Evaluation of Features with MMF

In Table 1, we can find the results of the analysis using the MMF. As can be seen, some of the features significantly differed across the groups, whereas some others not. In addition to that, no significant results (thus excluded from Table 1) were observed in almost any combination of features, when using the mean values, medians or standard deviations (although p-values were between 0.15–0.28), which highlights the superiority of the MMF, in which all the measurements are accounted for as measured.

Table 1 Mixed model analysis of variance results

Feature name	p-value ($\alpha = 0.05$)	F-value (dfn, dfe)[a]	Group means (SD) (pre-/post-DR)
Arteries widths	**0.01**	**6.53 (1, 299)**	**11.14 (2.20), 10.45 (1.93)**
Arteries angles	**0.022**	**5.24 (1, 99)**	**88.45 (8.74), 85.63 (6.93)**
Arteries BC	0.30	1.3 (1, 99)	1.24 (0.11), 1.29 (0.12)
Veins widths	**0.0005**	**16.95 (1, 299)**	**13.23 (2.81), 12.17 (2.28)**
Veins angles	0.62	0.24 (1, 99)	81.72 (6.9), 81.52 (6.62)
Veins BC	0.45	0.85 (1, 99)	1.1 2 (0.10), 1.12 (0.11)
Fractal Dim.	**0.024**	**6 (1, 24)**	**1.628 (0.06), 1.594 (0.06)**
Lacunarity	0.65	0.45 (1, 24)	0.22 (0.04), 0.22 (0.05)
Tortuosity (SD)	**0.021**	**5.79 (1, 24)**	**0.074 (0.013), 0.089 (0.02)**
CRVE	0.76	0.10 (1, 24)	29.13 (4.39), 28.01 (5.53)
CRAE	0.37	0.83 (1, 24)	20.21 (2.87), 19.74 (3)
AVR	0.81	0.07 (1, 24)	0.697 (0.10), 0.704 (0.14)

[a]*dfn* degree of freedom numerator, *dfe* degree of freedom error term

As can be seen in Table 1, arteries' widths and angles, veins' widths, arteries' angles, fractal dimension and tortuosity (standard deviation) are found to differ significantly between the two groups. The rest of them did not appear to do so, however, since all these features reflect functional changes, still remain useful for further investigation and possible inclusion in a classification system.

Interestingly enough, the arteries' widths have been decreased at the first year of DR by almost 6.5 % and the angles by 3.5 %. Similarly, but only for the widths, veins showed a decrease at the first year of DR by almost 8 %.

In Fig. 5, we can see two examples of how the differences between the post-DR and pre-DR measurements are correlated with the age of the patients, despite the fact that the data are limited for giving us a reliable result. However they can just be used as an indication or a trend of the data.

4.2 Classification with RRF

All the available features were initially included in the classifier, in order to evaluate their importance. In addition to the features that appear in Table 1, for selecting the feature subset, we included all the original features, including fractal-to-lacunarity ratio and Angle-to-BC ratio, as well as the descriptive statistics of them. In total 20 features were included, with 50 observations in total (25 for each class-balanced design), however fourteen of them were negatively affecting the performance. The classifier had a similar performance when all the initial values for arteries and veins were used, instead of the descriptive statistics, thus the aforementioned balanced structure was chosen.

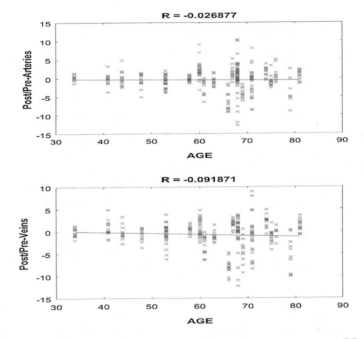

Fig. 5 The plot on the *top*, shows the differences between the measurements post-DR with the corresponding pre-DR measurements for the arteries. x axis: age, y axis: the individual differences. On the *bottom*, we find the same plot but for the veins. On *top* of them is the correlation coefficient parameter R

The final six selected features of our feature subset are (a) the mean of arteries' BCs, (b) Angle-to-BC ratio of veins, (c) Tortuosity, (d) Fractal dimension, (e) Vein SD and (f) Angle-to-BC ratio of arteries. In Fig. 6 we can see the importance for each of these features. The number of decision trees used for training the classifier was chosen at 5000, although it converged earlier. Choosing more trees than needed, does not affect the performance of the classifier. Larger number of trees produce more stable models and covariate importance estimates, but require more memory and a longer run time. The mtry parameter refers to the number of features available for splitting at each tree node and by default is set as the square root of the total number of features (rounded down).

Finally the performance of the classifier can be seen in Fig. 7 for the out of bag error and area under the ROC curve. As can be seen, the regularized random forests classifier achieved an OOB error of 22.5 % and AUC 0.7925 (Average over all the iterations of the cross-validation). Regarding accuracy, this was at 79.5 %.

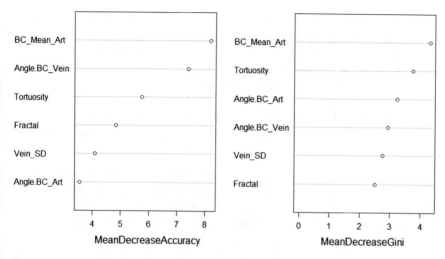

Fig. 6 Mean decrease accuracy shows how much the performance of the classifier will be affected if this feature is removed. A similar measure is the Gini index which is a measure of each feature's importance based on the Gini impurity index, used for the calculation of splits during training

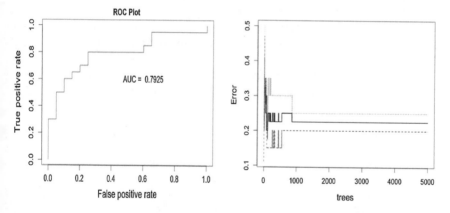

Fig. 7 On the *left* the ROC curve and the corresponding AUC value can be found. On the *right* the Out of Bag error for the whole training phase can be seen. The *red* and *green line* are the two classes and the *black* one is the average of them i.e. the final OBB error

4.3 Penalized Logistic Regression

As described in the previous section, when running a logistic regression model with elastic net penalty, a few factors have to be taken into account.

- Like in RRF the feature subset has to be selected. This occurs in two steps. The first step includes all the features under investigation. The second step is the final selection between the variables that had the best performance in the first step. In

both cases, a ten-fold cross-validation was used in order to calculate the mean
square error for the variables for different values of λ, and also of the penalty
parameter α, as the compromise between Lasso and ridge regression.

- Secondly, for the selected feature subset and the tuning parameter λ, we run the
  regression for varying penalties α, ranging from 0 to 1 with 0.1 step.
- After running the cross-validation for all the models, we evaluate which one fits
  best our data, and therefore define the optimum parameters for λ and α. Having
  these values set, we validate the performance by reporting the AUC, the accuracy
  and the ROC curve.

After running the relevant feature selection with the RRF, it was anticipated to obtain a
similar feature subset with the logistic regression, since the six selected features were
performing quite well. Indeed the same six features had the best score. In contrast,
the rest fourteen were all together deteriorating the performance of the classifier by
about 0.10 of the AUC, having extensive negative impact to the classifier. In Fig. 8,
the cross validation of the different features can be seen, which initially helps us
decide which features to discard and then work with the final ones. Secondly it can
be inferred how strong should the penalty be, after controlling for the λ parameter.
The best results were obtained for a penalty $\alpha = 0.2$.

Additionally, in Fig. 9, there is an informative illustration of how the coefficient
of each predictor changes along the different λ values. The optimum results were
obtained with $\lambda = 0.03$ as the tuning parameter that controls the overall strength of
the penalty.

Finally the logistic regression classifier had a similar performance with RRF, as
can be seen in Fig. 10, having an AUC $= 0.785$ and accuracy of 78 %.

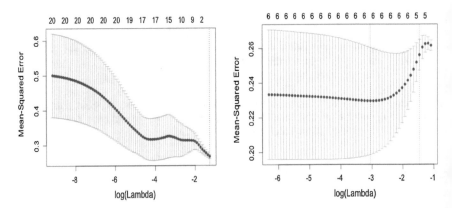

Fig. 8 On the *left* we can see the feature selection process for all the features, which leads us to the
right one, where we can see the final six features based on the their performance according to the
mean square error and for different values of λ. The *red dotted line* is the cross-validation curve,
together with the upper and lower standard deviation curves along the λ sequence

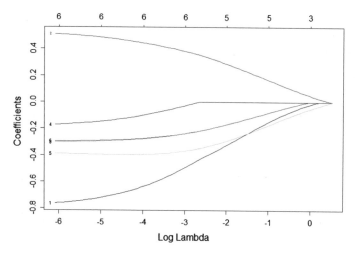

Fig. 9 Plot showing how the coefficients of all the features are adjusted according to the different values of λ that have been applied to each of them. For higher values of λ the predictors are starting moving towards zero. The x axis is the logarithm of λ

Fig. 10 ROC plot showing the Area under the ROC curve after a ten-fold cross validation. AUC in this case is 0.785. This value is the average over all the iterations of the cross validation

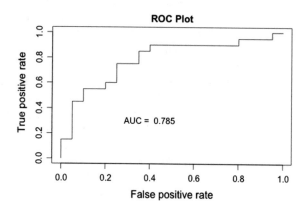

4.4 Discussion

Taking into account the limited amount of data, as well as the nature of the features, which represent the geometry of the retina and not any other image information, the performance of both classifiers is good enough to let us keep investigating those as well as additional features even further.

Another useful metric of the performance of the classifier is the precision/recall plot Fig. 11. Precision is a metric that gives us the positive predictive value of the classifier, while recall give us the true positive rate. Both these metrics are useful for evaluating a classifier, together with accuracy, AUC and ROC plot.

Fig. 11 Precision-Recall plot for both logistic regression (*black line*) and the RRF classifier (*red line*). Precision is defined as the $\frac{TruePositive}{TruePositive+FalsePositive}$, whereas recall is the $\frac{TruePositive}{TruePositive+FalseNegative}$

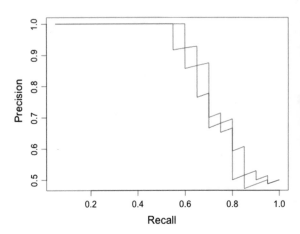

5 Conclussion and Discussion

Diabetes is a major disease, with millions of people being under medication in order to minimize its consequences. Identifying the changes in the vasculature during the progression of diabetes and measuring them is of paramount importance. Robust and reliable tools are needed for long term studies as well as properly designed experiments, in order to be able to discriminate over the different stages of progression. The alterations are so minor and in such a small scale that sometimes is very hard to measure and identify them. Hence novel tools for extracting information and analyzing data in a larger scale, are crucial for identifying the progression and also create reliable models with valid and robust biomarkers.

In this study, a comprehensive analysis was presented, using many different retinal geometric features and methods. To our best of knowledge, it is the first time that all these features together like CRVE/CRAE, tortuosity, fractal dimension, BC etc. were evaluated and/or utilized inside a classification system, yielding that performance, which is an improvement of approximately 2% from the previous study [20].

As aforementioned, it is a challenging task to extract all these features accurately, evaluate them and more importantly, associate any changes with the progression of diabetes. More data are always needed, in order to identify and investigate all of the possible underlying conditions and variations that occur as the disease progresses. The results of this study give us the boost to extend our investigation in more intervals of diabetes, by including even more data and features. Our immediate next work will include but not limited to building a multiclass system beyond the binary level for different periods of diabetes. Moreover, specific regions inside the retina will be investigated, focusing also on the bifurcations and the branching patterns of the vasculature.

Acknowledgments This research study was supported by a Marie Sklodowska-Curie grant from the European Commission in the framework of the REVAMMAD ITN (Initial Training Research network), Project number 316990.

References

1. Al-Diri, B., Hunter, A., Steel, D., Habib, M.: Manual measurement of retinal bifurcation features. In: 2010 Annual International Conference of the IEEE Engineering in Medicine and Biology Society (EMBC), pp. 4760–4764 (2010)
2. Annunziata, R., Garzelli, A., Ballerini, L., Mecocci, A., Trucco, E.: Leveraging multiscale hessian-based enhancement with a novel exudate inpainting technique for retinal vessel segmentation. IEEE J. Biomed. Health Inf. (2015)
3. Antonetti, D.A., Barber, A.J., Bronson, S.K., Freeman, W.M., Gardner, T.W., Jefferson, L.S., Simpson, I.A.: Diabetic retinopathy seeing beyond glucose-induced microvascular disease. Diabetes **55**(9), 2401–2411 (2006)
4. Avakian, A., Kalina, R.E., Helene Sage, E., Rambhia, A.H., Elliott, K.E., Chuang, E.L., Clark, J.I., Chuang, E.L., Parsons-Wingerter, P.: Fractal analysis of region-based vascular change in the normal and non-proliferative diabetic retina. Current Eye Res. **24**(4), 274–280 (2002)
5. Bankhead, P., Scholfield, C.N., McGeown, J.G., Curtis, T.M.: Fast retinal vessel detection and measurement using wavelets and edge location refinement. PloS One **7**(3), e32435 (2012)
6. Breiman, L.: Random forests. Mach. Learn. **45**(1), 5–32 (2001)
7. Cheung, N., McNab, A.A.: Venous anatomy of the orbit. Investig. Ophthalmol. Vis. Sci. **44**(3), 988–995 (2003)
8. Deng, H., Runger, G.: Feature selection via regularized trees. In: The 2012 International Joint Conference on IEEE Neural Networks (IJCNN), pp. 1–8 (2012)
9. Family, F., Masters, B.R., Platt, D.E.: Fractal pattern formation in human retinal vessels. Phys. D: Nonlinear Phenom. **38**(1), 98–103 (1989)
10. Friedman, J., Hastie, T., Tibshirani, R.: Regularization paths for generalized linear models via coordinate descent. J. Stat. Softw. **33**(1), 1–22 (2010)
11. Guo, Y., Logan, H.L., Glueck, D.H., Muller, K.E.: Selecting a sample size for studies with repeated measures. BMC Med. Res. Methodol. **13**(1), 100 (2013)
12. Habib, M.S., Al-Diri, B., Hunter, A., Steel, D.H.: The association between retinal vascular geometry changes and diabetic retinopathy and their role in prediction of progression-an exploratory study. BMC Ophthalmol. **14**(1), 89 (2014)
13. Hart, W.E., Goldbaum, M., Ct, B., Kube, P., Nelson, M.R.: Measurement and classification of retinal vascular tortuosity. Int. J. Med. Inf. **53**(2), 239–252 (1999)
14. Hunter, A., Lowell, J., Ryder, R., Basu, A., Steel, D.: Tram-line filtering for retinal vessel segmentation. In: Proceedings of the 3rd European Medical and Biological Engineering Conference (2005)
15. Jorgensen, C.M., Hardarson, S.H., Bek, T.: The oxygen saturation in retinal vessels from diabetic patients depends on the severity and type of vision threatening retinopathy. Acta Ophthalmol. **92**(1), 34–39 (2014)
16. Knudtson, M.D., Lee, K.E., Hubbard, L.D., Wong, T.Y., Klein, R., Klein, B.E.: Revised formulas for summarizing retinal vessel diameters. Curr. Eye Res. **27**(3), 143–149 (2013)
17. Leontidis, G., Al-Diri, B., Hunter, A.: Diabetic retinopathy: current and future methods for early screening from a retinal hemodynamic and geometric approach.Expert. Rev. Ophthalmol. **9**(5), 431–442 (2014)
18. Leontidis, G., Al-Diri, B., Hunter, A.: Study of the retinal vascular changes in the transition from diabetic to diabetic retinopathy eye. In: 2014 36th Annual International Conference of the IEEE Engineering in Medicine and Biology Society (EMBC), 26–30 August (2014)

19. Leontidis, G., Al-Diri, B., Hunter, A.: Retinal vascular geometry: examination of the changes between the early stages of diabetes and first year of diabetic retinopathy. In: Science and Information Conference (SAI), pp. 709–713 (2015). doi:10.1109/SAI.2015.7237220

20. Leontidis, G., Al-Diri, B., Wigdahl, J., Hunter, A.: Evaluation of geometric features as biomarkers of diabetic retinopathy for characterizing the retinal vascular changes during the progression of diabetes. In: 2015 37th Annual International Conference of the IEEE Engineering in Medicine Biology Society (EMBC), 25–29 August (2015)

21. Leontidis, G., Caliva, F., Al-Diri, B., Hunter, A.: Study of the retinal vascular changes between the early stages of diabetes and first year of diabetic retinopathy. Investig. Ophthalmol. Vis. Sci. **56**(7) (2015)

22. Leontidis, G., Wigdahl, J., Al-Diri, B., Ruggeri, A., Hunter, A.: Evaluating tortuosity in retinal fundus images of diabetic patients who progressed to diabetic retinopathy. In: 2015 37th Annual International Conference of the IEEE Engineering in Medicine and Biology Society (EMBC), 25–29 August (2015)

23. Li, J., Du, Q., Sun, C.: An improved box-counting method for image fractal dimension estimation. Pattern Recognit. **42**(11), 2460–2469 (2009)

24. Mandelbrot, B.B.: The fractal geometry of nature. Macmillan **173** (1983)

25. Nguyen, T.T., Wong, T.Y.: Retinal vascular changes and diabetic retinopathy. Curr. Diabetes Rep. **9**(4), 227–283 (2009)

26. Parr, J.C., Spears, G.F.S.: General caliber of the retinal arteries expressed as the equivalent width of the central retinal artery. Am. J. Pphthalmol. **77**(4), 472–477 (1974)

27. Shimizu, K., Kobayashi, Y., Muraoka, K.: Midperipheral fundus involvement in diabetic retinopathy. Ophthalmology **88**(7), 601–612 (1981)

28. Tolle, C.R., McJunkin, T.R., Gorsich, D.J.: An efficient implementation of the gliding box lacunarity algorithm. Phys. D: Nonlinear Phenom. **237**(3), 306–315 (2008)

29. Zhao, Y., Rada, L., Chen, K., Harding, S., Zheng, Y.: Automated vessel segmentation using infinite perimeter active contour model with hybrid region information with application to retinal images. IEEE Trans. Med. Imaging **34**(9), 1797–1807 (2015)

A New Method for Improving the Detection Capability of RADAR in the Presence of Noise

Md Saiful Islam, Jung-Chul Lee, Kabju Hwang and Uipil Chong

Abstract A RADAR system deals with many different and diverse problems for the last few decades. The detection capability of radar is one of the most important factors. The main objective of radar target detection is to improve probability of detection while reducing probability of a false alarm at the same time. To improve the probability of detection of moving target, a new approach is proposed in this paper using wavelet and Hough transforms. The wavelet de-noising technique is used to remove noise from received signal. Then the image processing technique of the Hough transform is used to detect moving target. To reduce the noises form received signal, we propose a new wavelet threshold function that reduces constant error of soft thresholding and improves the discontinuity of hard thresholding. We present performances of our method on a basis of the new thresholding technique and compare with traditional method. It is shown that detection performance of proposed method is superior to that obtained through traditional method.

Keywords Moving targets · Image denoising · Detection probability · False alarm · Radar

1 Introduction

Radar is an object-detection system that uses radio waves to determine the distance, direction, position or speed of targets. Radar that promises fast detection is the principal source of air defense system. Target detection is one of the most significant

M.S. Islam · J.-C. Lee · K. Hwang · U. Chong (✉)
School of Electrical Engineering, University of Ulsan, Ulsan, South Korea
e-mail: upchong@ulsan.ac.kr

M.S. Islam
e-mail: Saiful05eee@yahoo.com

J.-C. Lee
e-mail: jungclee@ulsan.ac.kr

© Springer International Publishing Switzerland 2016
L. Chen et al. (eds.), *Emerging Trends and Advanced Technologies for Computational Intelligence*, Studies in Computational Intelligence 647,
DOI 10.1007/978-3-319-33353-3_21

applications of radar systems. Detection performance in sever noise is one of the important abilities of radar network system.

Detection and tracking a moving target is very challenging because of two independent motions: the motion of the platform itself and the motion of moving objects to be tracked in the environment. Generally, moving target detection cannot achieve properly based on conventional detection system [1] due to severe noise and fast velocity. For small target detection, many methods have been proposed [2–4] based on matched Fourier transform, range stretching and joint time frequency processing, distinct optical properties, pixel based fusion algorithm and Keystone transform. For multiple targets [5, 6] a particle filter and high-order spectrum correlation are used for target detection and tracking. In the references [7–9], these methods are widely used for radar detection. However, the experimental evidence has shown various limitations of those methods [10]. To improve this shortcoming, image processing [11, 12] specifically Hough transform [13] are used in the area of radar detection to detect the targets in strong noise.

For finding the trajectory of a target, Carlson et al. [14–16] first applied the Hough transform(HT) using ρ and θ in search radar. After Carlson, the HT has been widely used in radar such as SAR [17, 18] and ground penetrating radar [19, 20]. Since all the points in data space are mapped to parameter space one by one, computational burden of this method is very heavy that limits applications of HT in radar applications. To reduce computational complexity, Liu et al. [21] propose a modified HT algorithm by shifting the parameter space cells for radar detection. However this approach allows the limited effect of noise. In this paper, we proposed a new method for multiple moving targets detection based on wavelet and HT. This approach is based on wavelet and HT. Firstly, the wavelet de-noising technique is used to remove noise from the received signal. We propose a new wavelet threshold function [22] that reduces constant error of soft thresholding and improves the discontinuity of hard thresholding. Secondly, the image processing technique of the HT is used.

The rest of the paper is organized as follows. In Sect. 2, literature of HT regarding to our proposed method are presented briefly. In Sect. 3, we describe the noise reduction technique based on wavelet de-noising technique and thoroughly explain our proposed improved threshold function. Section 4 is devoted to the performance assessment of the proposed method. The performances of proposed method have been presented in compare to the traditional method to detect moving targets. Finally, some conclusions are given in Sect. 5.

2 Problem Statement and Proposed Method

Hough transform is an efficient tool to find line in an image [23]. An arbitrary straight line can be represented by several data points in the range-time data space. The straight line can be defined by the algebraic distance ρ from the origin and the angle $\emptyset$ of its perpendicular from the origin. The equation of line corresponding to this geometry

$$\rho = r \cos \phi + t \sin \phi \tag{1}$$

If we define the value of $\emptyset$ from $0°$ to $180°$, then parameters are unique for each line. This specifies that every line in the r-t space corresponds to a unique point in the Hough space. Essentially, a line segment in a range-time space is a set of pixels. According to Eq. (1), each pixel is projected to a sine curve in the Hough parameter space.

Equation (1) is actually used for the mapping from range-time plane to Hough plane. According to Eq. (1), a simple matrix multiplication can be used for mapping from range-time space to Hough parameter space. A data matrix R represents the all pixel of range-time space.

$$R = \begin{bmatrix} P_1 & P_2 & P_3 & \ldots & P_m \\ t_1 & t_2 & t_3 & \ldots & t_m \end{bmatrix} \tag{2}$$

where, the columns are the range and time values of all pixels. P_1 is the pixel value at corresponding time t_1, P_2 is another pixel value at time t_2 and so on. According to Eq. (1), a transformation matrix T, composed of sines and cosines can be defined as

$$T = \begin{bmatrix} \sin \phi_1 & \cos \phi_1 \\ \sin \phi_2 & \cos \phi_2 \\ \ldots & \ldots \\ \ldots & \ldots \\ \sin \phi_n & \cos \phi_n \end{bmatrix} \tag{3}$$

where, $\emptyset$ has n discrete value between $0°$ to $180°$.

The multiplication of R and T produces an $n \times m$ matrix, H, containing the required value of ρ. Each column of H contains the ρ values for one of the parameter space sinusoids.

$$H = TR = \begin{bmatrix} \rho_{1,\phi_1} & \rho_{2,\phi_1} & \rho_{3,\phi_1} & \ldots & \rho_{m,\phi_1} \\ \rho_{1,\phi_2} & \rho_{2,\phi_2} & \rho_{3,\phi_2} & \ldots & \rho_{m,\phi_2} \\ \rho_{1,\phi_3} & \rho_{2,\phi_3} & \rho_{3,\phi_3} & \ldots & \rho_{m,\phi_3} \\ . & . & . & & . \\ . & . & . & \ldots & . \\ . & . & . & & . \\ \rho_{1,\phi_n} & \rho_{2,\phi_n} & \rho_{3,\phi_n} & \ldots & \rho_{m,\phi_n} \end{bmatrix} \tag{4}$$

Figure 1 shows the parameters for a line in r-t space. Figure 2 represents a view of the Hough transform of the points in Fig. 1. In the Fig. 2, Hough space shows four sinusoids correspond to four data points of Fig. 1. Through the four data points, the point of intersection defines ρ and Φ of line in data space. The point of intersection of all of the mapped sinusoids in Hough parameter space indicates all points in range-time space does exist on a line.

Fig. 1 The parameters for a
line in the *r-t* space

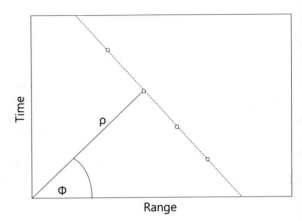

Fig. 2 Hough space of
Fig. 1. Each sinusoid
represents the one point of
Fig. 1

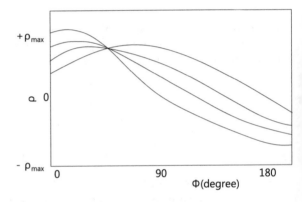

Carlson et al. [14–16] applied the Hough Transform in search radar for finding
the trajectory of a target. Carlson uses two thresholds named as primary threshold
and secondary threshold to declare detections.

In this paper, we employ the same method as [14] to analyze the detection performance except we use wavelet denoising technique to remove noise instead of primary
threshold. To reduce the noise from radar echo, we propose a new wavelet threshold
function. At first, the wavelet de-noising technique is used to remove noise from
received signal. Then, the image processing technique of the HT is used to detect
the trajectory of moving target. The new wavelet threshold function reduces constant
error of soft thresholding and improves the discontinuity of hard thresholding.

3 Noise Reduction by Wavelet

3.1 Wavelet Denoising

Digital images can be effected by different types of noises. Hence, noise reduction technique is essential. For removing noise and extracting signal from any data, Wavelet analysis is one of the most important methods. The Wavelets de-noising application has been used in spectrum cleaning of the atmospheric signals. There are different types of wavelets available like Morlet, Coiflets, Mexican Hat, Symlets, and Biorthogonal Haar, which have their own specifications such as filter coefficients and, reconstruction filter coefficients. In this paper, to eliminate noise embedded in the radar signal "sym8," wavelets have been used. The goal of this study is to denoise the radar signal. One often encounters the term 'de-noising' in recent wavelet literature, described in an informal way with various schemes that attempt to reject noise by damping or thresholding in the wavelet domain [24, 25]. The threshold of wavelet coefficient has near optimal noise reduction for different kinds of signals. Wavelets have many advantages over fast Fourier transform. Fourier analysis has a major drawback, which is that time information is lost, when transforming to the frequency domain. Thus, it is impossible to tell when a particular event took place under Fourier analysis. Wavelet analysis is capable of revealing aspects of data that other signal analysis technique, aspects such trends, breakdown points, discontinuities in higher derivatives, and self-similarity are unable to reveal. Wavelet analysis can often denoise a signal without appreciable degradation. The Wavelet transform performs a correlation analysis. Therefore, the output is expected to be maximal when the input signal most resembles the mother wavelet.

3.2 Wavelet Transform

According to definition of Wavelet transform [26], for function $f(t)$, Wavelet transform coefficient $W_f(a, \tau)$

$$W_f(a, \tau) = \langle f(t), \psi_{(a,\tau)}(t) \rangle = \frac{1}{\sqrt{a}} \int f(t) \psi^* (\frac{t - \tau}{a}) dt \qquad (5)$$

Here, $f(t), \psi_{(a,\tau)}(t)$ is the wavelet basis function, $\psi^*(\frac{t-\tau}{a})$ is a conjugate of wavelet basis function, τ is the amount of shift and a is scale.

3.3 Wavelet Denoising

The Wavelet de-noising procedure proceeds in three steps:

Step 1 Signal Decomposing: choose the wavelet basis function and to determine the decomposition level N, to get the coarse and detail coefficients by DWT.

Step 2 Threshold detail coefficients: For each level from 1 to N, Compare the detail coefficient and threshold value.

Step 3 Reconstructing the signal: reconstruct the denoised signal based on the original approximation coefficients of level N and the modified detail coefficients of levels from 1 to N.

3.4 Traditional Threshold Function

The major signal information mainly concentrates in the low frequency sub-band of wavelet transform domain. Noise equally distributes in all wavelet coefficient, so the wavelet transform factor should be bigger than the wavelet transform factor of the noises after wavelet decomposition. Therefore, the selection of wavelet threshold is an important step which directly impacts on the effect of noise reduction. Different methods have been proposed to choose the threshold. The standard and frequently used thresholding of wavelet coefficients is governed mainly by either soft or hard thresholding function, proposed by Donoho [32]. The soft thresholding is generally referred to as wavelet shrinkage, since it "shrinks" the coefficients with high amplitude towards zero, whereas the hard thresholding is commonly referred to simply as wavelet thresholding. Given that d_{jk} indicates the value of wavelet coefficient, $\tilde{d}_{jk}$ implies the value of d_{jk} after thresholding function, and T is the threshold value.

The soft thresholding function is defined as

$$\widehat{d}_{jk} = \begin{cases} 0, & if \ |d_{jk}| \leq T \\ d_{jk} - T, & if \ |d_{jk}| > T \\ d_{jk} + T, & if \ |d_{jk}| < -T \end{cases} \tag{6}$$

T is the threshold and generally can be a function of j and k. The hard thresholding function is defined as

$$\widehat{d}_{jk} = \begin{cases} 0, & if \ |d_{jk}| \leq T \\ d_{jk} & if \ |d_{jk}| > T \end{cases} \tag{7}$$

The soft and hard thresholding are depicted in Fig. 3. Soft thresholding provides smoother results in comparison with the hard thresholding whereas thresholding technique provides better edge preservation in comparison with the soft thresholding technique.

Soft thresholding and hard thresholding have some limitations in denoising of signal [28, 29]. The Eq. (6) indicates that the reconstructed signal faces oscillation,

Fig. 3 Thresholding
functions with threshold
value λ **a** linear, **b** soft and **c**
hard

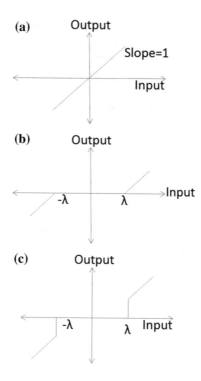

(a) Output

Slope=1

Input

(b) Output

-λ   λ   Input

(c) Output

-λ   λ   Input

since the estimated wavelet coefficients are not continuous at position $\pm T$ [30, 31]. Although the estimated wavelet coefficients of Eq. (7) have good continuity, these coefficients include constant errors [30, 32], which directly influence the accuracy of the reconstructed signal.

3.5 An Improved Threshold Function

To overcome the limitations of hard thresholding and soft thresholding de-noising methods, we proposed an improved thresholding [22]: Given that d_{jk} indicates the value of wavelet coefficient, $\widehat{d_{jk}}$ implies the value of d_{jk} after thresholding function, t is time and T is the threshold value.

$$\widehat{d_{Jk}} \begin{cases} 0, & \text{if } |d_{jk}| < T \\ d_{jk} + \frac{T(e^{-t}-1)}{m(e^{-t}+1)} & \text{if } |d_{jk}| \geq T \end{cases} \tag{8}$$

Here, value range of m is $1 < m \leq 10$, except this range, it is close to either soft threshold or hard threshold. When $\frac{1}{m} = 0$, the above equation will refer hard thresholding, while when $\frac{1}{m} = 1$, this equation will refer soft thresholding. We prefer

$m = 2$ and Eq. (8) become the combination of both soft and hard threshold and ensures $\widehat{d_{jk}}$ in a range between hard and soft threshold. This proposed method not only overcomes the shortcoming of hard thresholding and soft thresholding but also has many advantages. A more accurate wavelet coefficient can be achieved, since it takes advantage of both hard and soft thresholding. It improves the reconstruction precision, since it reduces the constant errors. Hence, it enhances the denoising effect. This threshold function also assures the continuity of estimated wavelet coefficients.

4   Simulation and Analysis

At the first in our simulation, we present performance of traditional method [14] and the improvements of our proposed approach in Sect. 4.1. Then in Sect. 4.2, we compare the signal to noise ratio (SNR) of our method with that of conventional Hough transforms. In this paper target detection decisions are based on threshold detection decision. We assume radar with 256 range gates and a time history composed of 128 scans with the search frame time being 0.375 s and total time history of 48 s.

4.1   Detection Performance by Proposed Method

A radar echo signal from moving target with noise is simulated as shown in grayscale plot of range-time data space in Fig. 4. For this simulation, we consider single moving target that is an approaching target with radial velocity. From Fig. 4, we cannot identify the trajectory of target due to the presence of severe noise in radar echo signal. Figure 5 represents the Hough parameter space where the target is dissemble with noise as well as ghost/false targets. It is impossible to identify the real target from this figure due to ghost/false targets with sever noise. It is worthy to mention that these false target appear due to noise in the echo signal. If we take the detection decision from these data, we have to decide 5 targets as shown in Fig. 6. Here, the 4 targets among the 5 targets are false target. To improve the probability of detection,

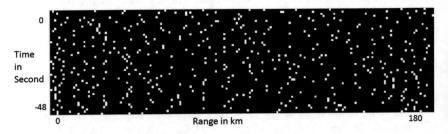

Fig. 4 Trajectory of moving target in the range time space from radar echo

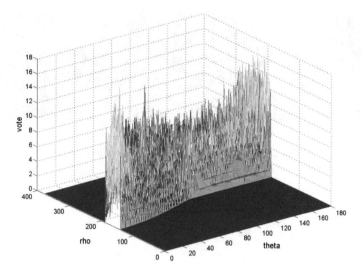

Fig. 5 Hough parameter space for radar echo

Fig. 6 Performance of detection decision with traditional method

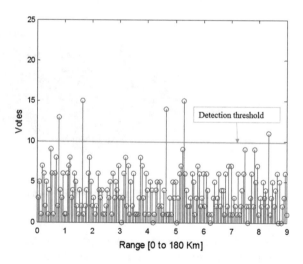

we reduce the noise from the echo signal using the proposed wavelet de-noising technique.

Trajectory of the target is very clear when noises are reduced by proposed method, shown in gray-scale plot in Fig. 7. Also, target can be now identified undoubtedly shown in the Hough space in Fig. 8. Trajectory of moving target in Fig. 7 and peak of target in Fig. 8 are very clear since much of the noise has been reduced by proposed wavelet de-noising technique. Figure 9 indicates the performance of detection decision with proposed method. In this figure only real target is appear and the other ghost/false targets are removed significantly.

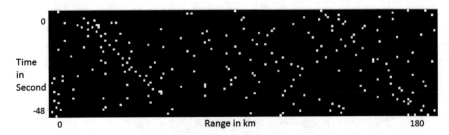

Fig. 7 Trajectory of moving target in the range time space from radar echo when noises are reduced by proposed method

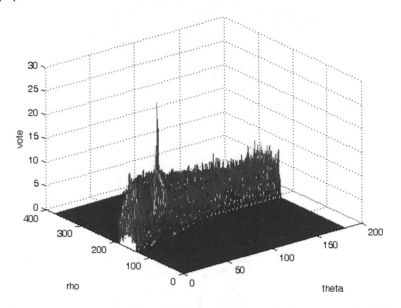

Fig. 8 Hough parameter space for radar echo when noises are reduced by proposed method

Let N be the total number of range-time cells that can contribute to a given parameter space cell and ξ be the threshold in parameter space. The probability of detection follows as [21].

$$P_D = \frac{\Gamma\left(N, \frac{\xi}{1+s}\right)}{\Gamma\left(N\right)} \tag{9}$$

where, s is the signal to noise ratio. Figure 10 shows the relation between probability of detection and signal to noise ratio.

Fig. 9 Performance of detection decision with proposed method

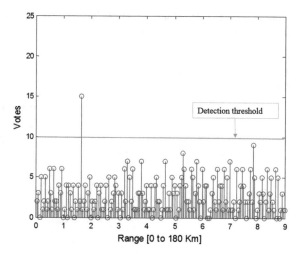

Fig. 10 Probability of detection (P_D) as a function of SNR

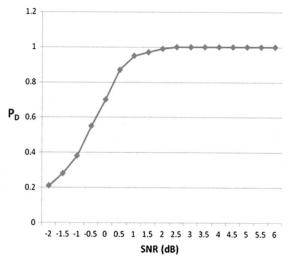

4.2 Performance Analysis

Performance analyses are done in this section to evaluate proposed approach. We compare the SNR of our method with that of conventional HT [14], shown in the Fig. 11. It is very obvious from Fig. 11 that proposed method with traditional wavelet provides higher SNR in compare to traditional HT which indicates our method is more effective in compare to traditional HT used to detect moving target and removing ghost/false target from the radar echo. Moreover, we compare our proposed threshold function with traditional wavelet threshold function. Our proposed wavelet transform provide higher SNR than the conventional wavelet transform shown in Fig. 11. This implies that our proposed threshold function is more effective than the traditional soft and hard threshold function.

Fig. 11 Plots of SNR versus detection range

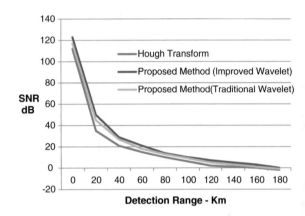

5 Conclusion

The paper presents a new method of improving the detection capability of radar in presence of noise based on Hough transform and wavelet denoising. A theoretical analysis is presented on Hough transform and wavelet denoising technique. An important contribution of this paper is to improve the detection capability of radar system for moving targets. We present the performances of moving target detection with their limitations based on Hough transform. To improve the detection ability, wavelet de-noising technique has been proposed with Hough transform for moving target detection. Experimental results demonstrate that the proposed method has a robust detection performance on moving target detection, and simultaneously can effectively get rid of false targets arising from the strong noise. Our proposed new wavelet threshold function reduces constant error of soft thresholding and improves the discontinuity of hard thresholding. It can be noticed that our proposed threshold function is more effective than the traditional soft and hard threshold functions.

References

1. Youlin, G.X., Niu, J., Zhang, R., Wu, Y., Fang, S., Hu, Y.: The study of moving target detection algorithm based on wind field detected by Lidar. In: Second International Conference on Intelligent Human-Machine Systems and Cybernetics, 26–28 Aug, Nanjing, Jiangsu (2010). doi:10.1109/IHMSC.2010.20

2. Zhang, S.S., Zeng, T., Long, T., Yuan, H.P.: Dim target detection based on Keystone transform. In: IEEE International Radar Conference, pp. 889– 894 (2005). doi:10.1109/RADAR.2005.1435953
3. Wang, J., Zhang, S.H., Bao, Z.: On motion compensation for weak radar reflected signal detection. In: The Sixth International Conference on Signal Processing, vol. 1, pp. 1445–1448 (2002). doi:10.1109/ICOSP.2002.1180065
4. Chen, J.J., Chen, J., Wang, S.L.: Detection of ultra-high speed moving target based on matched Fourier transform. In: CIE International Conference on Radar, pp. 1– 4 (2006). doi:10.1109/ICR.2006.343350
5. Morelande, M.R., Kreucher, C.M., Kastella, K.: A Bayesian approach to multiple target detection and tracking. IEEE Trans. Signal Process **55**, 1589–1604 (2007). doi:10.1109/TSP.2006.889470
6. Liou, R.-J., Azimi-Sadjadi, M.R.: Multiple target detection using modified high order correlations. IEEE Trans. Aerosp. Electron. Syst. **34**(2), 553–568 (1998). doi:10.1109/7.670336
7. He, Q., Lehmann, N.H., Blum, R.S., Haimovich, A.M.: MIMO radar moving target detection in homogenous clutter. IEEE Trans. Aerosp. Electr. Syst. **46**(3), 1290–1301 (2006). doi:10.1109/TAES.2010.5545189
8. Wang, P., Li, H., Himed, B.: Moving target detection using distributed MIMO radar in clutter with non-homogeneous power. IEEE Trans. Signal Process. **59**(10), 4809–4820 (2011). doi:10.1109/TSP.2011.2160861
9. He, Q., Lehmann, N.H., Blum, R.S., Haimovich, A.: MIMO radar moving target detection in homogeneous clutter. IEEE Trans. Aerosp. Electron. Syst. **46**(3), 1290–1301 (2010). doi:10.1109/TAES.2010.5545189
10. Li, N., Cui, G., Kong, L., Zhang, T., Liu, Q.H.: MIMO radar moving target detection in compound-Gaussian clutter. In: IEEE Radar Conference (2014). doi:10.1109/RADAR.2014.6875574
11. Mao, Q., Bharanitharan, K., Chang, C.-C.: Edge direction automatic control point selection algorithm for image morphing. IETE Tech. Rev. **30**(4), 343 (2013). doi:10.4103/0256-4602.116723
12. Atawneh, S., Almomani, A., Sumari, P.: Steganography in digital images: common approaches and tools. IETE Tech. Rev. **30**(4), 344–358 (2013). doi:10.4103/0256-4602.116724
13. Hough, P.V.C.: Methods and means for recognizing complex patterns. U.S. Patent 3,069,654 (1962)
14. Carlson, B.D., Evans, E.D., Wilson, S.L.: Search radar detection and track with the hough transform, part I: system concept. IEEE Trans. Aerosp. Electr. Syst. **30**, 102–108 (1994). doi:10.1109/7.250410
15. Carlson, B.D., Evans, E.D., Wilson, S.L.: Search radar detection and track with the Hough transform, part II: detection statistics. IEEE Trans. Aerosp. Electr. Syst. **30**, 109–115 (1994). doi:10.1109/7.250411
16. Carlson, B.D., Evans, E.D., Wilson, S.L.: Search radar detection and track with the Hough transform, part III: detection performance with binary integration. IEEE Trans. Aerosp. Electr. Syst. **30**, 116–125 (1994). doi:10.1109/7.250412
17. Dell'Acqua, F., Gamla, P.: Detection of urban structures in SAR images by robust fuzzy clustering algorithms: the example of street tracking. IEEE Trans. Geosci. Remote Sens. **39**(10), 2287–2297 (2001). doi:10.1109/36.957292
18. Amberg, V., et al.: Structure extraction from high resolution SAR data on urban areas. In: Proceedings of 2004 IEEE International Geoscience and Remote Sensing Symposium, vol. 3, pp. 1784–1787 (2004). doi:10.1109/IGARSS.2004.1370680
19. Tantum, S.L., et al.: Comparison of algorithms for land mine detection and discrimination using ground penetrating radar. In: Proceedings of the Society for Photo-Instrumentation Engineers, vol. 4742, pp. 728–735 (2002). doi:10.1117/12.479146
20. Long, K., et al.: Image processing of ground penetrating radar data for landmine detection. In: Proceedings of the Society for Photo-Instrumentation Engineers, vol. 6217, p. 62172R (2006). doi:10.1117/12.664038

21. Liu, H., He, Z., Zeng, J.: An improved radar detection algorithm based on hough transform. Springer Science+Business Media, LLC 2008, Sens Imaging, vol. 9, pp. 1–7 (2008). doi:10. 1007/s11220-008-0039-1
22. Islam, M.S., Chong, U.: Noise reduction of continuous wave radar and pulse radar using matched filter and wavelets. EURASIP J. Image Video Process. (2014). doi:10.1186/1687-5281-2014-43
23. Duda, R.O., Hart, P.E.: Use of the Hough transformation to detect lines and curves in pictures. Commun. ACM 15(1), 11–15 (1972). doi:10.1145/361237.361242
24. Li, Yuan: Wavelet Analysis for Change Points and Nonlinear Wavelet Estimates in Time Series. China Statistics Press, Beijing (2001)
25. Yinfeng, D., Li, Y., Xiao, M., Lai, M.: Analysis of earthquake ground motions using an improved Hilbert–Huang transform. Soil Dyn. Earthq. Engi. 28(1), 7–19 (2008). doi:10.1016/j.soildyn. 2007.05.002
26. Zhi-qiang, Z., Guo-wei, Z., Yu, P., Wei, S., Cheng, L., Jin-zhao, L.: Study on pulse wave signal noise reduction and feature point identification. J. Converg. Inf. Technol. (JCIT) 8(9), 953–960 (2013). doi:10.4156/jcit
27. Luisier, F., Blu, T., Unser, M.: A new SURE approach to image denoising: interscale ortho-normal wavelet thresholding. IEEE Trans. Image Proces. 16(3), 593–607 (2007). doi:10.1109/ TIP.2007.891064
28. Coifman, R.R., Donoho, D.L.: Translation-invariant de-noising. In: Wavelets and Statistics, Springer Lecture Notes in Statistics, vol. 103, pp. 125–150. Springer, New York (1994). doi:10. 1007/978-1-4612-2544-7_9
29. Qin, S., Yang, C., Tang, B., Tan, S.: The denoise based on translation invariance wavelet transform and its applications. In: A Conference on Structural Dynamics, Los Angeles, vol. 1, pp. 783–787 (2002)
30. Zhao, X., Cao, G.: A novel de-noising method for heart sound signal using improved threshold-ing function in wavelet domain. In: International Conference on Future BioMedical Information Engineering, FBIE, vol. 2009, pp. 65–68 (2009). doi:10.1109/FBIE.2009.5405795
31. Ray, P., Maitra, A.K., Basuray, A.: A new threshold function for de-noising partial discharge signal based on wavelet transform. In: International Conference on Signal Processing, Image Processing and Pattern Recognition, Bangalore, India (2013)
32. Li, Z., Fan, Q., Chang, L., Yang, X.: Improved wavelet threshold denoising method for MEMS gyroscope. In: 11th IEEE International Conference on Control and Automation (ICCA), Taichung, Taiwan, pp. 530–534 (2014). doi:10.1109/ICCA.2014.6870975

Printed in the United States
By Bookmasters